SCIENCE OF EVERYDAY THINGS

SCIENCE OF EVERYDAY THINGS

VOLUME 1: REAL-LIFE CHEMISTRY

EDITED BY **NEIL SCHLAGER**
WRITTEN BY **JUDSON KNIGHT**

A SCHLAGER INFORMATION GROUP BOOK

GALE GROUP
THOMSON LEARNING

Detroit • New York • San Diego • San Francisco
Boston • New Haven, Conn. • Waterville, Maine
London • Munich

A Schlager Information Group Book
Neil Schlager, Editor
Written by Judson Knight

Gale Group Staff

Kimberley A. McGrath, *Senior Editor*

Maria Franklin, *Permissions Manager*

Margaret A. Chamberlain, *Permissions Specialist*

Shalice Shah-Caldwell, *Permissions Associate*

Mary Beth Trimper, *Manager, Composition and Electronic Prepress*

Evi Seoud, *Assistant Manager, Composition and Electronic Prepress*

Dorothy Maki, *Manufacturing Manager*

Rita Wimberley, *Buyer*

Michelle DiMercurio, *Senior Art Director*

Barbara J. Yarrow, *Manager, Imaging and Multimedia Content*

Robyn V. Young, *Project Manager, Imaging and Multimedia Content*

Leitha Etheridge-Sims, Mary K. Grimes, and David G. Oblender, *Image Catalogers*

Pam A. Reed, *Imaging Coordinator*

Randy Bassett, *Imaging Supervisor*

Robert Duncan, *Senior Imaging Specialist*

Dan Newell, *Imaging Specialist*

ISBN   0-7876-5631-3 (set)
           0-7876-5632-1 (vol. 1)   0-7876-5634-8 (vol. 3)
           0-7876-5633-X (vol. 2)   0-7876-5635-6 (vol. 4)

Printed in the United States of America
10 9 8 7 6 5 4 3 2 1

Library of Congress Cataloging-in-Publication Data

Knight, Judson.
    Science of everyday things / written by Judson Knight, Neil Schlager, editor.
       p. cm.
    Includes bibliographical references and indexes.
    Contents: v. 1. Real-life chemistry – v. 2 Real-life physics.
    ISBN 0-7876-5631-3 (set : hardcover) – ISBN 0-7876-5632-1 (v. 1) – ISBN
    0-7876-5633-X (v. 2)
    1. Science–Popular works. I. Schlager, Neil, 1966-II. Title.

Q162.K678 2001
500–dc21
                                           2001050121

CONTENTS

INTRODUCTION

OVERVIEW OF THE SERIES

Welcome to *Science of Everyday Things*. Our aim is to explain how scientific phenomena can be understood by observing common, real-world events. From luminescence to echolocation to buoyancy, the series will illustrate the chief principles that underlay these phenomena and explore their application in everyday life. To encourage cross-disciplinary study, the entries will draw on applications from a wide variety of fields and endeavors.

Science of Everyday Things initially comprises four volumes:

Volume 1: *Real-Life Chemistry*
Volume 2: *Real-Life Physics*
Volume 3: *Real-Life Biology*
Volume 4: *Real-Life Earth Science*

Future supplements to the series will expand coverage of these four areas and explore new areas, such as mathematics.

ARRANGEMENT OF REAL-LIFE PHYSICS

This volume contains 40 entries, each covering a different scientific phenomenon or principle. The entries are grouped together under common categories, with the categories arranged, in general, from the most basic to the most complex. Readers searching for a specific topic should consult the table of contents or the general subject index.

Within each entry, readers will find the following rubrics:

- **Concept** Defines the scientific principle or theory around which the entry is focused.

- **How It Works** Explains the principle or theory in straightforward, step-by-step language.
- **Real-Life Applications** Describes how the phenomenon can be seen in everyday events.
- **Where to Learn More** Includes books, articles, and Internet sites that contain further information about the topic.

Each entry also includes a "Key Terms" section that defines important concepts discussed in the text. Finally, each volume includes numerous illustrations, graphs, tables, and photographs.

In addition, readers will find the comprehensive general subject index valuable in accessing the data.

ABOUT THE EDITOR, AUTHOR, AND ADVISORY BOARD

Neil Schlager and Judson Knight would like to thank the members of the advisory board for their assistance with this volume. The advisors were instrumental in defining the list of topics, and reviewed each entry in the volume for scientific accuracy and reading level. The advisors include university-level academics as well as high school teachers; their names and affiliations are listed elsewhere in the volume.

NEIL SCHLAGER is the president of Schlager Information Group Inc., an editorial services company. Among his publications are *When Technology Fails* (Gale, 1994); *How Products Are Made* (Gale, 1994); the *St. James Press Gay and Lesbian Almanac* (St. James Press, 1998); *Best Literature By and About Blacks* (Gale,

2000); *Contemporary Novelists, 7th ed.* (St. James Press, 2000); and *Science and Its Times* (7 vols., Gale, 2000-2001). His publications have won numerous awards, including three RUSA awards from the American Library Association, two Reference Books Bulletin/Booklist Editors' Choice awards, two New York Public Library Outstanding Reference awards, and a CHOICE award for best academic book.

Judson Knight is a freelance writer, and author of numerous books on subjects ranging from science to history to music. His work on science titles includes *Science, Technology, and Society, 2000 B.C.-A.D. 1799* (U*X*L, 2002), as well as extensive contributions to Gale's seven-volume *Science and Its Times* (2000-2001). As a writer on history, Knight has published *Middle Ages Reference Library* (2000), *Ancient Civilizations* (1999), and a volume in U*X*L's *African American Biography* series (1998). Knight's publications in the realm of music include *Parents Aren't Supposed to Like It* (2001), an overview of contemporary performers and genres, as well as *Abbey Road to Zapple Records: A Beatles Encyclopedia* (Taylor, 1999). His wife, Deidre Knight, is a literary agent and president of the Knight Agency. They live in Atlanta with their daughter Tyler, born in November 1998.

COMMENTS AND SUGGESTIONS

Your comments on this series and suggestions for future editions are welcome. Please write: The Editor, *Science of Everyday Things*, Gale Group, 27500 Drake Road, Farmington Hills, MI 48331.

ADVISORY BOARD

William E. Acree, Jr.
Professor of Chemistry, University of North Texas

Russell J. Clark
Research Physicist, Carnegie Mellon University

Maura C. Flannery
Professor of Biology, St. John's University, New
  York

John Goudie
Science Instructor, Kalamazoo (MI) Area
  Mathematics and Science Center

Cheryl Hach
Science Instructor, Kalamazoo (MI) Area
  Mathematics and Science Center

Michael Sinclair
Physics instructor, Kalamazoo (MI) Area
  Mathematics and Science Center

Rashmi Venkateswaran
Senior Instructor and Lab Coordinator,
  University of Ottawa
  Ottawa, Ontario, Canada

MEASUREMENT

MEASUREMENT

TEMPERATURE AND HEAT

MASS, DENSITY, AND VOLUME

MEASUREMENT

CONCEPT

Measurement seems like a simple subject, on the surface at least; indeed, all measurements can be reduced to just two components: number and unit. Yet one might easily ask, "What numbers, and what units?"—a question that helps bring into focus the complexities involved in designating measurements. As it turns out, some forms of numbers are more useful for rendering values than others; hence the importance of significant figures and scientific notation in measurements. The same goes for units. First, one has to determine what is being measured: mass, length, or some other property (such as volume) that is ultimately derived from mass and length. Indeed, the process of learning how to measure reveals not only a fundamental component of chemistry, but an underlying—if arbitrary and manmade—order in the quantifiable world.

HOW IT WORKS

Numbers

In modern life, people take for granted the existence of the base-10, of decimal numeration system—a name derived from the Latin word *decem*, meaning "ten." Yet there is nothing obvious about this system, which has its roots in the ten fingers used for basic counting. At other times in history, societies have adopted the two hands or arms of a person as their numerical frame of reference, and from this developed a base-2 system. There have also been base-5 systems relating to the fingers on one hand, and base-20 systems that took as their reference point the combined number of fingers and toes.

Obviously, there is an arbitrary quality underlying the modern numerical system, yet it works extremely well. In particular, the use of decimal fractions (for example, 0.01 or 0.235) is particularly helpful for rendering figures other than whole numbers. Yet decimal fractions are a relatively recent innovation in Western mathematics, dating only to the sixteenth century. In order to be workable, decimal fractions rely on an even more fundamental concept that was not always part of Western mathematics: place-value.

PLACE-VALUE AND NOTATION SYSTEMS

Place-value is the location of a number relative to others in a sequence, a location that makes it possible to determine the number's value. For instance, in the number 347, the 3 is in the hundreds place, which immediately establishes a value for the number in units of 100. Similarly, a person can tell at a glance that there are 4 units of 10, and 7 units of 1.

Of course, today this information appears to be self-evident—so much so that an explanation of it seems tedious and perfunctory—to almost anyone who has completed elementary-school arithmetic. In fact, however, as with almost everything about numbers and units, there is nothing obvious at all about place-value; otherwise, it would not have taken Western mathematicians thousands of years to adopt a place-value numerical system. And though they did eventually make use of such a system, Westerners did not develop it themselves, as we shall see.

ROMAN NUMERALS. Numeration systems of various kinds have existed since at least 3000 B.C., but the most important number

STANDARDIZATION IS CRUCIAL TO MAINTAINING STABILI-
TY IN A SOCIETY. DURING THE GERMAN INFLATIONARY
CRISIS OF THE 1920S, HYPERINFLATION LED TO AN
ECONOMIC DEPRESSION AND THE RISE OF ADOLF
HITLER. HERE, TWO CHILDREN GAZE UP AT A STACK OF
100,000 GERMAN MARKS—THE EQUIVALENT AT THE
TIME TO ONE U.S. DOLLAR. *(© Bettmann/Corbis.)*

system in the history of Western civilization prior to the late Middle Ages was the one used by the Romans. Rome ruled much of the known world in the period from about 200 B.C. to about A.D. 200, and continued to have an influence on Europe long after the fall of the Western Roman Empire in A.D. 476—an influence felt even today. Though the Roman Empire is long gone and Latin a dead language, the impact of Rome continues: thus, for instance, Latin terms are used to designate species in biology. It is therefore easy to understand how Europeans continued to use the Roman numeral system up until the thirteenth century A.D.—despite the fact that Roman numerals were enormously cumbersome.

The Roman notation system has no means of representing place-value: thus a relatively large number such as 3,000 is shown as MMM, whereas a much smaller number might use many more "places": 438, for instance, is rendered as CDXXXVIII. Performing any sort of calculations with these numbers is a nightmare. Imagine, for

instance, trying to multiply these two. With the number system in use today, it is not difficult to multiply 3,000 by 438 in one's head. The problem can be reduced to a few simple steps: multiply 3 by 400, 3 by 30, and 3 by 8; add these products together; then multiply the total by 1,000—a step that requires the placement of three zeroes at the end of the number obtained in the earlier steps.

But try doing this with Roman numerals: it is essentially impossible to perform this calculation without resorting to the much more practical place-value system to which we're accustomed. No wonder, then, that Roman numerals have been relegated to the sidelines, used in modern life for very specific purposes: in outlines, for instance; in ordinal titles (for example, Henry VIII); or in designating the year of a motion picture's release.

HINDU-ARABIC NUMERALS. The system of counting used throughout much of the world—1, 2, 3, and so on—is the Hindu-Arabic notation system. Sometimes mistakenly referred to as "Arabic numerals," these are most accurately designated as Hindu or Indian numerals. They came from India, but because Europeans discovered them in the Near East during the Crusades (1095-1291), they assumed the Arabs had invented the notation system, and hence began referring to them as Arabic numerals.

Developed in India during the first millennium B.C., Hindu notation represented a vast improvement over any method in use up to or indeed since that time. Of particular importance was a number invented by Indian mathematicians: zero. Until then, no one had considered zero worth representing since it was, after all, nothing. But clearly the zeroes in a number such as 2,000,002 stand for something. They perform a place-holding function: otherwise, it would be impossible to differentiate between 2,000,002 and 22.

USES OF NUMBERS IN SCIENCE

SCIENTIFIC NOTATION. Chemists and other scientists often deal in very large or very small numbers, and if they had to write out these numbers every time they discussed them, their work would soon be encumbered by lengthy numerical expressions. For this purpose, they use scientific notation, a method for writing extremely large or small numbers by representing

THE UNITED STATES NAVAL OBSERVATORY IN WASHINGTON, D.C., IS AMERICA'S PREEMINENT STANDARD FOR THE EXACT TIME OF DAY. *(Richard T. Nowitz/Corbis. Reproduced by permission.)*

them as a number between 1 and 10 multiplied by a power of 10.

Instead of writing 75,120,000, for instance, the preferred scientific notation is $7.512 \cdot 10^7$. To interpret the value of large multiples of 10, it is helpful to remember that the value of 10 raised to any power n is the same as 1 followed by that number of zeroes. Hence 10^{25}, for instance, is simply 1 followed by 25 zeroes.

Scientific notation is just as useful—to chemists in particular—for rendering very small numbers. Suppose a sample of a chemical compound weighed 0.0007713 grams. The preferred scientific notation, then, is $7.713 \cdot 10^{-4}$. Note that for numbers less than 1, the power of 10 is a negative number: 10^{-1} is 0.1, 10^{-2} is 0.01, and so on.

Again, there is an easy rule of thumb for quickly assessing the number of decimal places where scientific notation is used for numbers less than 1. Where 10 is raised to any power −n, the decimal point is followed by n places. If 10 is raised to the power of −8, for instance, we know at a glance that the decimal is followed by 7 zeroes and a 1.

SIGNIFICANT FIGURES. In making measurements, there will always be a degree of uncertainty. Of course, when the standards of calibration (discussed below) are very high, and the measuring instrument has been properly calibrated, the degree of uncertainty will be very small. Yet there is bound to be uncertainty to some degree, and for this reason, scientists use significant figures—numbers included in a measurement, using all certain numbers along with the first uncertain number.

Suppose the mass of a chemical sample is measured on a scale known to be accurate to 10^{-5} kg. This is equal to 1/100,000 of a kilo, or 1/100 of a gram; or, to put it in terms of place-value, the scale is accurate to the fifth place in a decimal fraction. Suppose, then, that an item is placed on the scale, and a reading of 2.13283697 kg is obtained. All the numbers prior to the 6 are significant figures, because they have been obtained with certainty. On the other hand, the 6 and the numbers that follow are not significant figures because the scale is not known to be accurate beyond 10^{-5} kg.

Thus the measure above should be rendered with 7 significant figures: the whole number 2, and the first 6 decimal places. But if the value is given as 2.132836, this might lead to inaccuracies at some point when the measurement is factored into other equations. The 6, in fact, should be "rounded off" to a 7. Simple rules apply to the rounding off of significant figures: if the digit following the first uncertain number is less than 5, there is no need to round off. Thus, if the measurement had been 2.13283627 kg (note that the 9 was changed to a 2), there is no need to round off, and in this case, the figure of 2.132836 is correct. But since the number following the 6 is in fact a 9, the correct significant figure is 7; thus the total would be 2.132837.

FUNDAMENTAL STANDARDS OF MEASURE

So much for numbers; now to the subject of units. But before addressing systems of measurement, what are the properties being measured? All forms of scientific measurement, in fact, can be reduced to expressions of four fundamental properties: length, mass, time, and electric current. Everything can be expressed in terms of these properties: even the speed of an electron spinning around the nucleus of an atom can be shown as "length" (though in this case, the measurement of space is in the form of a circle or even more complex shapes) divided by time.

Of particular interest to the chemist are length and mass: length is a component of volume, and both length and mass are elements of density. For this reason, a separate essay in this book is devoted to the subject of Mass, Density, and Volume. Note that "length," as used in this most basic sense, can refer to distance along any plane, or in any of the three dimensions—commonly known as length, width, and height—of the observable world. (Time is the fourth dimension.) In addition, as noted above, "length" measurements can be circular, in which case the formula for measuring space requires use of the coefficient π, roughly equal to 3.14.

REAL-LIFE APPLICATIONS

STANDARDIZED UNITS OF MEASURE: WHO NEEDS THEM?

People use units of measure so frequently in daily life that they hardly think about what they are doing. A motorist goes to the gas station and pumps 13 gallons (a measure of volume) into an automobile. To pay for the gas, the motorist uses dollars—another unit of measure, economic rather than scientific—in the form of paper money, a debit card, or a credit card.

This is simple enough. But what if the motorist did not know how much gas was in a gallon, or if the motorist had some idea of a gallon that differed from what the gas station management determined it to be? And what if the value of a dollar were not established, such that the motorist and the gas station attendant had to haggle over the cost of the gasoline just purchased? The result would be a horribly confused situation: the motorist might run out of gas, or money, or both, and if such confusion were multiplied by millions of motorists and millions of gas stations, society would be on the verge of breakdown.

THE VALUE OF STANDARDIZATION TO A SOCIETY. Actually, there have been times when the value of currency was highly unstable, and the result was near anarchy. In Germany during the early 1920s, for instance, rampant inflation had so badly depleted the value of the mark, Germany's currency, that employees demanded to be paid every day so that they could cash their paychecks before the value went down even further. People made jokes about the situation: it was said, for instance, that when a woman went into a store and left a basket containing several million marks out front, thieves ran by and stole the basket—but left the money. Yet there was nothing funny about this situation, and it paved the way for the nightmarish dictatorship of Adolf Hitler and the Nazi Party.

It is understandable, then, that standardization of weights and measures has always been an important function of government. When Ch'in Shih-huang-ti (259-210 B.C.) united China for the first time, becoming its first emperor, he set about standardizing units of measure as a means of providing greater unity to the country—thus making it easier to rule. On the other hand, the Russian Empire of the late nineteenth century failed to adopt standardized systems that would have tied it more closely to the industrialized nations of Western Europe. The width of railroad tracks in Russia was different than in Western Europe, and Russia used the old Julian calendar, as opposed to the Gregorian calendar adopted throughout much of Western Europe after 1582. These and other factors made economic exchanges between Russia and Western Europe extremely difficult, and the Russian Empire remained cut off from the rapid progress of the West. Like Germany a few decades later, it became ripe for the establishment of a dictatorship—in this case under the Communists led by V. I. Lenin.

Aware of the important role that standardization of weights and measures plays in the governing of a society, the U.S. Congress in 1901 established the Bureau of Standards. Today it is known as the National Institute of Standards and Technology (NIST), a nonregulatory agency within the Commerce Department. As will be discussed at the conclusion of this essay, the NIST maintains a wide variety of standard definitions regarding mass, length, temperature and so forth, against which other devices can be calibrated.

THE VALUE OF STANDARD-IZATION TO SCIENCE. What if a nurse, rather than carefully measuring a quantity of medicine before administering it to a patient, simply gave the patient an amount that "looked right"? Or what if a pilot, instead of calculating fuel, distance, and other factors carefully before taking off from the runway, merely used a "best estimate"? Obviously, in either case, disastrous results would be likely to follow. Though neither nurses or pilots are considered scientists, both use science in their professions, and those disastrous results serve to highlight the crucial matter of using standardized measurements in science.

Standardized measurements are necessary to a chemist or any scientist because, in order for an experiment to be useful, it must be possible to duplicate the experiment. If the chemist does not know exactly how much of a certain element he or she mixed with another to form a given compound, the results of the experiment are useless. In order to share information and communicate the results of experiments, then, scientists need a standardized "vocabulary" of measures.

This "vocabulary" is the International System of Units, known as SI for its French name, Système International d'Unités. By international agreement, the worldwide scientific community adopted what came to be known as SI at the 9th General Conference on Weights and Measures in 1948. The system was refined at the 11th General Conference in 1960, and given its present name; but in fact most components of SI belong to a much older system of weights and measures developed in France during the late eighteenth century.

SI VS. THE ENGLISH SYSTEM

The United States, as almost everyone knows, is the wealthiest and most powerful nation on Earth. On the other hand, Brunei—a tiny nation-state on the island of Java in the Indonesian archipelago—enjoys considerable oil wealth, but is hardly what anyone would describe as a superpower. Yemen, though it is located on the Arabian peninsula, does not even possess significant oil wealth, and is a poor, economically developing nation. Finally, Burma in Southeast Asia can hardly be described even as a "developing" nation: ruled by an extremely repressive military regime, it is one of the poorest nations in the world.

So what do these four have in common? They are the only nations on the planet that have failed to adopt the metric system of weights and measures. The system used in the United States is called the English system, though it should more properly be called the American system, since England itself has joined the rest of the world in "going metric." Meanwhile, Americans continue to think in terms of gallons, miles, and pounds; yet American scientists use the much more convenient metric units that are part of SI.

HOW THE ENGLISH SYSTEM WORKS (OR DOES NOT WORK). Like methods of counting described above, most systems of measurement in premodern times were modeled on parts of the human body. The foot is an obvious example of this, while the inch originated from the measure of a king's first thumb joint. At one point, the yard was defined as the distance from the nose of England's King Henry I to the tip of his outstretched middle finger.

Obviously, these are capricious, downright absurd standards on which to base a system of measure. They involve things that change, depending for instance on whose foot is being used as a standard. Yet the English system developed in this willy-nilly fashion over the centuries; today, there are literally hundreds of units—including three types of miles, four kinds of ounces, and five kinds of tons, each with a different value.

What makes the English system particularly cumbersome, however, is its lack of convenient conversion factors. For length, there are 12 inches in a foot, but 3 feet in a yard, and 1,760 yards in a mile. Where volume is concerned, there are 16 ounces in a pound (assuming one is talking about an avoirdupois ounce), but 2,000 pounds in a ton. And, to further complicate matters, there are all sorts of other units of measure developed to address a particular property: horsepower, for instance, or the British thermal unit (Btu).

THE CONVENIENCE OF THE METRIC SYSTEM. Great Britain, though it has long since adopted the metric system, in 1824 established the British Imperial System, aspects of which are reflected in the system still used in America. This is ironic, given the desire of early Americans to distance themselves psychologically from the empire to which their nation had once belonged. In any case, England's great worldwide influence during the nineteenth century brought about widespread adoption of the English or British system in colonies such as Australia and Canada. This acceptance had everything to do with British power and tradition, and nothing to do with convenience. A much more usable standard had actually been embraced 25 years before in a land that was then among England's greatest enemies: France.

During the period leading up to and following the French Revolution of 1789, French intellectuals believed that every aspect of existence could and should be treated in highly rational, scientific terms. Out of these ideas arose much folly, particularly during the Reign of Terror in 1793, but one of the more positive outcomes was the metric system. This system is decimal—that is, based entirely on the number 10 and powers of 10, making it easy to relate one figure to another. For instance, there are 100 centimeters in a meter and 1,000 meters in a kilometer.

PREFIXES FOR SIZES IN THE METRIC SYSTEM. For designating smaller values of a given measure, the metric system uses principles much simpler than those of the English system, with its irregular divisions of (for instance) gallons, quarts, pints, and cups. In the metric system, one need only use a simple Greek or Latin prefix to designate that the value is multiplied by a given power of 10. In general, the prefixes for values greater than 1 are Greek, while Latin is used for those less than 1. These prefixes, along with their abbreviations and respective values, are as follows. (The symbol μ for "micro" is the Greek letter mu.)

The Most Commonly Used Prefixes in the Metric System

- giga (G) = 10^9 (1,000,000,000)
- mega (M) = 10^6 (1,000,000)
- kilo (k) == 10^3 (1,000)
- deci (d) = 10^{-1} (0.1)
- centi (c) = 10^{-2} (0.01)
- milli (m) = 10^{-3} (0.001)
- micro (μ) = 10^{-6} (0.000001)
- nano (n) = 10^{-9} (0.000000001)

The use of these prefixes can be illustrated by reference to the basic metric unit of length, the meter. For long distances, a kilometer (1,000 m) is used; on the other hand, very short distances may require a centimeter (0.01 m) or a millimeter (0.001 m) and so on, down to a nanometer (0.000000001 m). Measurements of length also provide a good example of why SI includes units that are not part of the metric system, though they are convertible to metric units. Hard as it may be to believe, scientists often measure lengths even smaller than a nanometer—the width of an atom, for instance, or the wavelength of a light ray. For this purpose, they use the angstrom (Å or A), equal to 0.1 nanometers.

CALIBRATION AND SI UNITS

THE SEVEN BASIC SI UNITS. The SI uses seven basic units, representing length, mass, time, temperature, amount of substance, electric current, and luminous intensity. The first four parameters are a part of everyday life, whereas the last three are of importance only to scientists. "Amount of substance" is the number of elementary particles in matter. This is measured by the mole, a unit discussed in the essay on Mass, Density, and Volume. Luminous intensity, or the brightness of a light source, is measured in candelas, while the SI unit of electric current is the ampere.

The other four basic units are the meter for length, the kilogram for mass, the second for time, and the degree Celsius for temperature. The last of these is discussed in the essay on Temperature; as for meters, kilograms, and seconds, they will be examined below in terms of the means used to define each.

CALIBRATION. Calibration is the process of checking and correcting the performance of a measuring instrument or device against the accepted standard. America's preeminent standard for the exact time of day, for instance, is the United States Naval Observatory in Washington, D.C. Thanks to the Internet, people all over the country can easily check the exact time, and calibrate their clocks accordingly—though, of course, the resulting accuracy is subject to factors such as the speed of the Internet connection.

There are independent scientific laboratories responsible for the calibration of certain instruments ranging from clocks to torque wrenches, and from thermometers to laser-beam power analyzers. In the United States, instruments or devices with high-precision applications—that is, those used in scientific studies, or by high-tech industries—are calibrated according to standards established by the NIST.

The NIST keeps on hand definitions, as opposed to using a meter stick or other physical model. This is in accordance with the methods of calibration accepted today by scientists: rather than use a standard that might vary—for instance, the meter stick could be bent imperceptibly—unvarying standards, based on specific behaviors in nature, are used.

METERS AND KILOGRAMS. A meter, equal to 3.281 feet, was at one time defined in terms of Earth's size. Using an imaginary line drawn from the Equator to the North Pole through Paris, this distance was divided into 10 million meters. Later, however, scientists came to the realization that Earth is subject to geological changes, and hence any measurement calibrated to the planet's size could not ultimately be reliable. Today the length of a meter is calibrated according to the amount of time it takes light to travel through that distance in a vacuum (an area of space devoid of air or other matter). The official definition of a meter, then, is the distance traveled by light in the interval of 1/299,792,458 of a second.

One kilogram is, on Earth at least, equal to 2.21 pounds; but whereas the kilogram is a unit of mass, the pound is a unit of weight, so the correspondence between the units varies depending on the gravitational field in which a pound is measured. Yet the kilogram, though it represents a much more fundamental property of the physical world than a pound, is still a somewhat arbi-

trary form of measure in comparison to the meter as it is defined today.

Given the desire for an unvarying standard against which to calibrate measurements, it would be helpful to find some usable but unchanging standard of mass; unfortunately, scientists have yet to locate such a standard. Therefore, the value of a kilogram is calibrated much as it was two centuries ago. The standard is a bar of platinum-iridium alloy, known as the International Prototype Kilogram, housed near Sèvres in France.

SECONDS. A second, of course, is a unit of time as familiar to non-scientifically trained Americans as it is to scientists and people schooled in the metric system. In fact, it has nothing to do with either the metric system or SI. The means of measuring time on Earth are not "metric": Earth revolves around the Sun approximately every 365.25 days, and there is no way to turn this into a multiple of 10 without creating a situation even more cumbersome than the English units of measure.

The week and the month are units based on cycles of the Moon, though they are no longer related to lunar cycles because a lunar year would soon become out-of-phase with a year based on Earth's rotation around the Sun. The continuing use of weeks and months as units of time is based on tradition—as well as the essential need of a society to divide up a year in some way.

A day, of course, is based on Earth's rotation, but the units into which the day is divided—hours, minutes, and seconds—are purely arbitrary, and likewise based on traditions of long standing. Yet scientists must have some unit of time to use as a standard, and, for this purpose, the second was chosen as the most practical. The SI definition of a second, however, is not simply one-sixtieth of a minute or anything else so strongly influenced by the variation of Earth's movement.

Instead, the scientific community chose as its standard the atomic vibration of a particular isotope of the metal cesium, cesium-133. The vibration of this atom is presumed to be unvarying, because the properties of elements—unlike the size of Earth or its movement—do not change. Today, a second is defined as the amount of time it takes for a cesium-133 atom to vibrate 9,192,631,770 times. Expressed in scientific notation, with significant figures, this is $9.19263177 \cdot 10^9$.

WHERE TO LEARN MORE

Gardner, Robert. *Science Projects About Methods of Measuring.* Berkeley Heights, N.J.: Enslow Publishers, 2000.

Long, Lynette. *Measurement Mania: Games and Activities That Make Math Easy and Fun.* New York: Wiley, 2001.

"*Measurement*" (Web site). <http://www.dist214.k12.il.us/users/asanders/meas.html> (May 7, 2001).

"*Measurement in Chemistry*" (Web site). <http://bradley.edu/~campbell/lectnotes/149ch2/tsld001.htm> (May 7, 2001).

MegaConverter 2 (Web site). <http://www.megaconverter.com> (May 7, 2001).

Patilla, Peter. *Measuring.* Des Plaines, IL: Heinemann Library, 2000.

Richards, Jon. *Units and Measurements.* Brookfield, CT: Copper Beech Books, 2000.

Sammis, Fran. *Measurements.* New York: Benchmark Books, 1998.

Units of Measurement (Web site). <http://www.unc.edu/~rowlett/units/> (May 7, 2001).

Wilton High School Chemistry Coach (Web site). <http://www.chemistrycoach.com> (May 7, 2001).

TEMPERATURE AND HEAT

CONCEPT

Temperature, heat, and related concepts belong to the world of physics rather than chemistry; yet it would be impossible for the chemist to work without an understanding of these properties. Thermometers, of course, measure temperature according to one or both of two well-known scales based on the freezing and boiling points of water, though scientists prefer a scale based on the virtual freezing point of all matter. Also related to temperature are specific heat capacity, or the amount of energy required to change the temperature of a substance, and also calorimetry, the measurement of changes in heat as a result of physical or chemical changes. Although these concepts do not originate from chemistry but from physics, they are no less useful to the chemist.

HOW IT WORKS

Energy

The area of physics known as thermodynamics, discussed briefly below in terms of thermodynamics laws, is the study of the relationships between heat, work, and energy. Work is defined as the exertion of force over a given distance to displace or move an object, and energy is the ability to accomplish work. Energy appears in numerous manifestations, including thermal energy, or the energy associated with heat.

Another type of energy—one of particular interest to chemists—is chemical energy, related to the forces that attract atoms to one another in chemical bonds. Hydrogen and oxygen atoms in water, for instance, are joined by chemical bond-

ing, and when those bonds are broken, the forces joining the atoms are released in the form of chemical energy. Another example of chemical energy release is combustion, whereby chemical bonds in fuel, as well as in oxygen molecules, are broken and new chemical bonds are formed. The total energy in the newly formed chemical bonds is less than the energy of the original bonds, but the energy that makes up the difference is not lost; it has simply been released.

Energy, in fact, is never lost: a fundamental law of the universe is the conservation of energy, which states that in a system isolated from all other outside factors, the total amount of energy remains the same, though transformations of energy from one form to another take place. When a fire burns, then, some chemical energy is turned into thermal energy. Similar transformations occur between these and other manifestations of energy, including electrical and magnetic (sometimes these two are combined as electromagnetic energy), sound, and nuclear energy. If a chemical reaction makes a noise, for instance, some of the energy in the substances being mixed has been dissipated to make that sound. The overall energy that existed before the reaction will be the same as before; however, the energy will not necessarily be in the same place as before.

Note that chemical and other forms of energy are described as "manifestations," rather than "types," of energy. In fact, all of these can be described in terms of two basic types of energy: kinetic energy, or the energy associated with movement, and potential energy, or the energy associated with position. The two are inversely related: thus, if a spring is pulled back to its maximum point of tension, its potential energy is

also at a maximum, while its kinetic energy is zero. Once it is released and begins springing through the air to return to the position it maintained before it was stretched, it begins gaining kinetic energy and losing potential energy.

HEAT

Thermal energy is actually a form of kinetic energy generated by the movement of particles at the atomic or molecular level: the greater the movement of these particles, the greater the thermal energy. When people use the word "heat" in ordinary language, what they are really referring to is "the quality of hotness"—that is, the thermal energy internal to a system. In scientific terms, however, heat is internal thermal energy that flows from one body of matter to another—or, more specifically, from a system at a higher temperature to one at a lower temperature.

Two systems at the same temperature are said to be in a state of thermal equilibrium. When this state exists, there is no exchange of heat. Though in everyday terms people speak of "heat" as an expression of relative warmth or coldness, in scientific terms, heat exists only in transfer between two systems. Furthermore, there can never be a transfer of "cold"; although coldness is a recognizable sensory experience in human life, in scientific terms, cold is simply the absence of heat.

If you grasp a snowball in your hand, the hand of course gets cold. The mind perceives this as a transfer of cold from the snowball, but in fact exactly the opposite has happened: heat has moved from your hand to the snow, and if enough heat enters the snowball, it will melt. At the same time, the departure of heat from your hand results in a loss of internal energy near the surface of the hand, experienced as a sensation of coldness.

UNDERSTANDING TEMPERATURE

Just as heat does not mean the same thing in scientific terms as it does in ordinary language, so "temperature" requires a definition that sets it apart from its everyday meaning. Temperature may be defined as a measure of the average internal energy in a system. Two systems in a state of thermal equilibrium have the same temperature; on the other hand, differences in temperature determine the direction of internal energy flow between two systems where heat is being transferred.

This can be illustrated through an experience familiar to everyone: having one's temperature taken with a thermometer. If one has a fever, the mouth will be warmer than the thermometer, and therefore heat will be transferred to the thermometer from the mouth. The thermometer, discussed in more depth later in this essay, measures the temperature difference between itself and any object with which it is in contact.

TEMPERATURE AND THERMODYNAMICS

One might pour a kettle of boiling water into a cold bathtub to heat it up; or one might put an ice cube in a hot cup of coffee "to cool it down." In everyday experience, these seem like two very different events, but from the standpoint of thermodynamics, they are exactly the same. In both cases, a body of high temperature is placed in contact with a body of low temperature, and in both cases, heat passes from the high-temperature body to the low-temperature body.

The boiling water warms the tub of cool water, and due to the high ratio of cool water to boiling water in the bathtub, the boiling water

BECAUSE OF WATER'S HIGH SPECIFIC HEAT CAPACITY, CITIES LOCATED NEXT TO LARGE BODIES OF WATER TEND TO STAY WARMER IN THE WINTER AND COOLER IN THE SUMMER. DURING THE EARLY SUMMER MONTHS, FOR INSTANCE, CHICAGO'S LAKEFRONT STAYS COOLER THAN AREAS FURTHER INLAND. THIS IS BECAUSE THE LAKE IS COOLED FROM THE WINTER'S COLD TEMPERATURES AND SNOW RUNOFF. *(Farrell Grehan/Corbis. Reproduced by permission.)*

expends all its energy raising the temperature in the bathtub as a whole. The greater the ratio of very hot water to cool water, of course, the warmer the bathtub will be in the end. But even after the bath water is heated, it will continue to lose heat, assuming the air in the room is not warmer than the water in the tub—a safe assumption. If the water in the tub is warmer than the air, it will immediately begin transferring thermal energy to the lower-temperature air until their temperatures are equalized.

As for the coffee and the ice cube, what happens is opposite to the explanation ordinarily given. The ice does not "cool down" the coffee: the coffee warms up, and presumably melts, the ice. However, it expends at least some of its thermal energy in doing so, and, as a result, the coffee becomes cooler than it was.

THE LAWS OF THERMODY-NAMICS. These situations illustrate the second of the three laws of thermodynamics. Not only do these laws help to clarify the relationship between heat, temperature, and energy, but they also set limits on what can be accomplished in the world. Hence British writer and scientist C. P. Snow (1905-1980) once described the thermody-

namics laws as a set of rules governing an impossible game.

The first law of thermodynamics is essentially the same as the conservation of energy: because the amount of energy in a system remains constant, it is impossible to perform work that results in an energy output greater than the energy input. It could be said that the conservation of energy shows that "the glass is half full": energy is never lost. By contrast, the first law of thermodynamics shows that "the glass is half empty": no system can ever produce more energy than was put into it. Snow therefore summed up the first law as stating that the game is impossible to win.

The second law of thermodynamics begins from the fact that the natural flow of heat is always from an area of higher temperature to an area of lower temperature—just as was shown in the bathtub and coffee cup examples above. Consequently, it is impossible for any system to take heat from a source and perform an equivalent amount of work: some of the heat will always be lost. In other words, no system can ever be perfectly efficient: there will always be a degree of breakdown, evidence of a natural tendency called entropy.

Snow summed up the second law of thermodynamics, sometimes called "the law of entropy," thus: not only is it impossible to win, it is impossible to break even. In effect, the second law compounds the "bad news" delivered by the first with some even worse news. Though it is true that energy is never lost, the energy available for work output will never be as great as the energy put into a system.

The third law of thermodynamics states that at the temperature of absolute zero—a phenomenon discussed later in this essay—entropy also approaches zero. This might seem to counteract the second law, but in fact the third states in effect that absolute zero is impossible to reach. The French physicist and engineer Sadi Carnot (1796-1832) had shown that a perfectly efficient engine is one whose lowest temperature was absolute zero; but the second law of thermodynamics shows that a perfectly efficient engine (or any other perfect system) cannot exist. Hence, as Snow observed, not only is it impossible to win or break even; it is impossible to get out of the game.

REAL-LIFE APPLICATIONS

EVOLUTION OF THE THERMOMETER

A thermometer is a device that gauges temperature by measuring a temperature-dependent property, such as the expansion of a liquid in a sealed tube. The Greco-Roman physician Galen (c. 129-c. 199) was among the first thinkers to envision a scale for measuring temperature, but development of a practical temperature-measuring device—the thermoscope—did not occur until the sixteenth century.

The great physicist Galileo Galilei (1564-1642) may have invented the thermoscope; certainly he constructed one. Galileo's thermoscope consisted of a long glass tube planted in a container of liquid. Prior to inserting the tube into the liquid—which was usually colored water, though Galileo's thermoscope used wine—as much air as possible was removed from the tube. This created a vacuum (an area devoid of matter, including air), and as a result of pressure differences between the liquid and the interior of the thermoscope tube, some of the liquid went into the tube.

But the liquid was not the thermometric medium—that is, the substance whose temperature-dependent property changes were measured by the thermoscope. (Mercury, for instance, is the thermometric medium in many thermometers today; however, due to the toxic quality of mercury, an effort is underway to remove mercury thermometers from U.S. schools.) Instead, the air was the medium whose changes the thermoscope measured: when it was warm, the air expanded, pushing down on the liquid; and when the air cooled, it contracted, allowing the liquid to rise.

EARLY THERMOMETERS: THE SEARCH FOR A TEMPERATURE SCALE. The first true thermometer, built by Ferdinand II, Grand Duke of Tuscany (1610-1670) in 1641, used alcohol sealed in glass. The latter was marked with a temperature scale containing 50 units, but did not designate a value for zero. In 1664, English physicist Robert Hooke (1635-1703) created a thermometer with a scale divided into units equal to about 1/500 of the volume of the thermometric medium. For the zero point, Hooke chose the temperature at which water freezes, thus establishing a standard still used today in the Fahrenheit and Celsius scales.

Olaus Roemer (1644-1710), a Danish astronomer, introduced another important standard. Roemer's thermometer, built in 1702, was based not on one but two fixed points, which he designated as the temperature of snow or crushed ice on the one hand, and the boiling point of water on the other. As with Hooke's use of the freezing point, Roemer's idea of designating the freezing and boiling points of water as the two parameters for temperature measurements has remained in use ever since.

TEMPERATURE SCALES

THE FAHRENHEIT SCALE. Not only did he develop the Fahrenheit scale, oldest of the temperature scales still used in Western nations today, but in 1714, German physicist Daniel Fahrenheit (1686-1736) built the first thermometer to contain mercury as a thermometric medium. Alcohol has a low boiling point, whereas mercury remains fluid at a wide range of temperatures. In addition, it expands and contracts at a very constant rate, and tends not to stick to glass. Furthermore, its silvery color makes a mercury thermometer easy to read.

Fahrenheit also conceived the idea of using "degrees" to measure temperature. It is no mistake that the same word refers to portions of a circle, or that exactly 180 degrees—half the number of degrees in a circle—separate the freezing and boiling points for water on Fahrenheit's thermometer. Ancient astronomers first divided a circle into 360 degrees, as a close approximation of the ratio between days and years, because 360 has a large quantity of divisors. So, too, does 180—a total of 16 whole-number divisors other than 1 and itself.

Though today it might seem obvious that 0 should denote the freezing point of water, and 180 its boiling point, such an idea was far from obvious in the early eighteenth century. Fahrenheit considered a 0-to-180 scale, but also a 180-to-360 one, yet in the end he chose neither—or rather, he chose not to equate the freezing point of water with zero on his scale. For zero, he chose the coldest possible temperature he could create in his laboratory, using what he described as "a mixture of sal ammoniac or sea salt, ice, and water." Salt lowers the melting point of ice (which is why it is used in the northern United States to melt snow and ice from the streets on cold winter days), and thus the mixture of salt and ice produced an extremely cold liquid water whose temperature he equated to zero.

On the Fahrenheit scale, the ordinary freezing point of water is 32°, and the boiling point exactly 180° above it, at 212°. Just a few years after Fahrenheit introduced his scale, in 1730, a French naturalist and physicist named Rene Antoine Ferchault de Reaumur (1683-1757) presented a scale for which 0° represented the freezing point of water and 80° the boiling point. Although the Reaumur scale never caught on to the same extent as Fahrenheit's, it did include one valuable addition: the specification that temperature values be determined at standard sea-level atmospheric pressure.

THE CELSIUS SCALE. With its 32° freezing point and its 212° boiling point, the Fahrenheit system lacks the neat orderliness of a decimal or base-10 scale. Thus when France adopted the metric system in 1799, it chose as its temperature scale not the Fahrenheit but the Celsius scale. The latter was created in 1742 by Swedish astronomer Anders Celsius (1701-1744).

Like Fahrenheit, Celsius chose the freezing and boiling points of water as his two reference points, but he determined to set them 100, rather than 180, degrees apart. The Celsius scale is sometimes called the centigrade scale, because it is divided into 100 degrees, cent being a Latin root meaning "hundred." Interestingly, Celsius planned to equate 0° with the boiling point, and 100° with the freezing point; only in 1750 did fellow Swedish physicist Martin Strömer change the orientation of the Celsius scale. In accordance with the innovation offered by Reaumur, Celsius's scale was based not simply on the boiling and freezing points of water, but specifically on those points at normal sea-level atmospheric pressure.

In SI, a scientific system of measurement that incorporates units from the metric system along with additional standards used only by scientists, the Celsius scale has been redefined in terms of the triple point of water. (Triple point is the temperature and pressure at which a substance is at once a solid, liquid, and vapor.) According to the SI definition, the triple point of water—which occurs at a pressure considerably below normal atmospheric pressure—is exactly 0.01°C.

THE KELVIN SCALE. French physicist and chemist J. A. C. Charles (1746-1823), who is credited with the gas law that bears his name (see below), discovered that at 0°C, the volume of gas at constant pressure drops by 1/273 for every Celsius degree drop in temperature. This suggested that the gas would simply disappear if cooled to -273°C, which of course made no sense.

The man who solved the quandary raised by Charles's discovery was William Thompson, Lord Kelvin (1824-1907), who, in 1848, put forward the suggestion that it was the motion of molecules, and not volume, that would become zero at −273°C. He went on to establish what came to be known as the Kelvin scale. Sometimes known as the absolute temperature scale, the Kelvin scale is based not on the freezing point of water, but on absolute zero—the temperature at which molecular motion comes to a virtual stop. This is −273.15°C (−459.67°F), which, in the Kelvin scale, is designated as 0K. (Kelvin measures do not use the term or symbol for "degree.")

Though scientists normally use metric units, they prefer the Kelvin scale to Celsius because the absolute temperature scale is directly related to average molecular translational energy, based on the relative motion of molecules. Thus if the Kelvin temperature of an object is doubled, this means its average molecular translational energy has doubled as well. The same cannot be said if the temperature were doubled from, say, 10°C to 20°C, or from 40°C to 80°F, since neither the Celsius nor the Fahrenheit scale is based on absolute zero.

CONVERSIONS BETWEEN SCALES. The Kelvin scale is closely related to the Celsius scale, in that a difference of one degree measures the same amount of temperature in both. Therefore, Celsius temperatures can be converted to Kelvins by adding 273.15. Conversion between Celsius and Fahrenheit figures, on the other hand, is a bit trickier.

To convert a temperature from Celsius to Fahrenheit, multiply by 9/5 and add 32. It is important to perform the steps in that order, because reversing them will produce a wrong figure. Thus, 100°C multiplied by 9/5 or 1.8 equals 180, which, when added to 32 equals 212°F. Obviously, this is correct, since 100°C and 212°F each represent the boiling point of water. But if one adds 32 to 100°, then multiplies it by 9/5, the result is 237.6°F—an incorrect answer.

For converting Fahrenheit temperatures to Celsius, there are also two steps involving multiplication and subtraction, but the order is reversed. Here, the subtraction step is performed before the multiplication step: thus 32 is subtracted from the Fahrenheit temperature, then the result is multiplied by 5/9. Beginning with 212°F, when 32 is subtracted, this equals 180. Multiplied by 5/9, the result is 100°C—the correct answer.

One reason the conversion formulae use simple fractions instead of decimal fractions (what most people simply call "decimals") is that 5/9 is a repeating decimal fraction (0.55555....) Furthermore, the symmetry of 5/9 and 9/5 makes memorization easy. One way to remember the formula is that Fahrenheit is multiplied by a fraction—since 5/9 is a real fraction, whereas 9/5 is actually a mixed number, or a whole number plus a fraction.

MODERN THERMOMETERS

MERCURY THERMOMETERS. For a thermometer, it is important that the glass tube be kept sealed; changes in atmospheric pressure contribute to inaccurate readings, because they influence the movement of the thermometric medium. It is also important to have a reliable thermometric medium, and, for this reason, water—so useful in many other contexts—was quickly discarded as an option.

Water has a number of unusual properties: it does not expand uniformly with a rise in temperature, or contract uniformly with a lowered temperature. Rather, it reaches its maximum density at 39.2°F (4°C), and is less dense both above and below that temperature. Therefore alcohol, which responds in a much more uniform fashion to changes in temperature, soon took the place of water, and is still used in many thermometers today. But for the reasons mentioned earlier, mercury is generally considered preferable to alcohol as a thermometric medium.

In a typical mercury thermometer, mercury is placed in a long, narrow sealed tube called a capillary. The capillary is inscribed with figures for a calibrated scale, usually in such a way as to allow easy conversions between Fahrenheit and Celsius. A thermometer is calibrated by measuring the difference in height between mercury at the freezing point of water, and mercury at the boiling point of water. The interval between these two points is then divided into equal increments—180, as we have seen, for the Fahrenheit scale, and 100 for the Celsius scale.

VOLUME GAS THERMOMETERS. Whereas most liquids and solids expand at an irregular rate, gases tend to follow a fairly regular pattern of expansion in response to increases in temperature. The predictable behavior of gases in these situations has led to the development of the volume gas thermometer, a highly reliable instrument against which other thermometers—including those containing mercury—are often calibrated.

In a volume gas thermometer, an empty container is attached to a glass tube containing mercury. As gas is released into the empty container; this causes the column of mercury to move upward. The difference between the earlier position of the mercury and its position after the introduction of the gas shows the difference between normal atmospheric pressure and the pressure of the gas in the container. It is then possible to use the changes in the volume of the gas as a measure of temperature.

ELECTRIC THERMOMETERS. All matter displays a certain resistance to electric current, a resistance that changes with temperature; because of this, it is possible to obtain temperature measurements using an electric thermometer. A resistance thermometer is equipped with a fine wire wrapped around an insulator: when a change in temperature occurs, the resistance in the wire changes as well. This allows much quicker temperature readings than those offered by a thermometer containing a traditional thermometric medium.

Resistance thermometers are highly reliable, but expensive, and primarily are used for very precise measurements. More practical for everyday use is a thermistor, which also uses the principle of electric resistance, but is much simpler and less expensive. Thermistors are used for providing measurements of the internal temperature of food, for instance, and for measuring human body temperature.

Another electric temperature-measurement device is a thermocouple. When wires of two different materials are connected, this creates a small level of voltage that varies as a function of temperature. A typical thermocouple uses two junctions: a reference junction, kept at some constant temperature, and a measurement junction. The measurement junction is applied to the item whose temperature is to be measured, and any temperature difference between it and the reference junction registers as a voltage change, measured with a meter connected to the system.

OTHER TYPES OF THERMOMETER. A pyrometer also uses electromagnetic properties, but of a very different kind. Rather than responding to changes in current or voltage, the pyrometer is gauged to respond to visible and infrared radiation. As with the thermocouple, a pyrometer has both a reference element and a measurement element, which compares light readings between the reference filament and the object whose temperature is being measured.

Still other thermometers, such as those in an oven that register the oven's internal temperature, are based on the expansion of metals with heat. In fact, there are a wide variety of thermometers, each suited to a specific purpose. A pyrometer, for instance, is good for measuring the temperature of a object with which the thermometer itself is not in physical contact.

MEASURING HEAT

The measurement of temperature by degrees in the Fahrenheit or Celsius scales is a part of daily life, but measurements of heat are not as familiar to the average person. Because heat is a form of energy, and energy is the ability to perform work, heat is therefore measured by the same units as work. The principal SI unit of work or energy is the joule (J). A joule is equal to 1 newton-meter (N • m)—in other words, the amount of energy required to accelerate a mass of 1 kilogram at the rate of 1 meter per second squared across a distance of 1 meter.

The joule's equivalent in the English system is the foot-pound: 1 foot-pound is equal to 1.356 J, and 1 joule is equal to 0.7376 ft • lbs. In the British system, Btu, or British thermal unit, is another measure of energy, though it is primarily used for machines. Due to the cumbersome nature of the English system, contrasted with the convenience of the decimal units in the SI system, these English units of measure are not used by chemists or other scientists for heat measurement.

SPECIFIC HEAT CAPACITY

Specific heat capacity (sometimes called specific heat) is the amount of heat that must be added to, or removed from, a unit of mass for a given substance to change its temperature by 1°C. Typically, specific heat capacity is measured in units of J/g • °C (joules per gram-degree Celsius).

The specific heat capacity of water is measured by the calorie, which, along with the joule, is an important SI measure of heat. Often another unit, the kilocalorie—which, as its name suggests—is 1,000 calories—is used. This is one of the few confusing aspects of SI, which is much simpler than the English system. The dietary Calorie (capital C) with which most people are familiar is not the same as a calorie (lowercase c)—rather, a dietary Calorie is the same as a kilocalorie.

COMPARING SPECIFIC HEAT CAPACITIES. The higher the specific heat capacity, the more resistant the substance is to changes in temperature. Many metals, in fact, have a low specific heat capacity, making them easy to heat up and cool down. This contributes to the tendency of metals to expand when heated, and thus affects their malleability. On the other hand, water has a high specific heat capacity, as discussed below; indeed, if it did not, life on Earth would hardly be possible.

One of the many unique properties of water is its very high specific heat capacity, which is easily derived from the value of a kilocalorie: it is 4.184, the same number of joules required to equal a calorie. Few substances even come close to this figure. At the low end of the spectrum are lead, gold, and mercury, with specific heat capacities of 0.13, 0.13, and 0.14 respectively. Aluminum has a specific heat capacity of 0.89, and ethyl alcohol of 2.43. The value for concrete, one of the highest for any substance other than water, is 2.9.

As high as the specific heat capacity of concrete is, that of water is more than 40% higher. On the other hand, water in its vapor state (steam) has a much lower specific heat capacity—2.01. The same is true for solid water, or ice, with a specific heat capacity of 2.03. Nonetheless, water in its most familiar form has an astoundingly high specific heat capacity, and this has several effects in the real world.

EFFECTS OF WATER'S HIGH SPECIFIC HEAT CAPACITY. For instance, water is much slower to freeze in the winter than most substances. Furthermore, due to other unusual aspects of water—primarily the fact that it actually becomes less dense as a solid—the top of a lake or other body of water freezes first. Because ice is a poor medium for the conduction of heat (a consequence of its specific heat capacity), the ice at the top forms a layer that protects the still-liquid water below it from losing heat. As a result, the water below the ice layer does not freeze, and the animal and plant life in the lake is preserved.

Conversely, when the weather is hot, water is slow to experience a rise in temperature. For this reason, a lake or swimming pool makes a good place to cool off on a sizzling summer day. Given the high specific heat capacity of water, combined with the fact that much of Earth's surface is composed of water, the planet is far less susceptible than other bodies in the Solar System to variations in temperature.

The same is true of another significant natural feature, one made mostly of water: the human body. A healthy human temperature is 98.6°F (37°C), and, even in cases of extremely high fever, an adult's temperature rarely climbs by more than 5°F (2.7°C). The specific heat capacity of the human body, though it is of course lower than that of water itself (since it is not entirely made of water), is nonetheless quite high: 3.47.

CALORIMETRY

The measurement of heat gain or loss as a result of physical or chemical change is called calorimetry (pronounced kal-or-IM-uh-tree). Like the word "calorie," the term is derived from a Latin root word meaning "heat." The foundations of calorimetry go back to the mid-nineteenth century, but the field owes much to the work of scientists about 75 years prior to that time.

In 1780, French chemist Antoine Lavoisier (1743-1794) and French astronomer and mathematician Pierre Simon Laplace (1749-1827) had used a rudimentary ice calorimeter for measuring heat in the formations of compounds. Around the same time, Scottish chemist Joseph Black (1728-1799) became the first scientist to make a clear distinction between heat and temperature.

By the mid-1800s, a number of thinkers had come to the realization that—contrary to prevailing theories of the day—heat was a form of energy, not a type of material substance. (The belief that heat was a material substance, called "phlogiston," and that phlogiston was the part of a substance that burned in combustion, had originated in the seventeenth century. Lavoisier was the first scientist to successfully challenge the phlogiston theory.) Among these were American-British physicist Benjamin Thompson, Count Rumford (1753-1814) and English chemist James Joule (1818-1889)—for whom, of course, the joule is named.

Calorimetry as a scientific field of study actually had its beginnings with the work of French chemist Pierre-Eugene Marcelin Berthelot (1827-1907). During the mid-1860s, Berthelot became intrigued with the idea of measuring heat, and, by 1880, he had constructed the first real calorimeter.

CALORIMETERS. Essential to calorimetry is the calorimeter, which can be any device for accurately measuring the temperature of a substance before and after a change occurs. A calorimeter can be as simple as a styrofoam cup. Its quality as an insulator, which makes styrofoam ideal both for holding in the warmth of coffee and protecting the human hand from scalding, also makes styrofoam an excellent material for calorimetric testing. With a styrofoam calorimeter, the temperature of the substance inside the cup is measured, a reaction is allowed to take place, and afterward, the temperature is measured a second time.

The most common type of calorimeter used is the bomb calorimeter, designed to measure the heat of combustion. Typically, a bomb calorimeter consists of a large container filled with water, into which is placed a smaller container, the combustion crucible. The crucible is made of metal, with thick walls into which is cut an opening to allow the introduction of oxygen. In addition, the combustion crucible is designed to be connected to a source of electricity.

In conducting a calorimetric test using a bomb calorimeter, the substance or object to be studied is placed inside the combustion crucible and ignited. The resulting reaction usually occurs so quickly that it resembles the explosion of a bomb—hence the name "bomb calorimeter." Once the "bomb" goes off, the resulting transfer of heat creates a temperature change in the water, which can be readily gauged with a thermometer.

To study heat changes at temperatures higher than the boiling point of water, physicists use substances with higher boiling points. For experiments involving extremely large temperature ranges, an aneroid (without liquid) calorimeter may be used. In this case, the lining of the combustion crucible must be of a metal, such as copper, with a high coefficient or factor of thermal conductivity—that is, the ability to conduct heat from molecule to molecule.

TEMPERATURE IN CHEMISTRY

THE GAS LAWS. A collection of statements regarding the behavior of gases, the gas laws are so important to chemistry that a separate essay is devoted to them elsewhere. Several of the gas laws relate temperature to pressure and volume for gases. Indeed, gases respond to changes in temperature with dramatic changes in volume; hence the term "volume," when used in reference to a gas, is meaningless unless pressure and temperature are specified as well.

Among the gas laws, Boyle's law holds that in conditions of constant temperature, an inverse relationship exists between the volume and pressure of a gas: the greater the pressure, the less the volume, and vice versa. Even more relevant to the subject of thermal expansion is Charles's law, which states that when pressure is kept constant, there is a direct relationship between volume and absolute temperature.

CHEMICAL EQUILIBRIUM AND CHANGES IN TEMPERATURE. Just as two systems that exchange no heat are said to be in a state of thermal equilibrium, chemical equilibrium describes a dynamic state in which the concentration of reactants and products remains constant. Though the concentrations of reactants and products do not change, note that chemical equilibrium is a dynamic state—in other words, there is still considerable molecular activity, but no net change.

Calculations involving chemical equilibrium make use of a figure called the equilibrium constant (K). According to Le Châtelier's principle, named after French chemist Henri Le Châtelier (1850-1936), whenever a stress or change is imposed on a chemical system in equilibrium, the system will adjust the amounts of the various substances in such a way as to reduce the impact of that stress. An example of a stress is a change in temperature, which changes the equilibrium equation by shifting K (itself dependant on temperature).

Using Le Châtelier's law, it is possible to determine whether K will change in the direction of the forward or reverse reaction. In an exothermic reaction (a reaction that produces heat), K will shift to the left, or in the direction of the forward reaction. On the other hand, in an endothermic reaction (a reaction that absorbs heat), K will shift to the right, or in the direction of the reverse reaction.

TEMPERATURE AND REACTION RATES. Another important function of temperature in chemical processes is its function of speeding up chemical reactions. An increase in the concentration of reacting molecules, naturally, leads to a sped-up reaction, because there are simply more molecules colliding with one another. But it is also possible to speed up the reaction without changing the concentration.

By definition, wherever a temperature increase is involved, there is always an increase in average molecular translational energy. When temperatures are high, more molecules are col-

KEY TERMS

ABSOLUTE ZERO: The temperature, defined as 0K on the Kelvin scale, at which the motion of molecules in a solid virtually ceases. The third law of thermodynamics establishes the impossibility of actually reaching absolute zero.

CALORIE: A measure of specific heat capacity in the SI or metric system, equal to the heat that must be added to or removed from 1 gram of water to change its temperature by 1°C. The dietary Calorie (capital C), with which most people are familiar, is the same as the kilocalorie.

CALORIMETRY: The measurement of heat gain or loss as a result of physical or chemical change.

CELSIUS SCALE: The metric scale of temperature, sometimes known as the centigrade scale, created in 1742 by Swedish astronomer Anders Celsius (1701-1744). The Celsius scale establishes the freezing and boiling points of water at 0° and 100° respectively. To convert a temperature from the Celsius to the Fahrenheit scale, multiply by 9/5 and add 32. Though the worldwide scientific community uses the metric or SI system for most measurements, scientists prefer the related Kelvin scale of absolute temperature.

CONSERVATION OF ENERGY: A law of physics which holds that within a system isolated from all other outside factors, the total amount of energy remains the same, though transformations of energy from one form to another take place. The first law of thermodynamics is the same as the conservation of energy.

ENERGY: The ability to accomplish work—that is, the exertion of force over a given distance to displace or move an object.

ENTROPY: The tendency of natural systems toward breakdown, and specifically the tendency for the energy in a system to be dissipated. Entropy is closely related to the second law of thermodynamics.

FAHRENHEIT SCALE: The oldest of the temperature scales still in use, created in 1714 by German physicist Daniel Fahrenheit (1686-1736). The Fahrenheit scale establishes the freezing and boiling points of water at 32° and 212° respectively. To convert a temperature from the Fahrenheit to the Celsius scale, subtract 32 and multiply by 5/9.

FIRST LAW OF THERMODYNAMICS: A law which states the amount of energy in a system remains constant, and therefore it is impossible to perform work that results in an energy output greater than the energy input. This is the same as the conservation of energy.

HEAT: Internal thermal energy that flows from one body of matter to another.

JOULE: The principal unit of energy—and thus of heat—in the SI or metric system, corresponding to 1 newton-meter (N • m). A joule (J) is equal to 0.7376 foot-pounds in the English system.

KELVIN SCALE: Established by William Thompson, Lord Kelvin (1824-1907), the Kelvin scale measures temperature in relation to absolute zero, or 0K. (Units in the Kelvin system, known as Kelvins, do not include the word or symbol

KEY TERMS CONTINUED

for degree.) The Kelvin scale, which is the system usually favored by scientists, is directly related to the Celsius scale; hence Celsius temperatures can be converted to Kelvins by adding 273.15.

KILOCALORIE: A measure of specific heat capacity in the SI or metric system, equal to the heat that must be added to or removed from 1 kilogram of water to change its temperature by 1°C. As its name suggests, a kilocalorie is 1,000 calories. The dietary Calorie (capital C) with which most people are familiar is the same as the kilocalorie.

KINETIC ENERGY: The energy that an object possesses by virtue of its motion.

MOLECULAR TRANSLATIONAL ENERGY: The kinetic energy in a system produced by the movement of molecules in relation to one another. Thermal energy is a manifestation of molecular translational energy.

SECOND LAW OF THERMODYNAMICS: A law of thermodynamics which states that no system can simply take heat from a source and perform an equivalent amount of work. This is a result of the fact that the natural flow of heat is always from a high-temperature reservoir to a low-temperature reservoir. In the course of such a transfer, some of the heat will always be lost—an example of entropy. The second law is sometimes referred to as "the law of entropy."

SPECIFIC HEAT CAPACITY: The amount of heat that must be added to, or removed from, a unit of mass of a given substance to change its temperature by 1°C. It is typically measured in J/g · °C (joules per gram-degree Celsius). A calorie is the specific heat capacity of 1 gram of water.

SYSTEM: In chemistry and other sciences, the term "system" usually refers to any set of interactions isolated from the rest of the universe. Anything outside of the system, including all factors and forces irrelevant to a discussion of that system, is known as the environment.

THERMAL ENERGY: Heat energy resulting from internal kinetic energy.

THERMAL EQUILIBRIUM: A situation in which two systems have the same temperature. As a result, there is no exchange of heat between them.

THERMODYNAMICS: The study of the relationships between heat, work, and energy.

THERMOMETER: A device that gauges temperature by measuring a temperature-dependent property, such as the expansion of a liquid in a sealed tube, or resistance to electric current.

THERMOMETRIC MEDIUM: A substance whose physical properties change with temperature. A mercury or alcohol thermometer measures such changes.

THIRD LAW OF THERMODYNAMICS: A law of thermodynamics stating that at the temperature of absolute zero, entropy also approaches zero. Zero entropy contradicts the second law of thermodynamics, meaning that absolute zero is therefore impossible to reach.

liding, and the collisions that occur are more energetic. The likelihood is therefore increased that any particular collision will result in the energy necessary to break chemical bonds, and thus bring about the rearrangements in molecules needed for a reaction.

WHERE TO LEARN MORE

"*About Temperature*" (Web site). <http://www.unidata.ucar.edu/staff/blynds/tmp.html> (April 18, 2001).

"*Basic Chemical Thermodynamics*" (Web site). <http://www.sunderland.ac.uk/~hs0bcl/td1.htm> (May 8, 2001).

"*Chemical Thermodynamics*" (Web site). <http://www.shodor.org/UNChem/advanced/thermo/> (May 8, 2001).

Ebbing, Darrell D.; R. A. D. Wentworth; and James P. Birk. *Introductory Chemistry.* Boston: Houghton Mifflin, 1995.

Gardner, Robert. *Science Projects About Methods of Measuring.* Berkeley Heights, NJ: Enslow Publishers, 2000.

MegaConverter 2 (Web site). <http://www.megaconverter.com> (May 7, 2001).

Royston, Angela. *Hot and Cold.* Chicago: Heinemann Library, 2001.

Santrey, Laurence. *Heat.* Illustrated by Lloyd Birmingham. Mahwah, N.J.: Troll Associates, 1985.

Suplee, Curt. *Everyday Science Explained.* Washington, D.C.: National Geographic Society, 1996.

Zumdahl, Steven S. *Introductory Chemistry: A Foundation,* 4th ed. Boston: Houghton Mifflin, 2000.

MASS, DENSITY, AND VOLUME

CONCEPT

Among the physical properties studied by chemists and other scientists, mass is one of the most fundamental. All matter, by definition, has mass. Mass, in turn, plays a role in two properties important to the study of chemistry: density and volume. All of these—mass, density, and volume—are simple concepts, yet in order to work in chemistry or any of the other hard sciences, it is essential to understand these types of measurement. Measuring density, for instance, aids in determining the composition of a given substance, while volume is a necessary component to using the gas laws.

HOW IT WORKS

FUNDAMENTAL PROPERTIES IN RELATION TO VOLUME AND DENSITY

Most qualities of the world studied by scientists can be measured in terms of one or more of four properties: length, mass, time, and electric charge. The volume of a cube, for instance, is a unit of length cubed—that is, length multiplied by "width," which is then multiplied by "height."

Width and height are not, for the purposes of science, distinct from length: they are simply versions of it, distinguished by their orientation in space. Length provides one dimension, while width provides a second perpendicular to the third. Height, perpendicular both to length and width, makes the third spatial dimension—yet all of these are merely expressions of length differentiated according to direction.

VOLUME AND DENSITY DEFINED. Volume, then, is measured in terms of length, and can be defined as the amount of three-dimensional space an object occupies. Volume is usually expressed in cubic units of length—for example, the milliliter (mL), also known as the cubic centimeter (cc), is equal to $6.10237 \cdot 10^{-2}$ in^3. As its name implies, there are 1,000 milliliters in a liter.

Density is the ratio of mass to volume—or, to put its definition in terms of fundamental properties, of mass divided by cubed length. Density can also be viewed as the amount of matter within a given area. In the SI system, density is typically expressed as grams per cubic centimeter (g/cm^3), equivalent to 62.42197 pounds per cubic foot in the English system.

MASS

MASS DEFINED. Though length is easy enough to comprehend, mass is more involved. In his second law of motion, Sir Isaac Newton (1642-1727) defined mass as the ratio of force to gravity. This, of course, is a statement that belongs to the realm of physics; for a chemist, it is more useful—and also accurate—to define mass as the quantity of matter that an object contains.

Matter, in turn, can be defined as physical substance that occupies space; is composed of atoms (or in the case of subatomic particles, is part of an atom); is convertible into energy—and has mass. The form or state of matter itself is not important: on Earth it is primarily observed as a solid, liquid, or gas, but it can also be found (particularly in other parts of the universe) in a fourth state, plasma.

ASTRONAUTS NEIL ARMSTRONG AND BUZZ ALDRIN, THE FIRST MEN TO WALK ON THE MOON, WEIGHED LESS ON THE MOON THAN ON EARTH. THE REASON IS BECAUSE WEIGHT DIFFERS AS A RESPONSE TO THE GRAVITATIONAL PULL OF THE PLANET, MOON, OR OTHER BODY ON WHICH IT IS MEASURED. THUS, A PERSON WEIGHS LESS ON THE MOON, BECAUSE THE MOON POSSESSES LESS MASS THAN EARTH AND EXERTS LESS GRAVITATIONAL FORCE. *(NASA/Roger Ressmeyer/Corbis. Reproduced by permission.)*

The more matter an object contains, the more mass. To refer again to the laws of motion, in his first law, Newton identified inertia: the tendency of objects in motion to remain in motion, or of objects at rest to remain at rest, in a constant velocity unless they are acted upon by some outside force. Mass is a measure of inertia, meaning that the more mass something contains, the more difficult it is to put it into motion, or to stop it from moving.

MASS VS. WEIGHT. Most people who are not scientifically trained tend to think that mass and weight are the same thing, but this is like saying that apples and apple pies are the same. Of course, an apple is an ingredient in an apple pie, but the pie contains something else—actually, a number of other things, such as flour and sugar. In this analogy, mass is equivalent to the apple, and weight the pie, while the acceleration due to gravity is the "something else" in weight.

It is understandable why people confuse mass with weight, since most weight scales provide measurements in both pounds and kilograms. However, the pound (lb) is a unit of weight in the English system, whereas a kilogram

TO DETERMINE WHETHER A PIECE OF GOLD IS GENUINE OR FAKE, ONE MUST MEASURE THE DENSITY OF THE SUB-
STANCE. HERE, A MAN EVALUATES GOLD PIECES IN HONG KONG. *(Christophe Loviny/Corbis. Reproduced by permission.)*

(kg) is a unit of mass in the metric and SI systems. Though the two are relatively convertible
on Earth (1 lb = 0.4536 kg; 1 kg = 2.21 lb), they
are actually quite different.

Weight is a measure of force, which Newton's second law of motion defined as the product of mass multiplied by acceleration. The acceleration component of weight is a result of Earth's
gravitational pull, and is equal to 32 ft (9.8 m)
per second squared. Thus a person's weight varies
according to gravity, and would be different if
measured on the Moon; mass, on the other hand,
is the same throughout the universe. Given its
invariable value, scientists typically speak in
terms of mass rather than weight.

Weight differs as a response to the gravitational pull of the planet, moon, or other body on
which it is measured. Hence a person weighs less
on the Moon, because the Moon possesses less
mass than Earth, and, thus, exerts less gravitational force. Therefore, it would be easier on the
Moon to lift a person from the ground, but it
would be no easier to move that person from a
resting position, or to stop him or her from
moving. This is because the person's mass, and
hence his or her resistance to inertia, has not
changed.

REAL-LIFE APPLICATIONS

ATOMIC MASS UNITS

Chemists do not always deal in large units of
mass, such as the mass of a human body—which,
of course, is measured in kilograms. Instead,
the chemist's work is often concerned with
measurements of mass for the smallest types
of matter: molecules, atoms, and other elementary particles. To measure these even in
terms of grams (0.001 kg) is absurd: a single
atom of carbon, for instance, has a mass of
$1.99 \cdot 10^{-23}$ g. In other words, a gram is about
50,000,000,000,000,000,000,000 times larger
than a carbon atom—hardly a usable comparison.

Instead, chemists use an atom mass unit
(abbreviated amu), which is equal to $1.66 \cdot 10^{-24}$
g. Even so, is hard to imagine determining the
mass of single atoms on a regular basis, so
chemists make use of figures for the average
atomic mass of a particular element. The average
atomic mass of carbon, for instance, is 12.01
amu. As is the case with any average, this means
that some atoms—different isotopes of carbon—
may weigh more or less, but the figure of 12.01

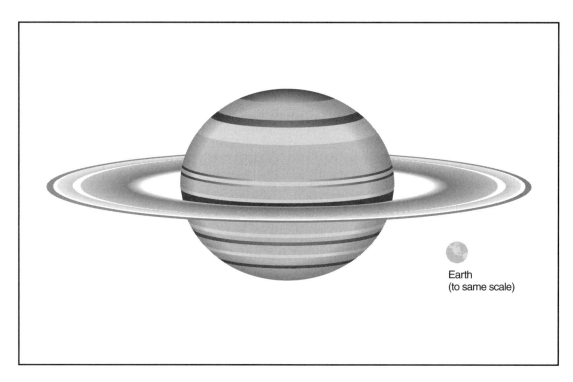

ALTHOUGH SATURN IS MUCH LARGER THAN EARTH, IT IS MUCH LESS DENSE.

amu is still reliable. Some other average atomic mass figures for different elements are as follows:

- Hydrogen (H): 1.008 amu
- Helium (He): 4.003 amu
- Lithium (Li): 6.941 amu
- Nitrogen (N): 14.01 amu
- Oxygen (O): 16.00
- Aluminum (Al): 26.98
- Chlorine (Cl): 35.46 amu
- Gold (Au): 197.0 amu
- Hassium (Hs): [265 amu]

The figure for hassium, with an atomic number of 108, is given in brackets because this number is the mass for the longest-lived isotope. The average value of mass for the molecules in a given compound can also be rendered in terms of atomic mass units: water (H_2O) molecules, for instance, have an average mass of 18.0153 amu. Molecules of magnesium oxide (MgO), which can be extracted from sea water and used in making ceramics, have an average mass much higher than for water: 40.304 amu.

These values are obtained simply by adding those of the atoms included in the molecule: since water has two hydrogen atoms and one oxygen, the average molecular mass is obtained by multiplying the average atomic mass of hydrogen by two, and adding it to the average atomic mass of oxygen. In the case of magnesium oxide, the oxygen is bonded to just one other atom—but magnesium, with an average atomic mass of 24.304, weighs much more than hydrogen.

MOLAR MASS

It is often important for a chemist to know exactly how many atoms are in a given sample, particularly in the case of a chemical reaction between two or more samples. Obviously, it is impossible to count atoms or other elementary particles, but there is a way to determine whether two items—regardless of the elements or compounds involved—have the same number of elementary particles. This method makes use of the figures for average atomic mass that have been established for each element.

If the average atomic mass of the substance is 5 amu, then there should be a very large number of atoms (if it is an element) or molecules (if it is a compound) of that substance having a total mass of 5 grams (g). Similarly, if the average atomic mass of the substance is 7.5 amu, then there should be a very large number of atoms or molecules of that substance having a total mass of 7.5 g. What is needed, clearly, is a very large number by which elementary particles must be multiplied in order to yield a mass whose value in grams is equal to the value, in amu, of its average atomic mass. This is known as Avogadro's number.

AVOGADRO'S NUMBER. The first scientist to recognize a meaningful distinction between atoms and molecules was Italian physicist Amedeo Avogadro (1776-1856). Avogadro maintained that gases consisted of particles—which he called molecules—that in turn consisted of one or more smaller particles. He further reasoned that one liter of any gas must contain the same number of particles as a liter of another gas.

In order to discuss the behavior of molecules, it was necessary to set a large quantity as a basic unit, since molecules themselves are very small. This led to the establishment of what is known as Avogadro's number, equal to 6.022137 $\times$ 10^{23} (more than 600 billion trillion.)

The magnitude of Avogadro's number is almost inconceivable. The same number of grains of sand would cover the entire surface of Earth at a depth of several feet. The same number of seconds, for instance, is about 800,000 times as long as the age of the universe (20 billion years). Avogadro's number—named after the man who introduced the concept of the molecule, but only calculated years after his death—serves a very useful purpose in computations involving molecules.

THE MOLE. To compare two substances containing the same number of atoms or molecules, scientists use the mole, the SI fundamental unit for "amount of substance." A mole (abbreviated mol) is, generally speaking, Avogadro's number of atoms or molecules; however, in the more precise SI definition, a mole is equal to the number of carbon atoms in 12.01 g (0.03 lb) of carbon. Note that, as stated earlier, carbon has an average atomic mass of 12.01 amu. This is no coincidence, of course: multiplication of the average atomic mass by Avogadro's number yields a figure in grams equal to the value of the average atomic mass in amu.

The term "mole" can be used in the same way we use the word "dozen." Just as "a dozen" can refer to twelve cakes or twelve chickens, so "mole" always describes the same number of molecules. Just as one liter of water, or one liter of mercury, has a certain mass, a mole of any given substance has its own particular mass,

expressed in grams. A mole of helium, for instance, has a mass of 4.003 g (0.01 lb), whereas a mole of iron is 55.85 g (0.12 lb) These figures represent the molar mass for each: that is, the mass of 1 mol of a given substance.

Once again, the value of molar mass in grams is the same as that of the average atomic mass in amu. Also, it should be clear that, given the fact that helium weighs much less than air—the reason why helium-filled balloons float—a quantity of helium with a mass of 4.003 g must be a great deal of helium. And indeed, as indicated earlier, the quantity of atoms or molecules in a mole is sufficiently great to make a sample that is large, but still usable for the purposes of study or comparison.

MEASURING VOLUME

Mass, because of its fundamental nature, is sometimes hard to comprehend, and density requires an explanation in terms of mass and volume. Volume, on the other hand, appears to be quite straightforward—and it is, when one is describing a solid of regular shape. In other situations, however, volume measurement is more complicated.

As noted earlier, the volume of a cube can be obtained simply by multiplying length by width by height. There are other means for measuring the volume of other straight-sided objects, such as a pyramid. Still other formulae, which make use of the constant π (roughly equal to 3.14) are necessary for measuring the volume of a cylinder, a sphere, or a cone.

For an object that is irregular in shape, however, one may have to employ calculus—but the most basic method is simply to immerse the object in water. This procedure involves measuring the volume of the water before and after immersion, and calculating the difference. Of course, the object being measured cannot be water-soluble; if it is, its volume must be measured in a non-water-based liquid such as alcohol.

LIQUID AND GAS VOLUME. Measuring liquid volumes is even easier than for solids, given the fact that liquids have no definite shape, and will simply take the shape of the container in which they are placed. Gases are similar to liquids in the sense that they expand to fit their container; however, measurement of gas volume is a more involved process than that used to measure either liquids or solids, because gases are highly responsive to changes in temperature and pressure.

If the temperature of water is raised from its freezing point to its boiling point—from 32°F (0°C) to 212°F (100°C)—its volume will increase by only 2%. If its pressure is doubled from 1 atm (defined as normal air pressure at sea level) to 2 atm, volume will decrease by only 0.01%. Yet if air were heated from 32° to 212°F, its volume would increase by 37%; if its pressure were doubled from 1 atm to 2, its volume would decrease by 50%.

Not only do gases respond dramatically to changes in temperature and pressure, but also, gas molecules tend to be non-attractive toward one another—that is, they tend not to stick together. Hence, the concept of "volume" in relation to a gas is essentially meaningless unless its temperature and pressure are known.

COMPARING DENSITIES

In the discussion of molar mass above, helium and iron were compared, and we saw that the mass of a mole of iron was about 14 times as great as that of a mole of helium. This may seem like a fairly small factor of difference between them: after all, helium floats on air, whereas iron (unless it is arranged in just the right way, for instance, in a tanker) sinks to the bottom of the ocean. But be careful: the comparison of molar mass is only an expression of the mass of a helium atom as compared to the mass of an iron atom. It makes no reference to density, which is the ratio of mass to volume.

Expressed in terms of the ratio of mass to volume, the difference between helium and iron becomes much more pronounced. Suppose, on the one hand, one had a gallon jug filled with iron. How many gallons of helium does it take to equal the mass of the iron? Fourteen? Try again: it takes more than 43,000 gallons of helium to equal the mass of the iron in one gallon jug! Clearly, what this shows is that the density of iron is much, much greater than that of helium.

This, of course, is hardly a surprising revelation; still, it is sometimes easy to get confused by comparisons of mass as opposed to comparisons of density. One might even get tricked by the old elementary-school brain-teaser that goes something like this: "Which is heavier, a ton of feathers or a ton of cannonballs?" Of course neither is heavier, but the trick element in the question

relates to the fact that it takes a much greater volume of feathers (measured in cubic feet, for instance) than of cannonballs to equal a ton.

One of the interesting things about density, as distinguished from mass and volume, is that it has nothing to do with the amount of material. A kilogram of iron differs from 10 kg of iron both in mass and volume, but the density of both samples is the same. Indeed, as discussed below, the known densities of various materials make it possible to determine whether a sample of that material is genuine.

COMPARING DENSITIES. As noted several times, the densities of numerous materials are known quantities, and can be easily compared. Some examples of density, all expressed in terms of grams per cubic centimeter, are listed below. These figures are measured at a temperature of 68°F (20°C), and for hydrogen and oxygen, the value was obtained at normal atmospheric pressure (1 atm):

Comparisons of Densities for Various Substances:

- Oxygen: 0.00133 g/cm³
- Hydrogen: 0.000084 g/cm³
- Ethyl alcohol: 0.79 g/cm³
- Ice: 0.920 g/cm³
- Water: 1.00 g/cm³
- Concrete: 2.3 g/cm³
- Iron: 7.87 g/cm³
- Lead: 11.34 g/cm³
- Gold: 19.32 g/cm³

SPECIFIC GRAVITY

IS IT REALLY GOLD? Note that pure water (as opposed to sea water, which is 3% more dense) has a density of 1.0 g per cubic centimeter. Water is thus a useful standard for measuring the specific gravity of other substances, or the ratio between the density of that substance and the density of water. Since the specific gravity of water is 1.00—also the density of water in g/cm³—the specific gravity of any substance (a number, rather than a number combined with a unit of measure) is the same as the value of its own density in g/cm³.

Comparison of densities make it possible to determine whether a piece of jewelry alleged to be solid gold is really genuine. To determine the answer, one must drop the sample in a beaker of water with graduated units of measure clearly

KEY TERMS

ATOMIC MASS UNIT: An SI unit (abbreviated amu), equal to $1.66 \cdot 10^{-24}$ g, for measuring the mass of atoms.

AVERAGE ATOMIC MASS: A figure used by chemists to specify the mass—in atomic mass units—of the average atom in a large sample. The average atomic mass of carbon, for instance, is 12.01 amu. If a substance is a compound, the average atomic mass of all atoms in a molecule of that substance must be added together to yield the average molecular mass of that substance.

AVOGADRO'S NUMBER: A figure, named after Italian physicist Amedeo Avogadro (1776-1856), equal to $6.022137 \times +0^{23}$. Avogadro's number indicates the number of atoms, molecules, or other elementary particles in a mole.

DENSITY: The ratio of mass to volume—in other words, the amount of matter within a given area. In the SI system, density is typically expressed as grams per cubic centimeter (g/cm^3), equal to 62.42197 pounds per cubic foot in the English system.

MASS: The amount of matter an object contains.

MATTER: Physical substance that occupies space, has mass, is composed of atoms (or in the case of subatomic particles, is part of an atom), and is convertible to energy.

MILLILITER: One of the most commonly used units of volume in the SI system of measures. The milliliter (abbreviated mL), also known as a cubic centimeter

(cc), is equal to $6.10237 \cdot 10^{-2}$ cubic inches in the English system. As the name implies, there are 1,000 milliliters in a liter.

MOLAR MASS: The mass, in grams, of 1 mole of a given substance. The value in grams of molar mass is always equal to the value, in atomic mass units, of the average atomic mass of that substance: thus, carbon has a molar mass of 12.01 g, and an average atomic mass of 12.01 amu.

MOLE: The SI fundamental unit for "amount of substance." A mole is, generally speaking, Avogadro's number of atoms, molecules, or other elementary particles; however, in the more precise SI definition, a mole is equal to the number of carbon atoms in 12.01 g of carbon.

SPECIFIC GRAVITY: The density of an object or substance relative to the density of water; or more generally, the ratio between the densities of two objects or substances. Since the specific gravity of water is 1.00—also the density of water in g/cm^3—the specific gravity of any substance is the same as the value of its own density in g/cm^3. Specific gravity is simply a number, without any unit of measure.

VOLUME: The amount of three-dimensional space an object occupies. Volume is usually expressed in cubic units of length—for instance, the milliliter.

WEIGHT: The product of mass multiplied by the acceleration due to gravity (32 ft or 9.8 m/sec^2). A pound is a unit of weight, whereas a kilogram is a unit of mass.

marked. Suppose the item has a mass of 10 g. The density of gold is 19 g/cm^3, and because density is equal to mass divided by volume, the volume of water displaced should be equal to the mass divided by the density. The latter figure is equal to 10 g divided by 19 g/cm^3, or 0.53 ml. Suppose that instead, the item displaced 0.88 ml of water. Clearly it is not gold, but what is it?

Given the figures for mass and volume, its density is equal to 11.34 g/cm^3—which happens to be the density of lead. If, on the other hand, the amount of water displaced were somewhere between the values for pure gold and pure lead, one could calculate what portion of the item was gold and which lead. It is possible, of course, that it could contain some other metal, but given the high specific gravity of lead, and the fact that its density is relatively close to that of gold, lead is a favorite gold substitute among jewelry counterfeiters.

SPECIFIC GRAVITY AND THE DENSITIES OF PLANETS. Most rocks near the surface of Earth have a specific gravity somewhere between 2 and 3, while the specific gravity of the planet itself is about 5. How do scientists know that the density of Earth is around 5 g/cm^3? The computation is fairly simple, given the fact that the mass and volume of the planet are known. And given the fact that most of what lies close to Earth's surface—sea water, soil, rocks—has a specific gravity well below 5, it is clear that Earth's interior must contain high-density materials, such as nickel or iron. In the same way, calculations regarding the density of other objects in the Solar System provide a clue as to their interior composition.

This brings the discussion back around to a topic raised much earlier in this essay, when comparing the weight of a person on Earth versus that person's weight on the Moon. It so happens that the Moon is smaller than Earth, but that is not the reason it exerts less gravitational pull: as noted earlier, the gravitational force a planet, moon, or other body exerts is related to its mass, not its size.

It so happens, too, that Jupiter is much larger than Earth, and that it exerts a gravitational pull much greater than that of Earth. This is because it has a mass many times as great as Earth's. But what about Saturn, the second-largest planet in the Solar System? In size it is only about 17% smaller than Jupiter, and both are much, much larger than Earth. Yet a person would weigh much less on Saturn than on Jupiter, because Saturn has a mass much smaller than Jupiter's. Given the close relation in size between the two planets, it is clear that Saturn has a much lower density than Jupiter, or in fact even Earth: the great ringed planet has a specific gravity of less than 1.

WHERE TO LEARN MORE

Chahrour, Janet. *Flash! Bang! Pop! Fizz!: Exciting Science for Curious Minds.* Illustrated by Ann Humphrey Williams. Hauppauge, NY: Barron's, 2000.

"Density and Specific Gravity" (Web site). <http://www.tpub.com/fluid/ch1e.htm> (March 27, 2001).

"Density, Volume, and Cola" (Web site). <http://student.biology.arizona.edu/sciconn/density/density_coke.html> (March 27, 2001).

Ebbing, Darrell D.; R. A. D. Wentworth; and James P. Birk. *Introductory Chemistry.* Boston: Houghton Mifflin, 1995.

"The Mass Volume Density Challenge" (Web site). <http://science-math-technology.com/mass_volume_density.html> (March 27, 2001).

"MegaConverter 2" (Web site). <http://www.megaconverter.com> (May 7, 2001).

"Metric Density and Specific Gravity" (Web site). <http://www.essex1.com/people/speer/density.html> (March 27, 2001).

Robson, Pam. *Clocks, Scales and Measurements.* New York: Gloucester Press, 1993.

"Volume, Mass, and Density" (Web site). <http://www.nyu.edu/pages/mathmol/modules/water/density_intro.html> (March 27, 2001).

Zumdahl, Steven S. *Introductory Chemistry: A Foundation,* 4th ed. Boston: Houghton Mifflin, 2000.

MATTER

PROPERTIES OF MATTER

GASES

PROPERTIES OF MATTER

CONCEPT

Matter is physical substance that occupies space, has mass, is composed of atoms—or, in the case of subatomic particles, is part of an atom—and is convertible to energy. On Earth, matter appears in three clearly defined forms—solid, liquid, and gas—whose varying structural characteristics are a function of the speeds at which its molecules move in relation to one another. A single substance may exist in any of the three phases: liquid water, for instance, can be heated to become steam, a vapor; or, when sufficient heat is removed from it, it becomes ice, a solid. These are merely physical changes, which do not affect the basic composition of the substance itself: it is still water. Matter, however, can and does undergo chemical changes, which (as with the various states or phases of matter) are an outcome of activity at the atomic and molecular level.

HOW IT WORKS

MATTER AND ENERGY

One of the characteristics of matter noted in its definition above is that it is convertible to energy. We rarely witness this conversion, though as Albert Einstein (1879-1955) showed with his Theory of Relativity, it occurs in a massive way at speeds approaching that of light.

Einstein's famous formula, $E = mc^2$, means that every item possesses a quantity of energy equal to its mass multiplied by the squared speed of light. Given the fact that light travels at 186,000 mi (299,339 km) per second, the quanti-

ties of energy available from even a tiny object traveling at that speed are enormous indeed. This is the basis for both nuclear power and nuclear weaponry, each of which uses some of the smallest particles in the known universe to produce results that are both amazing and terrifying.

Even in everyday life, it is still possible to observe the conversion of mass to energy, if only on a very small scale. When a fire burns—that is, when wood experiences combustion in the presence of oxygen, and undergoes chemical changes—a tiny fraction of its mass is converted to energy. Likewise, when a stick of dynamite explodes, it too experiences chemical changes and the release of energy. The actual amount of energy released is, again, very small: for a stick of dynamite weighing 2.2 lb (1 kg), the portion of its mass that "disappears" is be equal to 6 parts out of 100 billion.

Actually, none of the matter in the fire or the dynamite blast disappears: it simply changes forms. Most of it becomes other types of matter—perhaps new compounds, and certainly new mixtures of compounds. A very small part, as we have seen, becomes energy. One of the most fundamental principles of the universe is the conservation of energy, which holds that within a system isolated from all other outside factors, the total amount of energy remains the same, though transformations of energy from one form to another take place. In this situation, some of the energy remains latent, or "in reserve" as matter, while other components of the energy are released; yet the total amount of energy remains the same.

PHYSICAL AND CHEMICAL CHANGES

In discussing matter—as, for instance, in the context of matter transforming into energy—one may speak in physical or chemical terms, or both. Generally speaking, physicists study physical properties and changes, while chemists are concerned with chemical processes and changes.

A physicist views matter in terms of its mass, temperature, mechanical properties (for example, elasticity); electrical conductivity; and other structural characteristics. The chemical makeup of matter, on the other hand, is of little concern to a physicist. For instance, in analyzing a fire or an explosion, the physicist is not concerned with the interactions of combustible or explosive materials and oxygen. The physicist's interest, rather, is in questions such as the amount of heat in the fire, the properties of the sound waves emitted in the explosion of the dynamite, and so on.

The changes between different states or phases of matter, as they are discussed below, are physical changes. If water boils and vaporizes as steam, it is still water; likewise if it freezes to become solid ice, nothing has changed with regard to the basic chemical structure of the H_2O molecules that make up water. But if water reacts with another substance to form a new compound, it has undergone chemical change. Likewise, if water molecules experience electrolysis, a process in which electric current is used to decompose H_2O into molecules of H_2 and O_2, this is also a chemical change.

Similarly, a change from matter to energy, while it is also a physical change, typically involves some chemical or nuclear process to serve as "midwife" to that change. Yet physical and chemical changes have at least one thing in common: they can be explained in terms of behavior at the atomic or molecular level. This is true of many physical processes—and of all chemical ones.

ATOMS

In his highly readable *Six Easy Pieces*—a work that includes considerable discussion of chemistry as well as physics—the great American physicist Richard Feynman (1918-1988) asked, "If, in some cataclysm, all of scientific knowledge were to be destroyed, and only one sentence passed on to the next generations of creatures, what statement would contain the most information in the fewest words?"

The answer he gave was this: "I believe it is the atomic hypothesis (or the atomic fact, or whatever you wish to call it) that all things are made of atoms—little articles that move around in perpetual motion, attracting each other when they are a little distance apart, but repelling upon being squeezed into one another. In that sentence, you will see, there is an enormous amount of information about the world, if just a little imagination and thinking are applied."

Indeed, what Feynman called the "atomic hypothesis" is one of the most important keys to understanding both physical and chemical changes. The behavior of particles at the atomic level has a defining role in the shape of the world studied by the sciences, and an awareness of this behavior makes it easier to understand physical processes, such as changes of state between solid, liquid, and gas; chemical processes, such as the formation of new compounds; and other processes, such as the conversion of matter to energy, which involve both physical and chemical changes. Only when one comprehends the atomic structure of matter is it possible to move on to the chemical elements that are the most basic materials of chemistry.

STRUCTURE OF THE ATOM. As Feynman went on to note, atoms are so tiny that if an apple were magnified to the size of Earth, the atoms in it would each be about the size of a regular apple. Clearly, atoms and other atomic particles are far too small to be glimpsed even by the most highly powered optical microscope. Yet physicists and other scientists are able to study the behavior of atoms, and by doing so, they are able to form a picture of what occurs at the atomic level.

An atom is the fundamental particle in a chemical element. The atom is not, however, the smallest particle in the universe: atoms are composed of subatomic particles, including protons, neutrons, and electrons. These are distinguished from one another in terms of electric charge: as with the north and south poles of magnets, positive and negative charges attract one another, but like charges repel. (In fact, magnetism is simply a manifestation of a larger electromagnetic force that encompasses both electricity and magnetism.)

Clustered at the center, or nucleus, of the atom are protons, which are positively charged, and neutrons, which exert no charge. Spinning around the nucleus are electrons, which exert a negative charge. The vast majority of the atom's mass is made up by the protons and neutrons, which have approximately the same mass; that of the electron is much smaller. If an electron had a mass of 1—not a unit, but simply a figure used for comparison—the mass of the proton would be 1,836, and of the neutron 1,839.

Atoms of the same element always have the same number of protons, and since this figure is unique for a given element, each element is assigned an atomic number equal to the number of protons in its nucleus. Two atoms may have the same number of protons, and thus be of the same element, yet differ in their number of neutrons. Such atoms are called isotopes.

The number of electrons is usually the same as the number of protons, and thus atoms have a neutral charge. In certain situations, however, the atom may lose or gain one or more electrons and acquire a net charge, becoming an ion. But electric charge, like energy, is conserved, and the electrons are not "lost" when an atom becomes an ion: they simply go elsewhere.

It is useful, though far from precise, to compare the interior of an atom to a planet spinning very quickly around a sun. If the nucleus were our own Sun, then the electrons spinning at the edge of the atom would be on an orbit somewhere beyond Mars: in other words, the ratio between the size of the nucleus and the furthest edge of the atom is like that between the Sun's diameter and an orbital path about 80 million miles beyond Mars.

One of many differences between an atom and a solar system, however, is the fact that the electrons are spinning around the nucleus at a relative rate of motion much, much greater than any planet is revolving around the Sun. Furthermore, what holds the atom together is not gravitational force, as in the Solar System, but electromagnetic force. A final and critical difference is the fact that electrons move in much more complex orbital patterns than the elliptical paths that planets make in their movement around the Sun.

MOLECULES

Though an atom is the fundamental unit of matter, most of the substances people encounter in

A COAL GASIFICATION PLANT. COAL GASIFICATION MAKES IT POSSIBLE TO BURN "CLEAN" COAL. *(Roger Ressmeyer/Corbis. Reproduced by permission.)*

the world are not pure elements such as oxygen or iron. They are compounds in which atoms of more than one element join—usually in molecules. All molecules are composed of more than one atom, but not necessarily of more than one element: oxygen, for instance, generally appears in the form of molecules in which two oxygen atoms are bonded. Because of this, pure oxygen is represented by the chemical symbol O_2, as opposed to the symbol for the element oxygen, which is simply O.

One of the most well-known molecular forms in the world is water, or H_2O, composed of two hydrogen atoms and one oxygen atom. The arrangement is extremely precise and never varies: scientists know, for instance, that the two hydrogen atoms join the oxygen atom (which is much larger than the hydrogen atoms) at an angle of 105°3'. Since the oxygen atom is much larger than the two hydrogens, its shape can be compared to a basketball with two softballs attached.

Other molecules are much more complex than those of water, and some are much, much more complex, a fact reflected in the sometimes lengthy names and complicated symbolic representations required to identify their chemical components. On the other hand, not all materials are made up of molecules: salt, for instance, is an ionic solid, as discussed below.

QUANTIFYING ATOMS AND MOLECULES

The nucleus of an atom is about 10^{-13} cm in diameter, and the diameter of the entire atom is about 10^{-8} cm—about 0.0000003937 in. Obviously, special units are required for describing the size of atoms, and usually measurements are provided in terms of the angstrom, equal to 10^{-10} m, or 10^{-8} cm. To put this on some sort of imaginable scale, there are 10 million angstroms in a millimeter.

Measuring the spatial dimensions of an atom, however, is not as important as measuring its mass—and naturally, the mass of an atom is also almost inconceivably small. For instance, it takes about $5.0 \cdot 10^{23}$ carbon atoms to equal just one gram of mass. Again, the numbers boggle the mind, but the following may put this into perspective. We have already established just how tiny an angstrom is; now consider the following. If $5.0 \cdot 10^{23}$ angstrom lengths were laid end to end, they would stretch for a total of about 107,765 round trips from Earth to the Sun!

It is obvious, then, that an entirely different unit should be used for measuring the mass of an atom, and for this purpose, chemists and other scientists use an atom mass unit (abbreviated amu). The latter is equal to $1.66 \cdot 10^{-24}$ g. Even so, scientists can hardly be expected to be constantly measuring the mass of individual atoms; rather, they rely on figures determined for the average atomic mass of a particular element.

Average atomic mass figures range from 1.008 amu for hydrogen to over 250 amu for elements of very high atomic numbers. Figures for average atomic mass can be used to determine the average mass of a molecule as well, simply by combining the average atomic mass figures for each atom the molecule contains. A water molecule, for instance, has an average mass equal to the average atomic mass of hydrogen multiplied by two, and added to the average atomic mass of oxygen.

AVOGADRO'S NUMBER AND THE MOLE. Just as using average atomic mass is much more efficient than measuring the mass of individual atoms or molecules, scientists need a useful means for comparing atoms or molecules of different substances—and for doing so in such a way that they know they are analyzing equal numbers of particles. This cannot be done in terms of mass, because the number of atoms in each sample would vary: a gram of hydrogen, for instance, would contain about 12 times as many atoms as a gram of carbon, which has an average atomic mass of 12.01 amu. What is needed, instead, is a way to designate a certain number of atoms or molecules, such that accurate comparisons are possible.

In order to do this, scientists make use of a figure known as Avogadro's number. Named after Italian physicist Amedeo Avogadro (1776-1856), it is equal to 6.022137×10^{23} Earlier, we established the almost inconceivable scale represented by the figure $5.0 \cdot 10^{23}$; here we are confronted with a number 20% larger. But Avogadro's number, which is equal to 6,022,137 followed by 17 zeroes, is more than simply a mind-boggling series of digits.

In general terms, Avogadro's number designates the quantity of molecules (and sometimes atoms, if the substance in question is an element that, unlike oxygen, appears as single atoms) in a mole (abbreviated mol). A mole is the SI unit for "amount of substance," and is defined precisely as the number of carbon atoms in 12.01 g of carbon. It is here that the value of Avogadro's number becomes clear: as noted, carbon has an average atomic mass of 12.01 amu, and multiplication of the average atomic mass by Avogadro's number yields a figure in grams equal to the value of the average atomic mass in atomic mass units.

MOLAR MASS AND DENSITY. By comparison, a mole of helium has a molar mass of 4.003 g (0.01 lb) The molar mass of iron (that is, the mass of 1 mole of iron) is 55.85 g (0.12 lb) Note that there is not a huge ratio of difference between the molar mass of iron and that of helium: iron has a molar mass about 14 times greater. This, of course, seems very small in light of the observable differences between iron and helium: after all, who ever heard of a balloon filled with iron, or a skyscraper with helium girders?

The very striking differences between iron and helium, clearly, must come from something other than the molar mass differential between them. Of course, a mole of iron contains the same number of atoms as a mole of helium, but this says nothing about the relative density of the two substances. In terms of volume—that is, the amount of space that something occupies—the difference is much more striking: the volume of a mole of helium is about 43,000 times as large as that of a mole of iron.

What this tells us is that the densities of iron and helium—the amount of mass per unit of volume—are very different. This difference in density is discussed in the essay on Mass, Density, and Volume; here the focus is on a larger judgment that can be formed by comparing the two densities. Helium, of course, is almost always in the form of a gas: to change it to a solid requires a temperature near absolute zero. And iron is a solid, meaning that it only turns into a liquid at extraordinarily high temperatures. These differences in overall structure can, in turn, be attributed to the relative motion, attraction, and energy of the molecules in each.

Molecular Attraction and Motion

At the molecular level, every item of matter in the world is in motion, and the rate of that motion is a function of the attraction between molecules. Furthermore, the rate at which molecules move in relation to one another determines phase of matter—that is, whether a particular item can be described as solid, liquid, or gas. The movement of molecules generates kinetic energy, or the energy of movement, which is manifested as thermal energy—what people call "heat" in ordinary language. (The difference between thermal energy and heat is explained in the essay on Temperature and Heat.) In fact, thermal energy is the result of molecules' motion relative to one another: the faster they move, the greater the kinetic energy, and the greater the "heat."

When the molecules in a material move slowly—merely vibrating in place—they exert a strong attraction toward one another, and the material is called a solid. Molecules of liquid, by contrast, move at moderate speeds and exert a moderate attraction. A material substance whose molecules move at high speeds, and therefore exert little or no attraction, is known as a gas. In short, the weaker the attraction, the greater the rate of relative motion—and the greater the amount of thermal energy the object contains.

REAL-LIFE APPLICATIONS

Types of Solids

Particles of solids resist attempts to compress them, or push them together, and because of their close proximity, solid particles are fixed in an orderly and definite pattern. As a result, a solid usually has a definite volume and shape.

A crystalline solid is a type of solid in which the constituent parts are arranged in a simple, definite geometric pattern that is repeated in all directions. But not all crystalline solids are the same. Table salt is an example of an ionic solid: a form of crystalline solid that contains ions. When mixed with a solvent such as water, ions from the salt move freely throughout the solution, making it possible to conduct an electric current.

Regular table sugar (sucrose) is a molecular solid, or one in which the molecules have a neutral electric charge—that is, there are no ions present. Therefore, a solution of water and sugar would not conduct electricity. Finally, there are crystalline solids known as atomic solids, in which atoms of one element bond to one another. Examples include diamonds (made of pure carbon), silicon, and all metals.

Other solids are said to be amorphous, meaning that they possess no definite shape. Amorphous solids—an example of which is clay—either possess very tiny crystals, or consist of several varieties of crystal mixed randomly. Still other solids, among them glass, do not contain crystals.

Freezing and Melting

VIBRATIONS AND FREEZING. Because of their slow movement in relation to one another, solid particles exert strong attractions; yet as slowly as they move, solid particles do move—as is the case with all forms of matter at the atomic level. Whereas the particles in a liquid or gas move fast enough to be in relative motion with regard to one another, however, solid particles merely vibrate from a fixed position.

As noted earlier, the motion and attraction of particles in matter has a direct effect on thermal energy, and thus on heat and temperature. The cooler the solid, the slower and weaker the vibrations, and the closer the particles are to one another. Thus, most types of matter contract when freezing, and their density increases. Absolute zero, or 0K on the Kelvin scale of temperature—equal to –459.67°F (–273°C)—is the point at which vibration virtually ceases.

Note that the vibration virtually stops, but does not totally stop. In fact, as established in the third law of thermodynamics, absolute zero is impossible to achieve: thus, the relative motion of molecules never ceases. The lowest temperature actually achieved, at a Finnish nuclear laboratory in 1993, is $2.8 \cdot 10^{-10}$ K, or 0.00000000028K—still above absolute zero.

UNUSUAL CHARACTERISTICS OF SOLID AND LIQUID WATER.

The behavior of water when frozen is interesting and exceptional. Above 39.2°F (4°C) water, like most substances, expands when heated. In other words, the molecules begin moving further apart as expected, because—in this temperature range, at least—water behaves like other substances, becoming "less solid" as the temperature increases.

Between 32°F (0°C) and 39.2°F (4°C), however, water actually contracts. In this temperature range, it is very "cold" (that is, it has relatively little heat), but it is not frozen. The density of water reaches its maximum—in other words, water molecules are as closely packed as they can be—at 39.2°F; below that point, the density starts to decrease again. This is highly unusual: in most substances, the density continues to increase with lowered temperatures, whereas water is actually most dense slightly above the freezing point.

Below the freezing point, then, water expands, and therefore when water in pipes freezes, it may increase in volume to the point where it bursts the pipe. This is also the reason why ice floats on water: its weight is less than that of the water it has displaced, and thus it is buoyant. Additionally, the buoyant qualities of ice atop very cold water helps explain the behavior of lake water in winter; although the top of a lake may freeze, the entire lake rarely freezes solid—even in the coldest of inhabited regions.

Instead of freezing from the bottom up, as it would if ice were less buoyant than the water, the lake freezes from the top down—an important thing to remember when ice-fishing! Furthermore, water in general (and ice in particular) is a poor conductor of heat, and thus little of the heat from the water below it escapes. Therefore, the lake does not freeze completely—only a layer at the top—and this helps preserve animal and plant life in the body of water.

MELTING. When heated, particles begin to vibrate more and more, and therefore move further apart. If a solid is heated enough, it loses its rigid structure and becomes a liquid. The temperature at which a solid turns into a liquid is called the melting point, and melting points are different for different substances. The melting point of a substance, incidentally, is the same as its freezing point: the difference is a matter of orientation—that is, whether the process is one of a solid melting to become a liquid, or of a liquid freezing to become a solid.

The energy required to melt 1 mole of a solid substance is called the molar heat of fusion. It can be calculated by the formula $Q = sm\delta T$, where Q is energy, s is specific heat capacity, m is mass, and δT means change in temperature. (In the symbolic language often employed by scientists, the Greek letter δ, or delta, stands for "change in.") Specific heat capacity is measured in units of J/g • °C (joules per gram-degree Celsius), and energy in joules or kilojoules (kJ)—that is, 1,000 joules.

In melting, all the thermal energy in a solid is used in breaking up the arrangement of crystals, called a lattice. This is why water melted from ice does not feel any warmer than the ice did: the thermal energy has been expended, and there is none left over for heating the water. Once all the ice is melted, however, the absorbed energy from the particles—now moving at much greater speeds than when the ice was in a solid state—causes the temperature to rise.

For the most part, solids composed of particles with a higher average atomic mass require more energy—and hence higher temperatures—to induce the vibrations necessary for melting. Helium, with an average atomic mass of 4.003 amu, melts or freezes at an incredibly low temperature: –457.6°F (–272°C), or close to absolute zero. Water, for which, as noted earlier, the average atomic mass is the sum of the masses for its two hydrogen atoms and one oxygen atom, has an average molecular mass of 18.016 amu. Ice melts (or water freezes) at much higher temperatures than helium: 32°F (0°C). Copper, with an average atomic mass of 63.55 amu, melts at much, much higher temperatures than water: 1,985°F (1,085°C).

LIQUIDS

The particles of a liquid, as compared to those of a solid, have more energy, more motion, and—generally speaking—less attraction to one another. The attraction, however, is still fairly strong: thus, liquid particles are in close enough proximity that the liquid resists attempts at compression.

On the other hand, their arrangement is loose enough that the particles tend to move around one another rather than simply vibrate in place the way solid particles do. A liquid is therefore not definite in shape. Due to the fact that the particles in a liquid are farther apart than those of a solid, liquids tend to be less dense than solids. The liquid phase of a substance thus tends to be larger in volume than its equivalent in solid form. Again, however, water is exceptional in this regard: liquid water actually takes up less space than an equal mass of frozen water.

BOILING

When a liquid experiences an increase in temperature, its particles take on energy and begin to move faster and faster. They collide with one another, and at some point the particles nearest the surface of the liquid acquire enough energy to break away from their neighbors. It is at this point that the liquid becomes a gas or vapor.

As heating continues, particles throughout the liquid begin to gain energy and move faster, but they do not immediately transform into gas. The reason is that the pressure of the liquid, combined with the pressure of the atmosphere above the liquid, tends to keep particles in place. Those particles below the surface, therefore, remain where they are until they acquire enough energy to rise to the surface.

The heated particle moves upward, leaving behind it a hollow space—a bubble. A bubble is not an empty space: it contains smaller trapped particles, but its small mass, relative to that of the liquid it disperses, makes it buoyant. Therefore, a bubble floats to the top, releasing its trapped particles as gas or vapor. At that point, the liquid is said to be boiling.

THE EFFECT OF ATMOSPHERIC PRESSURE. The particles thus have to overcome atmospheric pressure as they rise, which means that the boiling point for any liquid depends in part on the pressure of the surrounding air. Normal atmospheric pressure (1 atm) is equal to 14 lb/in^2 (1.013 x 10^5 Pa), and is measured at sea level. The greater the altitude, the less the air pressure, because molecules of air—since air is a gas, and therefore its particles are fast-moving and non-attractive—respond less to Earth's gravitational pull. This is why airplanes require pressurized cabins to maintain an adequate oxygen supply; but even at altitudes much lower than the flight path of an airplane, differences in air pressure are noticeable.

It is for this reason that cooking instructions often vary with altitude. Atop Mt. Everest, Earth's highest peak at about 29,000 ft (8,839 m) above sea level, the pressure is approximately one-third of normal atmospheric pressure. Water boils at a much lower temperature on Everest than it does elsewhere: 158°F (70°C), as opposed to 212°F (100°C) at sea level. Of course, no one lives on the top of Mt. Everest—but people do live in Denver, Colorado, where the altitude is 5,577 ft (1,700 m) and the boiling point of water is 203°F (95°C).

Given the lower boiling point, one might assume that food would cook faster in Denver than in New York, Los Angeles, or in any city close to sea level. In fact, the opposite is true: because heated particles escape the water so much faster at high altitudes, they do not have time to acquire the energy needed to raise the temperature of the water. It is for this reason that a recipe may include a statement such as "at altitudes above XX feet, add XX minutes to cooking time."

If lowered atmospheric pressure means a lowered boiling point, what happens in outer space, where there is no atmospheric pressure? Liquids boil at very, very low temperatures. This is one of the reasons why astronauts have to wear pressurized suits: if they did not, their blood would boil—even though space itself is incredibly cold.

LIQUID TO GAS AND BACK AGAIN. Note that the process of changing a liquid to a gas is similar to that which occurs when a solid changes to a liquid: particles gain heat and therefore energy, begin to move faster, break free from one another, and pass a certain threshold into a new phase of matter. And just as the freezing and melting point for a given substance are the same temperature—the only difference being one of orientation—the boiling point of a liquid transforming into a gas is the same as the condensation point for a gas turning into a liquid.

The behavior of water in boiling and condensation makes possible distillation, one of the principal methods for purifying sea water in various parts of the world. First the water is boiled, then it is allowed to cool and condense, thus forming water again. In the process, the water separates from the salt, leaving it behind in the form of brine. A similar separation takes place when salt water freezes: because salt, like most crystalline solids, has a much lower freezing point than water, very little of it remains joined to the water in ice. Instead, the salt takes the form of a briny slush.

GASES

A liquid that is vaporized, or any substance that exists normally as a gas, is quite different in physical terms from a solid or a liquid. This is illustrated by the much higher energy component in the molar heat of vaporization, or the amount of energy required to turn 1 mole of a liquid into a gas.

Consider, for instance, what happens to water when it experiences phase changes. Assuming that heat is added at a uniform rate, when ice reaches its melting point, there is only a relatively small period of time when the H_2O is composed of both ice and liquid. But when the liquid reaches its boiling point, the water is present both as a liquid and a vapor for a much longer period of time. In fact, it takes almost seven times as much energy to turn liquid water into pure steam than it does to turn ice into purely liquid water. Thus, the molar heat of fusion for water is 6.02 kJ/mol, while the molar heat of vaporization is 40.6 kJ/mol.

Although liquid particles exert a moderate attraction toward one another, particles in a gas (particularly a substance that normally exists as a gas at ordinary temperatures on Earth) exert little to no attraction. They are thus free to move, and to move quickly. The overall shape and arrangement of gas is therefore random and indefinite—and, more importantly, the motion of gas particles provides much greater kinetic energy than is present in any other major form of matter on Earth.

The constant, fast, and random motion of gas particles means that they are regularly collid-

ing and thereby transferring kinetic energy back and forth without any net loss of energy. These collisions also have the overall effect of producing uniform pressure in a gas. At the same time, the characteristics and behavior of gas particles indicate that they tend not to remain in an open container. Therefore, in order to have any pressure on a gas—other than normal atmospheric pressure—it is necessary to keep it in a closed container.

THE PHASE DIAGRAM

The vaporization of water is an example of a change of phase—the transition from one phase of matter to another. The properties of any substance, and the points at which it changes phase, are plotted on what is known as a phase diagram. The phase diagram typically shows temperature along the x-axis, and pressure along the y-axis.

For simple substances, such as water and carbon dioxide (CO_2), the solid form of the substance appears at a relatively low temperature and at pressures anywhere from zero upward. The line between solids and liquids, indicating the temperature at which a solid becomes a liquid at any pressure above a certain level, is called the fusion curve. Though it appears to be a more or less vertical line, it is indeed curved, indicating that at high pressures, a solid well below the normal freezing point may be melted to create a liquid.

Liquids occupy the area of the phase diagram corresponding to relatively high temperatures and high pressures. Gases or vapors, on the other hand, can exist at very low temperatures, but only if the pressure is also low. Above the melting point for the substance, gases exist at higher pressures and higher temperatures. Thus, the line between liquids and gases often looks almost like a 45° angle. But it is not a straight line, as its name, the vaporization curve, implies. The curve of vaporization demonstrates that at relatively high temperatures and high pressures, a substance is more likely to be a gas than a liquid.

THE CRITICAL POINT. There are several other interesting phenomena mapped on a phase diagram. One is the critical point, found at a place of very high temperature and pressure along the vaporization curve. At the critical point, high temperatures prevent a liquid from remaining a liquid, no matter how high the pressure.

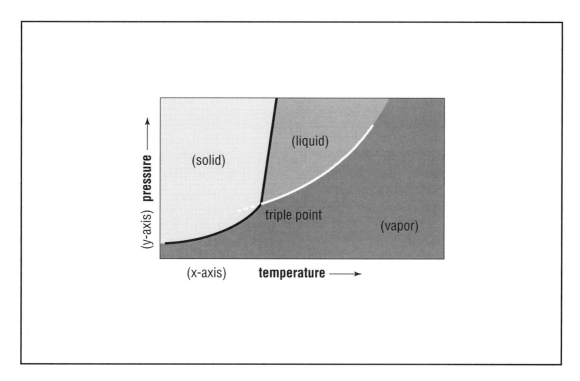

A PHASE DIAGRAM FOR WATER.

At the same time, the pressure causes gas beyond that point to become increasingly more dense, but due to the high temperatures, it does not condense into a liquid. Beyond the critical point, the substance cannot exist in anything other than the gaseous state. The temperature component of the critical point for water is 705.2°F (374°C)—at 218 atm, or 218 times ordinary atmospheric pressure. For helium, however, critical temperature is just a few degrees above absolute zero. This is, in part, why helium is rarely seen in forms other than a gas.

THE SUBLIMATION CURVE. Another interesting phenomenon is the sublimation curve, or the line between solid and gas. At certain very low temperatures and pressures, a substance may experience sublimation, meaning that a gas turns into a solid, or a solid into a gas, without passing through a liquid stage.

A well-known example of sublimation occurs when "dry ice," made of carbon dioxide, vaporizes at temperatures above (−78.5°C). Carbon dioxide is exceptional, however, in that it experiences sublimation at relatively high pressures that occur in everyday life: for most substances, the sublimation point transpires at such a low pressure point that it is seldom witnessed outside of a laboratory.

THE TRIPLE POINT. The phenomenon known as the triple point shows how an ordinary substance such as water or carbon dioxide can actually be a liquid, solid, and vapor—all at once. Most people associate water as a gas or vapor (that is, steam) with very high temperatures. Yet, at a level far below normal atmospheric pressure, water can be a vapor at temperatures as low as −4°F (−20 °C). (All of the pressure values in the discussion of water at or near the triple point are far below atmospheric norms: the pressure at which water turns into a vapor at −4°F, for instance, is about 0.001 atm.)

Just as water can exist as a vapor at low temperatures and low pressures, it is also possible for water at temperatures below freezing to remain liquid. Under enough pressure, ice melts and is thereby transformed from a solid to a liquid, at temperatures below its normal freezing point. On the other hand, if the pressure of ice falls below a very low threshold, it will sublimate.

The phase diagram of water shows a line between the solid and liquid states that is almost, but not quite, exactly perpendicular to the x-axis. But in fact, it is a true fusion curve: it slopes slightly upward to the left, indicating that solid ice turns into water with an increase of pressure. Below a certain level of pressure is the vaporization curve, and where the fusion curve intersects the vaporization curve, there is a place called the triple point. Just below freezing, in conditions equivalent to about 0.007 atm, water is a solid, liquid, and vapor all at once.

OTHER STATES OF MATTER

PLASMA. Principal among states of matter other than solid, liquid, and gas is plasma, which is similar to gas. (The term "plasma," when referring to the state of matter, has nothing to do with the word as it is often used, in reference to blood plasma.) As with gas, plasma particles collide at high speeds—but in plasma the speeds are even greater, and the kinetic energy levels even higher.

The speed and energy of these collisions is directly related to the underlying property that distinguishes plasma from gas. So violent are the collisions between plasma particles that electrons are knocked away from their atoms. As a result, plasma does not have the atomic structure typical of a gas; rather, it is composed of positive ions and electrons. Plasma particles are thus electrically charged, and therefore greatly influenced by electric and magnetic fields.

Formed at very high temperatures, plasma is found in stars. The reaction between plasma and atomic particles in the upper atmosphere is responsible for the aurora borealis, or "northern lights." Though found on Earth only in very small quantities, plasma—ubiquitous in other parts of the universe—may be the most plentiful of all the states of matter.

QUASI-STATES. Among the quasi-states of matter discussed by scientists are several terms describing the structure in which particles are joined, rather than the attraction and relative movement of those particles. Thus "crystalline," "amorphous," and "glassy" are all terms to describe what may be individual states of matter; so too is "colloidal."

A colloid is a structure intermediate in size between a molecule and a visible particle, and it has a tendency to be dispersed in another medium—the way smoke, for instance, is dispersed in air. Brownian motion describes the behavior of most colloidal particles. When one sees dust floating in a ray of sunshine through a window, the light reflects off colloids in the dust, which are driven back and forth by motion in the air otherwise imperceptible to the human senses.

DARK MATTER. The number of states or phases of matter is clearly not fixed, and it is quite possible that more will be discovered in outer space, if not on Earth. One intriguing candidate is called dark matter, so described because it neither reflects nor emits light, and is therefore invisible. In fact, luminous or visible matter may very well make up only a small fraction of the mass in the universe, with the rest being taken up by dark matter.

If dark matter is invisible, how do astronomers and physicists know it exists? By analyzing the gravitational force exerted on visible objects in such cases where there appears to be no visible object to account for that force. An example is the center of our galaxy, the Milky Way. It appears to be nothing more than a dark "halo," but in order to cause the entire galaxy to revolve around it—in the same way that planets revolve around the Sun, though on a vastly larger scale—it must contain a staggering quantity of invisible mass.

THE BOSE-EINSTEIN CONDENSATE

Physicists at the Joint Institute of Laboratory Astrophysics in Boulder, Colorado, in 1995 revealed a highly interesting aspect of atomic behavior at temperatures approaching absolute zero. Some 70 years before, Einstein had predicted that, at extremely low temperatures, atoms would fuse to form one large "superatom." This hypothesized structure was dubbed the Bose-Einstein Condensate (BEC) after Einstein and Satyendranath Bose (1894-1974), an Indian physicist whose statistical methods contributed to the development of quantum theory.

Cooling about 2,000 atoms of the element rubidium to a temperature just 170 billionths of a degree Celsius above absolute zero, the physicists succeeded in creating an atom 100 micrometers across—still incredibly small, but vast in comparison to an ordinary atom. The superatom, which lasted for about 15 seconds, cooled down all the way to just 20 billionths of a degree above absolute zero. The Colorado physicists won the Nobel Prize in physics in 1997 for their work.

In 1999, researchers in a lab at Harvard University also created a superatom of BEC, and used it to slow light to just 38 MPH (61.2 km/h)—about 0.02% of its ordinary speed.

Dubbed a "new" form of matter, the BEC may lead to a greater understanding of quantum mechanics, and may aid in the design of smaller, more powerful computer chips.

SOME UNUSUAL PHASE TRANSITIONS

At places throughout this essay, references have been made variously to "phases" and "states" of matter. This is not intended to confuse, but rather to emphasize a particular point. Solids, liquids, and gases are referred to as "phases" because many (though far from all) substances on Earth regularly move from one phase to another.

There is absolutely nothing incorrect in referring to "states of matter." But "phases of matter" is used in the present context as a means of emphasizing the fact that substances, at the appropriate temperature and pressure, can be solid, liquid, or gas. The phases of matter, in fact, can be likened to the phases of a person's life: infancy, babyhood, childhood, adolescence, adulthood, old age. The transition between these stages is indefinite, yet it is easy enough to say when a person is at a certain stage.

LIQUID CRYSTALS. A liquid crystal is a substance that, over a specific range of temperature, displays properties both of a liquid and a solid. Below this temperature range, it is unquestionably a solid, and above this range it is just as certainly a liquid. In between, however, liquid crystals exhibit a strange solid-liquid behavior: like a liquid, their particles flow, but like a solid, their molecules maintain specific crystalline arrangements.

The cholesteric class of liquid crystals is so named because the spiral patterns of light through the crystal are similar to those which appear in cholesterols. Depending on the physical properties of a cholesteric liquid crystal, only certain colors may be reflected. The response of liquid crystals to light makes them useful in liquid crystal displays (LCDs) found on laptop computer screens, camcorder views, and in other applications.

LIQUEFACTION OF GASES. One interesting and useful application of phase change is the liquefaction of gases, or the change of gas into liquid by the reduction in its molecu-

lar energy levels. Liquefied natural gas (LNG) and liquefied petroleum gas (LPG), the latter a mixture of by-products obtained from petroleum and natural gas, are among the examples of liquefied gas in daily use. In both cases, the volume of the liquefied gas is far less than it would be if the gas were in a vaporized state, thus enabling ease and economy of transport.

Liquefied gases are used as heating fuel for motor homes, boats, and homes or cabins in remote areas. Other applications of liquefied gases include liquefied oxygen and hydrogen in rocket engines; liquefied oxygen and petroleum used in welding; and a combination of liquefied oxygen and nitrogen used in aqualung devices. The properties of liquefied gases figure heavily in the science of producing and studying low-temperature environments. In addition, liquefied helium is used in studying the behavior of matter at temperatures close to absolute zero.

COAL GASIFICATION. Coal gasification, as one might discern from the name, is the conversion of coal to gas. Developed before World War II, it fell out of favor after the war, due to the lower cost of oil and natural gas. However, increasingly stringent environmental regulations imposed by the federal government on industry during the 1970s, combined with a growing concern for the environment on the part of the populace as a whole, led to a resurgence of interest in coal gasification.

Though widely used as a fuel in power plants, coal, when burned by ordinary means, generates enormous air pollution. Coal gasification, on the other hand, makes it possible to burn "clean" coal. Gasification involves a number of chemical reactions, some exothermic or heat-releasing, and some endothermic or heat-absorbing. At one point, carbon monoxide is released in an exothermic reaction, then mixed with hydrogen released from the coal to create a second exothermic reaction. The energy discharged in these first two reactions is used to initiate a third, endothermic, reaction.

The finished product of coal gasification is a mixture containing carbon monoxide, methane, hydrogen, and other substances, and this—rather than ordinary coal—is burned as a fuel. The composition of the gases varies according to the process used. Products range from coal synthesis gas and medium-Btu gas (both composed of car-

ABSOLUTE ZERO: The temperature, defined as 0K on the Kelvin scale, at which the motion of molecules in a solid virtually ceases. Absolute zero is equal to −459.67°F (−273.15°C).

ATOM: The smallest particle of an element that retains the chemical and physical properties of the element. An atom can exist either alone or in combination with other atoms in a molecule. Atoms are made up of protons, neutrons, and electrons. An atom that loses or gains one or more electrons, and thus has a net charge, is an ion. Atoms that have the same number of protons—that is, are of the same element—but differ in number of neutrons, are known as isotopes.

ATOMIC MASS UNIT: An SI unit (abbreviated amu), equal to $1.66 \cdot 10^{-24}$ g, for measuring the mass of atoms.

ATOMIC NUMBER: The number of protons in the nucleus of an atom. Since this number is different for each element, elements are listed on the periodic table in order of atomic number.

ATOMIC SOLID: A form of crystalline solid in which atoms of one element bond to one another. Examples include diamonds (made of pure carbon), silicon, and all metals.

AVERAGE ATOMIC MASS: A figure used by chemists to specify the mass—in atomic mass units—of the average atom in a large sample. If a substance is a compound, the average atomic mass of all atoms in a molecule of that substance must be added together to yield the average molecular mass of that substance.

AVOGADRO'S NUMBER: A figure, named after Italian physicist Amedeo Avogadro (1776-1856), equal to 6.022137×10^{23}. Avogadro's number indicates the number of atoms, molecules, or other elementary particles in a mole.

CHANGE OF PHASE: The transition from one phase of matter to another.

CRITICAL POINT: A coordinate, plotted on a phase diagram, above which a substance cannot exist in anything other than the gaseous state. Located at a position of very high temperature and pressure, the critical point marks the termination of the vaporization curve.

COMPOUND: A substance made up of atoms of more than one element. These atoms are usually, but not always, joined in molecules.

CRYSTALLINE SOLID: A type of solid in which the constituent parts have a simple and definite geometric arrangement that is repeated in all directions. Types of crystalline solids include atomic solids, ionic solids, and molecular solids.

ELEMENT: A substance made up of only one kind of atom.

ELECTRON: A negatively charged particle in an atom. Electrons, which spin around the protons and neutrons that make up the atom's nucleus, constitute a very small portion of the atom's mass. In most atoms, the number of electrons and protons is the same, thus canceling out one another. When an atom loses one or more electrons, however—thus becoming an ion—it acquires a net electric charge.

ENERGY: The ability to accomplish work—that is, the exertion of force over a given distance to displace or move an object.

FUSION CURVE: The boundary between solid and liquid for any given substance as plotted on a phase diagram.

GAS: A phase of matter in which molecules move at high speeds, and therefore exert little or no attraction toward one another.

ION: An atom or atoms that has lost or gained one or more electrons, and thus has a net electric charge.

IONIC SOLID: A form of crystalline solid that contains ions. When mixed with a solvent such as water, ions from table salt—an example of an ionic solid—move freely throughout the solution, making it possible to conduct an electric current.

ISOTOPES: Atoms that have an equal number of protons, and hence are of the same element, but differ in their number of neutrons.

KINETIC ENERGY: The energy that an object possesses by virtue of its movement.

LIQUID: A phase of matter in which molecules move at moderate speeds, and therefore exert moderate attractions toward one another.

MATTER: Physical substance that occupies space, has mass, is composed of atoms (or in the case of subatomic particles, is part of an atom), and is convertible to energy.

MOLAR HEAT OF FUSION: The amount of energy required to melt 1 mole of a solid substance.

MOLAR HEAT OF VAPORIZATION: The amount of energy required to turn 1 mole of a liquid into a gas. Because gases possess much more energy than liquids or solids, molar heat of vaporization is usually much higher than molar heat of fusion.

MOLAR MASS: The mass, in grams, of 1 mole of a given substance. The value in grams of molar mass is always equal to the value, in atomic mass units, of the average atomic mass of that substance: thus carbon has a molar mass of 12.01 g, and an average atomic mass of 12.01 amu.

MOLE: The SI fundamental unit for "amount of substance." A mole is, generally speaking, Avogadro's number of atoms, molecules, or other elementary particles; however, in the more precise SI definition, a mole is equal to the number of carbon atoms in 12.01 g of carbon.

MOLECULAR SOLID: A form of crystalline solid in which the molecules have a neutral electric charge, meaning that there are no ions present as there are in an ionic solid. Table sugar (sucrose) is an example of a molecular solid. When mixed with a solvent such as water, the resulting solution does not conduct electricity.

MOLECULE: A group of atoms, usually of more than one element, joined in a structure.

NEUTRON: A subatomic particle that has no electric charge. Neutrons are found at the nucleus of an atom, alongside protons.

NUCLEUS: The center of an atom, a region where protons and neutrons are located, and around which electrons spin.

PHASE DIAGRAM: A chart, plotted for any particular substance, identifying the particular phase of matter for that substance at a given temperature and pressure level. A phase diagram usually shows temperature along the x-axis, and pressure along the y-axis.

PHASES OF MATTER: The various forms of material substance (matter), which are defined primarily in terms of the behavior exhibited by their atomic or molecular structures. On Earth, three principal phases of matter exist, namely solid, liquid, and gas. Other forms of matter include plasma.

PLASMA: One of the phases of matter, closely related to gas. Plasma is found primarily in stars. Containing neither atoms nor molecules, plasma is made up of electrons and positive ions.

PROTON: A positively charged particle in an atom. Protons and neutrons, which together form the nucleus around which electrons spin, have approximately the

same mass—a mass that is many times greater than that of an electron. The number of protons in the nucleus of an atom is the atomic number of an element.

SOLID: A phase of matter in which molecules move slowly relative to one another, and therefore exert strong attractions toward one another.

SUBLIMATION CURVE: The boundary between solid and gas for any given substance, as plotted on a phase diagram.

SYSTEM: In chemistry and other sciences, the term "system" usually refers to any set of interactions isolated from the rest of the universe. Anything outside of the system, including all factors and forces irrelevant to a discussion of that system, is known as the environment.

THERMAL ENERGY: "Heat" energy, a form of kinetic energy produced by the relative motion of atoms or molecules. The greater the movement of these particles, the greater the thermal energy.

VAPORIZATION CURVE: The boundary between liquid and gas for any given substance as plotted on a phase diagram.

bon monoxide and hydrogen, though combined in different forms) to substitute natural gas, which consists primarily of methane.

Not only does coal gasification produce a clean-burning product, but it does so without the high costs associated with flue-gas desulfurization systems. The latter, often called "scrubbers," were originally recommended by the federal government to industry, but companies discovered that coal gasification could produce the same results for much less money. In addition, the waste products from coal gasification can be used

for other purposes. At the Cool Water Integrated Gasification Combined Cycle Plant, established in Barstow, California, in 1984, sulfur obtained from the reduction of sulfur dioxide is sold off for about $100 a ton.

THE CHEMICAL DIMENSION TO CHANGES OF PHASE

Throughout much of this essay, we have discussed changes of phase primarily in physical terms; yet clearly these changes play a significant role in chemistry. Furthermore, coal gasification serves to illustrate the impact chemical processes can have on changes of state.

Much earlier, figures were given for the melting points of copper, water, and helium, and these were compared with the average atomic mass of each. Those figures, again, are:

Average Atomic Mass and Melting Points of a Sample Gas, Liquid, and Solid

- Helium: 4.003 amu; −457.6°F (−210°C)
- Water: 40.304 amu 32°F (0°C)
- Copper: 63.55 amu; 1,985°F (1,085°C).

Something seems a bit strange about those comparisons: specifically, the differences in melting point appear to be much more dramatic than the differences in average atomic mass. Clearly, another factor is at work—a factor that relates to the difference in the attractions between molecules in each. Although the differences between solids, liquids, and gases are generally physical, the one described here—a difference between substances—is clearly chemical in nature.

To discuss this in the detail it deserves would require a lengthy digression on the chemical dimensions of intermolecular attraction. Nonetheless, it is possible here to offer at least a cursory answer to the question raised by these striking differences in response to temperature.

DIPOLES, ELECTRON SEAS, AND LONDON DISPERSION

Water molecules are polar, meaning that one area of a water molecule is positively charged, while another area has a negative charge. Thus the positive side of one molecule is drawn to the negative side of another, and vice versa, which gives water a much stronger intermolecular bond than, for instance, oil, in which the positive and negative charges are evenly distributed throughout the molecule.

Yet the intermolecular attraction between the dipoles (as they are called) in water is not nearly as strong as the bond that holds together a metal. Particles in copper or other metals "float" in a tightly packed "sea" of highly mobile electrons, which provide a bond that is powerful, yet lacking in a firm directional orientation. Thus metals are both strong and highly malleable (that is, they can be hammered very flat without breaking.)

Water, of course, appears most often as a liquid, and copper as a solid, precisely because water has a very high boiling point (the point at which it becomes a vapor) and copper has a very high melting point. But consider helium, which has the lowest freezing point of any element: just above absolute zero. Even then, a pressure equal to 25 times that of normal atmospheric pressure is required to push it past the freezing point.

Helium and other Group 8 or Group 18 elements, as well as non-polar molecules such as oils, are bonded by what is called London dispersion forces. The latter, as its name suggests, tends to keep molecules dispersed, and induces instantaneous dipoles when most of the electrons happen to be on one side of an atom. Of course, this happens only for an infinitesimal fraction of time, but it serves to create a weak attraction. Only at very low temperatures do London dispersion forces become strong enough to result in the formation of a solid.

WHERE TO LEARN MORE

Biel, Timothy L. *Atom: Building Blocks of Matter.* San Diego, CA: Lucent Books, 1990.

Feynman, Richard. *Six Easy Pieces: Essentials of Physics Explained by Its Most Brilliant Teacher.* New introduction by Paul Davies. Cambridge, MA: Perseus Books, 1995.

"High School Chemistry Table of Contents—Solids and Liquids" Homeworkhelp.com (Web site). <http://www.homeworkhelp.com/homeworkhelp/freemember/text/chem/hig h/topic09.htm> (April 10, 2001).

"Matter: Solids, Liquids, Gases." Studyweb (Web site). <http://www.studyweb.com/links/4880.html> (April 10, 2001).

"The Molecular Circus" (Web site). <http://www.cpo.com/Weblabs/circus.htm> (April 10, 2001).

Paul, Richard. *A Handbook to the Universe: Explorations of Matter, Energy, Space, and Time for Beginning Scientific Thinkers.* Chicago: Chicago Review Press, 1993.

"Phases of Matter" (Web site). <http://pc65.frontier.osrhe.edu/hs/science/pphase.htm> (April 10, 2001).

Royston, Angela. *Solids, Liquids, and Gasses.* Chicago: Heinemann Library, 2001.

Wheeler, Jill C. *The Stuff Life's Made Of: A Book About Matter.* Minneapolis, MN: Abdo & Daughters Publishing, 1996.

Zumdahl, Steven S. *Introductory Chemistry: A Foundation,* 4th ed. Boston: Houghton Mifflin, 2000.

GASES

CONCEPT

The number of elements that appear ordinarily in the form of a gas is relatively small: oxygen, hydrogen, fluorine, and chlorine in the halogen "family"; and a handful of others, most notably the noble gases in Group 8 of the periodic table. Yet many substances can exist in the form of a gas, depending on the relative attraction and motion of molecules in that substance. A simple example, of course, is water, or H_2O, which, though it appears as a liquid at room temperature, begins to vaporize and turn into steam at 212°F (100°C). In general, gases respond more dramatically to changes in pressure and temperature than do most other types of matter, and this allows scientists to predict gas behaviors under certain conditions. These predictions can explain mundane occurrences, such as the fact that an open can of soda will soon lose its fizz, but they also apply to more dramatic, life-and-death situations.

HOW IT WORKS

Molecular Motion and Phases of Matter

On Earth, three principal phases or states of matter exist: solid, liquid, and gas. The differences between these three are, on the surface at least, easily perceivable. Clearly water is a liquid, just as ice is a solid and steam a vapor or gas. Yet the ways in which various substances convert between phases are often complex, as are the interrelations between these phases. Ultimately, understanding of the phases depends on an awareness of what takes place at the molecular level.

All molecules are in motion, and the rate of that motion determines the attraction between them. The movement of molecules generates kinetic energy, or the energy of movement, which is manifested as thermal energy. In everyday language, thermal energy is what people mean when they say "heat"; but in scientific terms, heat has a different definition.

The force that attracts atoms to atoms, or molecules to molecules, is not the same as gravitational force, which holds the Moon in orbit around Earth, Earth in orbit around the Sun, and so on. By contrast, the force of interatomic and intermolecular attraction is electromagnetic. Just as the north pole of a magnet is attracted to the south pole of another magnet and repelled by that other magnet's north pole, so positive electric charges are attracted to negative charges, and negatives to positives. (In fact, electricity and magnetism are both manifestations of an electromagnetic interaction.)

The electromagnetic attractions between molecules are much more complex than this explanation makes it seem, and they play a highly significant role in chemical bonding. In simple terms, however, one can say that the greater the rate of motion for the molecules in relation to one another, the less the attraction between molecules. In addition, the kinetic energy, and hence the thermal energy, is greater in a substance whose molecules are relatively free to move.

When the molecules in a material move slowly in relation to one another, they exert a strong attraction, and the material is called a solid. Molecules of liquid, by contrast, move at

STATE	VOLUME FORM SHAPE COMPRESSIBILITY	ARRANGEMENT AND CLOSENESS OF PARTICLES	MOTION OF PARTICLES	ATTRACTION BETWEEN PARTICLES	BOILING POINT
GAS	No definite volume, form or shape Compressible	Random Far apart	Fast	Little to None	Lower than Room Temperature
LIQUID	Has a definite volume, but no definite form or shape Non compressive tendency	Random Close	Moderate	Moderate	Higher than Room Temperature
SOLID	Definite volume, has own shape or form Non compressible	Definite Close	Slow	Strong	Much higher than Room Temperature

A COMPARISON OF GAS, LIQUID, AND SOLID STATES.

moderate speeds and exert a moderate attraction. A material substance whose molecules move at high speeds, and therefore exert little or no attraction, is known as a gas.

COMPARISON OF GASES TO OTHER PHASES OF MATTER

WATER AND AIR COMPARED. Gases respond to changes in pressure and temperature in a manner remarkably different from that of solids or liquids. Consider the behavior of liquid water as compared with air—a combination of oxygen (O_2), nitrogen (N_2), and other gases—in response to experiments involving changes in pressure and temperature.

In the first experiment, both samples are subjected to an increase in pressure from 1 atm (that is, normal atmospheric pressure at sea level) to 2 atm. In the second, both experience an increase in temperature from 32°F (0°C) to 212°F (100°C). The differences in the responses of water and air are striking.

A sample of water that experiences an increase in pressure from 1 to 2 atm will decrease in volume by less than 0.01%, while a temperature increase from the freezing point to the boiling point will result in only a 2% increase in volume. For air, however, an equivalent pressure

increase will decrease the volume by a whopping 50%, and an equivalent temperature increase results in a volume increase of 37%.

Air and other gases, by definition, have a boiling point below room temperature. If they did not boil and thus become gas well below ordinary temperatures, they would not be described as substances that are in the gaseous state in most circumstances. The boiling point of water, of course, is higher than room temperature, and that of solids is much higher.

THE ARRANGEMENT OF PARTICLES. Solids possess a definite volume and a definite shape, and are relatively noncompressible: for instance, if one applies extreme pressure to a steel plate, it will bend, but not much. Liquids have a definite volume, but no definite shape, and tend to be noncompressible. Gases, on the other hand, possess no definite volume or shape, and are highly compressible.

At the molecular level, particles of solids tend to be definite in their arrangement and close in proximity—indeed, part of what makes a solid "solid," in the everyday meaning of that term, is the fact that its constituent parts are basically immovable. Liquid molecules, too, are close in proximity, though random in arrangement. Gas

THE DIVERS PICTURED HERE HAVE ASCENDED FROM A SUNKEN SHIP AND HAVE STOPPED AT THE 10-FT (3-METER) DECOMPRESSION LEVEL TO AVOID GETTING DECOMPRESSION SICKNESS, BETTER KNOWN AS THE "BENDS." *(Jonathan Blair/Corbis. Reproduced by permission)*

molecules are random in arrangement, but tend to be more widely spaced than liquid molecules.

PRESSURE

There are a number of statements, collectively known as the "gas laws," that describe and predict the behavior of gases in response to changes in temperature, pressure, and volume. Temperature and volume are discussed elsewhere in this book. However, the subject of pressure requires some attention before we can continue with a discussion of the gas laws.

When a force is applied perpendicular to a surface area, it exerts pressure on that surface. Hence the formula for pressure is $p = F/A$, where p is pressure, F force, and A the area over which the force is applied. The greater the force, and the smaller the area of application, the greater the pressure; conversely, an increase in area—even without a reduction in force—reduces the overall pressure.

Pressure is measured by a number of units in the English and SI systems. Because $p = F/A$, all

units of pressure represent some ratio of force to surface area.

UNITS OF PRESSURE. The principal SI unit of pressure is called a pascal (Pa), or 1 N/m². It is named for French mathematician and physicist Blaise Pascal (1623-1662), who is credited with Pascal's principle. The latter holds that the external pressure applied on a fluid—which, in the physical sciences, can mean either a gas or a liquid—is transmitted uniformly throughout the entire body of that fluid.

A newton (N), the SI unit of force, is equal to the force required to accelerate 1 kg of mass at a rate of 1 m/sec². Thus a Pascal is the pressure of 1 newton over a surface area of 1 m². In the English or British system, pressure is measured in terms of pounds per square inch, abbreviated as lbs./in². This is equal to $6.89 \cdot 10^3$ Pa, or 6,890 Pa.

Another important measure of pressure is the atmosphere (atm), which is the average pressure exerted by air at sea level. In English units, this is equal to 14.7 lb/in², and in SI units, to $1.013 \cdot 10^5$ Pa.

There are two other specialized units of pressure measurement in the SI system: the bar, equal to 10^5 Pa, and the torr, equal to 133 Pa. Meteorologists, scientists who study weather patterns, use the millibar (mb), which, as its name implies, is equal to 0.001 bars. At sea level, atmospheric pressure is approximately 1,013 mb.

The torr, also known as the millimeter of mercury (mm Hg), is the amount of pressure required to raise a column of mercury (chemical symbol Hg) by 1 mm. It is named for Italian physicist Evangelista Torricelli (1608-1647), who invented the barometer, an instrument for measuring atmospheric pressure.

THE BAROMETER. The barometer constructed by Torricelli in 1643 consisted of a long glass tube filled with mercury. The tube was open at one end, and turned upside down into a dish containing more mercury: the open end was submerged in mercury, while the closed end at the top constituted a vacuum—that is, an area devoid of matter, including air.

The pressure of the surrounding air pushed down on the surface of the mercury in the bowl, while the vacuum at the top of the tube provided an area of virtually no pressure into which the mercury could rise. Thus the height to which the

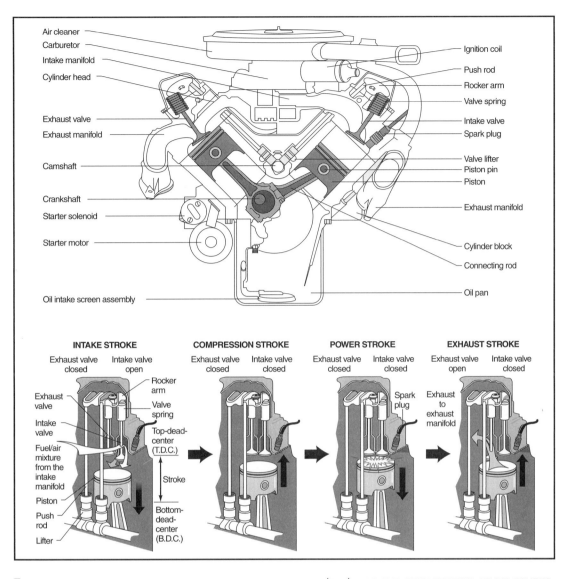

THE MAJOR COMPONENTS OF AN INTERNAL COMBUSTION ENGINE (TOP), AND THE FOUR STROKES OF ITS COMBUSTION SEQUENCE (BOTTOM).

mercury rose in the glass tube represented normal air pressure (that is, 1 atm.) Torricelli discovered that at standard atmospheric pressure, the column of mercury rose to 760 mm (29.92 in).

The value of 1 atm was thus established as equal to the pressure exerted on a column of mercury 760 mm high at a temperature of 0°C (32°F). In time, Torricelli's invention became a fixture both of scientific laboratories and of households. Since changes in atmospheric pressure have an effect on weather patterns, many home indoor-outdoor thermometers today also include a barometer.

REAL-LIFE APPLICATIONS

INTRODUCTION TO THE GAS LAWS

English chemist Robert Boyle (1627-1691), who made a number of important contributions to chemistry—including his definition and identification of elements—seems to have been influenced by Torricelli. If so, this is an interesting example of ideas passing from one great thinker to another: Torricelli, a student of Galileo Galilei (1564-1642), was no doubt influenced by Galileo's thermoscope.

Like Torricelli, Boyle conducted tests involving the introduction of mercury to a tube closed at the other end. The tube Boyle used was shaped like the letter J, and it was so long that he had to use the multi-story foyer of his house as a laboratory. At the tip of the curved bottom was an area of trapped gas, and into the top of the tube, Boyle introduced increasing quantities of mercury. He found that the greater the volume of mercury, the greater the pressure on the gas, and the less the volume of gas at the end of the tube. As a result, he formulated the gas law associated with his name.

The gas laws are not a set of government regulations concerning use of heating fuel; rather, they are a series of statements concerning the behavior of gases in response to changes in temperature, pressure, and volume. These were derived, beginning with Boyle's law, during the seventeenth, eighteenth, and nineteenth centuries by scientists whose work is commemorated through the association of their names with the laws they discovered. In addition to Boyle, these men include fellow English chemists John Dalton (1766-1844) and William Henry (1774-1836); French physicists and chemists J. A. C. Charles (1746-1823) and Joseph Gay-Lussac (1778-1850); and Italian physicist Amedeo Avogadro (1776-1856).

There is a close relationship between Boyle's, Charles's, and Gay-Lussac's laws. All of these treat one of three parameters—temperature, pressure, or volume—as fixed quantities in order to explain the relationship between the other two variables. Avogadro's law treats two of the parameters as fixed, thereby establishing a relationship between volume and the number of molecules in a gas. The ideal gas law sums up these four laws, and the kinetic theory of gases constitutes an attempt to predict the behavior of gases based on these laws. Finally, Dalton's and Henry's laws both relate to partial pressure of gases.

BOYLE'S, CHARLES'S, AND GAY-LUSSAC'S LAWS

BOYLE'S AND CHARLES'S LAWS. Boyle's law holds that in isothermal conditions (that is, a situation in which temperature is kept constant), an inverse relationship exists between the volume and pressure of a gas. (An inverse relationship is a situation involving two variables, in which one of the two increases in direct proportion to the decrease in the other.)

In this case, the greater the pressure, the less the volume and vice versa. Therefore, the product of the volume multiplied by the pressure remains constant in all circumstances.

Charles's law also yields a constant, but in this case the temperature and volume are allowed to vary under isobarometric conditions—that is, a situation in which the pressure remains the same. As gas heats up, its volume increases, and when it cools down, its volume reduces accordingly. Hence, Charles established that the ratio of temperature to volume is constant.

ABSOLUTE TEMPERATURE. In about 1787, Charles made an interesting discovery: that at 0°C (32°F), the volume of gas at constant pressure drops by 1/273 for every Celsius degree drop in temperature. This seemed to suggest that the gas would simply disappear if cooled to -273°C (-459.4°F), which, of course, made no sense. In any case, the gas would most likely become first a liquid, and then a solid, long before it reached that temperature.

The man who solved the quandary raised by Charles's discovery was born a year after Charles died. He was William Thomson, Lord Kelvin (1824-1907); in 1848, he put forward the suggestion that it was molecular translational energy—the energy generated by molecules in motion—and not volume, that would become zero at -273°C. He went on to establish what came to be known as the Kelvin scale of absolute temperature.

Sometimes known as the absolute temperature scale, the Kelvin scale is based not on the freezing point of water, but on absolute zero—the temperature at which molecular motion comes to a virtual stop. This is -273.15°C (-459.67°F). In the Kelvin scale, which uses neither the term nor the symbol for "degree," absolute zero is designated as 0K.

Scientists prefer the Kelvin scale to the Celsius, and certainly to the Fahrenheit, scales. If the Kelvin temperature of an object is doubled, its average molecular translational energy has doubled as well. The same cannot be said if the temperature were doubled from, say, 10°C to 20°C, or from 40°F to 80°F, since neither the Celsius nor the Fahrenheit scale is based on absolute zero.

GAY-LUSSAC'S LAW. From Boyle's and Charles's law, a pattern should be emerging: both treat one parameter (temperature in Boyle's, pressure in Charles's) as unvarying, while

two other factors are treated as variables. Both, in turn, yield relationships between the two variables: in Boyle's law, pressure and volume are inversely related, whereas in Charles's law, temperature and volume are directly related.

In Gay-Lussac's law, a third parameter, volume, is treated as a constant, and the result is a constant ratio between the variables of pressure and temperature. According to Gay-Lussac's law, the pressure of a gas is directly related to its absolute temperature.

AVOGADRO'S LAW

Gay-Lussac also discovered that the ratio in which gases combine to form compounds can be expressed in whole numbers: for instance, water is composed of one part oxygen and two parts hydrogen. In the language of modern chemistry, this is expressed as a relationship between molecules and atoms: one molecule of water contains one oxygen atom and two hydrogen atoms.

In the early nineteenth century, however, scientists had yet to recognize a meaningful distinction between atoms and molecules, and Avogadro was the first to achieve an understanding of the difference. Intrigued by the whole-number relationship discovered by Gay-Lussac, Avogadro reasoned that one liter of any gas must contain the same number of particles as a liter of another gas. He further maintained that gas consists of particles—which he called molecules—that in turn consist of one or more smaller particles.

In order to discuss the behavior of molecules, Avogadro suggested the use of a large quantity as a basic unit, since molecules themselves are very small. Avogadro himself did not calculate the number of molecules that should be used for these comparisons, but when that number was later calculated, it received the name "Avogadro's number" in honor of the man who introduced the idea of the molecule. Equal to $6.022137 \cdot 10^{23}$, Avogadro's number designates the quantity of atoms or molecules (depending on whether the substance in question is an element or a compound) in a mole.

Today the mole (abbreviated mol), the SI unit for "amount of substance," is defined precisely as the number of carbon atoms in 12.01 g of carbon. The term "mole" can be used in the same way we use the word "dozen." Just as "a dozen" can refer to twelve cakes or twelve chickens, so "mole" always describes the same number of molecules. The ratio of mass between a mole of Element A and Element B, or Compound A and Compound B, is the same as the ratio between the mass of Atom A and Atom B, or Molecule A and Molecule B. Avogadro's law describes the connection between gas volume and number of moles. According to Avogadro's law, if the volume of gas is increased under isothermal and isobarometric conditions, the number of moles also increases. The ratio between volume and number of moles is therefore a constant.

THE IDEAL GAS LAW

Once again, it is easy to see how Avogadro's law can be related to the laws discussed earlier. Like the other three, this one involves the parameters of temperature, pressure, and volume, but it also introduces a fourth—quantity of molecules (that is, number of moles). In fact, all the laws so far described are brought together in what is known as the ideal gas law, sometimes called the combined gas law.

The ideal gas law can be stated as a formula, $pV = nRT$, where p stands for pressure, V for volume, n for number of moles, and T for temperature. R is known as the universal gas constant, a figure equal to 0.0821 atm · liter/mole · K. (Like most figures in chemistry, this one is best expressed in metric rather than English units.)

Given the equation $pV = nRT$ and the fact that R is a constant, it is possible to find the value of any one variable—pressure, volume, number of moles, or temperature—as long as one knows the value of the other three. The ideal gas law also makes it possible to discern certain relationships: thus, if a gas is in a relatively cool state, the product of its pressure and volume is proportionately low; and if heated, its pressure and volume product increases correspondingly.

THE KINETIC THEORY OF GASES

From the preceding gas laws, a set of propositions known collectively as the kinetic theory of gases has been derived. Collectively, these put forth the proposition that a gas consists of numerous molecules, relatively far apart in space, which interact by colliding. These collisions are responsible for the production of thermal energy, because when the velocity of the molecules increases—as it does after collision—the temperature increases as well.

There are five basic postulates to the kinetic theory of gases:

- 1. Gases consist of tiny molecular or atomic particles.
- 2. The proportion between the size of these particles and the distances between them is so small that the individual particles can be assumed to have negligible volume.
- 3. These particles experience continual random motion. When placed in a container, their collisions with the walls of the container constitute the pressure exerted by the gas.
- 4. The particles neither attract nor repel one another.
- 5. The average kinetic energy of the particles in a gas is directly related to absolute temperature.

These observations may appear to resemble statements made earlier concerning the differences between gases, liquids, and solids in terms of molecular behavior. If so, that is no accident: the kinetic theory constitutes a generally accepted explanation for the reasons why gases behave as they do. Kinetic theories do not work as well for explaining the behaviors of solids and liquids; nonetheless, they do go a long way toward identifying the molecular properties inherent in the various phases of matter.

LAWS OF PARTIAL PRESSURE

In addition to all the gas laws so far discussed, two laws address the subject of partial pressure. When two or more gases are present in a container, partial pressure is the pressure that one of them exerts if it alone is in the container.

Dalton's law of partial pressure states that the total pressure of a gas is equal to the sum of its partial pressures. As noted earlier, air is composed mostly of nitrogen and oxygen. Along with these are small components, carbon dioxide, and gases collectively known as the rare or noble gases: argon, helium, krypton, neon, radon, and xenon. Hence, the total pressure of a given quantity of air is equal to the sum of the pressures exerted by each of these gases.

Henry's law states that the amount of gas dissolved in a liquid is directly proportional to the partial pressure of the gas above the surface of the solution. This applies only to gases such as oxygen and hydrogen that do not react chemically to liquids. On the other hand, hydrochloric acid will ionize when introduced to water: one or more of its electrons will be removed, and its atoms will convert to ions, which are either positive or negative in charge.

APPLICATIONS OF DALTON'S AND HENRY'S LAWS

PARTIAL PRESSURE: A MATTER OF LIFE AND POSSIBLE DEATH FOR SCUBA DIVERS. The gas laws are not just a series of abstract statements. Certainly, they do concern the behavior of ideal as opposed to real gases. Like all scientific models, they remove from the equation all outside factors, and treat specific properties in isolation. Yet, the behaviors of the ideal gases described in the gas laws provide a key to understanding the activities of real gases in the real world. For instance, the concept of partial pressure helps scuba divers avoid a possibly fatal sickness.

Imagine what would happen if a substance were to bubble out of one's blood like carbon dioxide bubbling out of a soda can, as described below. This is exactly what can happen to an undersea diver who returns to the surface too quickly: nitrogen rises up within the body, producing decompression sickness—known colloquially as "the bends." This condition may manifest as itching and other skin problems, joint pain, choking, blindness, seizures, unconsciousness, permanent neurological defects such as paraplegia, and possibly even death.

If a scuba diver descending to a depth of 150 ft (45.72 m) or more were to use ordinary air in his or her tanks, the results would be disastrous. The high pressure exerted by the water at such depths creates a high pressure on the air in the tank, meaning a high partial pressure on the nitrogen component in the air. The result would be a high concentration of nitrogen in the blood, and hence the bends.

Instead, divers use a mixture of helium and oxygen. Helium gas does not dissolve well in blood, and thus it is safer for a diver to inhale this oxygen-helium mixture. At the same time, the oxygen exerts the same pressure that it would normally—in other words, it operates in accordance with Dalton's observations concerning partial pressure.

OPENING A SODA CAN. Inside a can or bottle of carbonated soda is carbon dioxide gas (CO_2), most of which is dissolved in the drink itself. But some of it is in the space (sometimes referred to as "head space") that makes up the difference between the volume of the soft drink and the volume of the container.

At the bottling plant, the soda manufacturer adds high-pressure carbon dioxide (CO_2) to the head space in order to ensure that more CO_2 will be absorbed into the soda itself. This is in accordance with Henry's law: the amount of gas (in this case CO_2) dissolved in the liquid (soda) is directly proportional to the partial pressure of the gas above the surface of the solution—that is, the CO_2 in the head space. The higher the pressure of the CO_2 in the head space, the greater the amount of CO_2 in the drink itself; and the greater the CO_2 in the drink, the greater the "fizz" of the soda.

Once the container is opened, the pressure in the head space drops dramatically. Once again, Henry's law indicates that this drop in pressure will be reflected by a corresponding drop in the amount of CO_2 dissolved in the soda. Over a period of time, the soda will release that gas, and eventually, it will go "flat."

FIRE EXTINGUISHERS. A fire extinguisher consists of a long cylinder with an operating lever at the top. Inside the cylinder is a tube of carbon dioxide surrounded by a quantity of water, which creates pressure around the CO_2 tube. A siphon tube runs vertically along the length of the extinguisher, with one opening in the water near the bottom. The other end opens in a chamber containing a spring mechanism attached to a release valve in the CO_2 tube.

The water and the CO_2 do not fill the entire cylinder: as with the soda can, there is "head space," an area filled with air. When the operating lever is depressed, it activates the spring mechanism, which pierces the release valve at the top of the CO_2 tube. When the valve opens, the CO_2 spills out in the "head space," exerting pressure on the water. This high-pressure mixture of water and carbon dioxide goes rushing out of the siphon tube, which was opened when the release valve was depressed. All of this happens, of course, in a fraction of a second—plenty of time to put out the fire.

AEROSOL CANS. Aerosol cans are similar in structure to fire extinguishers, though with one important difference. As with the fire extinguisher, an aerosol can includes a nozzle that depresses a spring mechanism, which in turn allows fluid to escape through a tube. But instead of a gas cartridge surrounded by water, most of the can's interior is made up of the product (for instance, deodorant), mixed with a liquid propellant.

The "head space" of the aerosol can is filled with highly pressurized propellant in gas form, and, in accordance with Henry's law, a corresponding proportion of this propellant is dissolved in the product itself. When the nozzle is depressed, the pressure of the propellant forces the product out through the nozzle.

A propellant, as its name implies, propels the product itself through the spray nozzle when the nozzle is depressed. In the past, chlorofluorocarbons (CFCs)—manufactured compounds containing carbon, chlorine, and fluorine atoms—were the most widely used form of propellant. Concerns over the harmful effects of CFCs on the environment, however, has led to the development of alternative propellants, most notably hydrochlorofluorocarbons (HCFCs), CFC-like compounds that also contain hydrogen atoms.

APPLICATIONS OF BOYLE'S, CHARLES'S, AND GAY-LUSSAC'S LAWS

WHEN THE TEMPERATURE CHANGES. A number of interesting results occur when gases experience a change in temperature, some of them unfortunate and some potentially lethal. In these instances, it is possible to see the gas laws—particularly Boyle's and Charles's—at work.

There are numerous examples of the disastrous effects that result from an increase in the temperature of combustible gases, including natural gas and petroleum-based products. In addition, the pressure on the gases in aerosol cans makes the cans highly explosive—so much so that discarded cans at a city dump may explode on a hot summer day. Yet, there are other instances when heating a gas can produce positive effects.

A hot-air balloon, for instance, floats because the air inside it is not as dense than the air outside. According to Charles's law, heating a gas will increase its volume, and since gas molecules exert little attraction toward one another, they tend to "spread out" even further with an increase of volume. This, in turn, creates a significant difference in density between the air in the balloon and the air outside, and as a result, the balloon floats.

Although heating a gas can be beneficial, cooling a gas is not always a wise idea. If someone were to put a bag of potato chips into a freezer, thinking this would preserve their flavor, he would be in for a disappointment. Much of what maintains the flavor of the chips is the pressurization of the bag, which ensures a consistent internal environment so that preservative chemicals, added during the manufacture of the chips, can keep them fresh. Placing the bag in the freezer causes a reduction in pressure, as per Gay-Lussac's law, and the bag ends up a limp version of its former self.

Propane tanks and tires offer an example of the pitfalls that may occur by either allowing a gas to heat up or cool down by too much. Because most propane tanks are made according to strict regulations, they are generally safe, but it is not entirely inconceivable that the extreme heat of a summer day could cause a defective tank to burst. An increase in temperature leads to an increase in pressure, in accordance with Gay-Lussac's law, and could lead to an explosion.

Because of the connection between heat and pressure, propane trucks on the highways during the summer are subjected to weight tests to ensure that they are not carrying too much gas. On the other hand, a drastic reduction in temperature could result in a loss in gas pressure. If a propane tank from Florida were transported by truck during the winter to northern Canada, the pressure is dramatically reduced by the time it reaches its destination.

THE INTERNAL-COMBUSTION ENGINE.

In operating a car, we experience two applications of the gas laws. One of these is what makes the car run: the combustion of gases in the engine, which illustrates the interrelation of volume, pressure, and temperature expressed in the laws attributed to Boyle, Charles, and Gay-Lussac. The other is, fortunately, a less frequent phenomenon—but it can and does save lives. This is the operation of an airbag, which depends, in part, on the behaviors explained in Charles's law.

When the driver of a modern, fuel-injection automobile pushes down on the accelerator, this activates a throttle valve that sprays droplets of gasoline mixed with air into the engine. The mixture goes into the cylinder, where the piston moves up, compressing the gas and air. While the mixture is still at a high pressure, the electric spark plug produces a flash that ignites the gasoline-air mixture. The heat from this controlled explosion increases the volume of air, which forces the piston down into the cylinder. This opens an outlet valve, causing the piston to rise and release exhaust gases.

As the piston moves back down again, an inlet valve opens, bringing another burst of gasoline-air mixture into the chamber. The piston, whose downward stroke closed the inlet valve, now shoots back up, compressing the gas and air to repeat the cycle. The reactions of the gasoline and air to changes in pressure, temperature, and volume are what move the piston, which turns a crankshaft that causes the wheels to rotate.

THE AIRBAG.

So much for moving— what about stopping? Most modern cars are equipped with an airbag, which reacts to sudden impact by inflating. This protects the driver and front-seat passenger, who, even if they are wearing seatbelts, may otherwise be thrown against the steering wheel or dashboard.

In order to perform its function properly, the airbag must deploy within 40 milliseconds (0.04 seconds) of impact. Not only that, but it has to begin deflating before the body hits it. If a person's body, moving forward at speeds typical in an automobile accident, were to smash against a fully inflated airbag, it would feel like hitting concrete—with all the expected results.

The airbag's sensor contains a steel ball attached to a permanent magnet or a stiff spring. The spring or magnet holds the ball in place through minor mishaps when an airbag is not warranted—for instance, if a car were simply to be "tapped" by another in a parking lot. But in a case of sudden deceleration, the magnet or spring releases the ball, sending it down a smooth bore. The ball flips a switch, turning on an electrical circuit. This in turn ignites a pellet of sodium azide, which fills the bag with nitrogen gas.

At this point, the highly pressurized nitrogen gas molecules begin escaping through vents. Thus, as the driver's or rider's body hits the airbag, the deflation of the bag is moving it in the same direction that the body is moving—only

ABSOLUTE TEMPERATURE: Temperature in relation to absolute zero (-273.15°C or -459.67°F), as measured on the Kelvin scale. The Kelvin and Celsius scales are directly related; hence, Celsius temperatures can be converted to Kelvins (for which neither the word nor the symbol for "degree" are used) by adding 273.15.

ATMOSPHERE: A measure of pressure, abbreviated "atm" and equal to the average pressure exerted by air at sea level. In English units, this is equal to 14.7 lb/in^2, and in SI units, to 101,300 pascals.

AVOGADRO'S LAW: A statement, derived by the Italian physicist Amedeo Avogadro (1776-1856), which holds that as the volume of gas increases under isothermal and isobarometric conditions, the number of molecules (expressed in terms of mole number), increases as well. Thus, the ratio of volume to mole number is a constant.

BAROMETER: An instrument for measuring atmospheric pressure.

BOYLE'S LAW: A statement, derived by English chemist Robert Boyle (1627-1691), which holds that for gases in isothermal conditions, an inverse relationship exists between the volume and pressure of a gas. This means that the greater the pressure, the less the volume and vice versa, and therefore the product of pressure multiplied by volume yields a constant figure.

CHARLES'S LAW: A statement, derived by French physicist and chemist J. A. C. Charles (1746-1823), which holds that for gases in isobarometric conditions, the ratio between the volume and temperature of a gas is constant. This means that the greater the temperature, the greater the volume, and vice versa.

DALTON'S LAW OF PARTIAL PRESSURE: A statement, derived by the English chemist John Dalton (1766-1844), which holds that the total pressure of a gas is equal to the sum of its partial pressures—that is, the pressure exerted by each component of the gas mixture.

GAS: A phase of matter in which molecules move at high speeds, and therefore exert little or no attraction toward one another.

GAS LAWS: A series of statements concerning the behavior of gases in response to changes in temperature, pressure, and volume. The gas laws, developed by scientists during the seventeenth, eighteenth, and nineteenth centuries, include Avogadro's law, Boyle's law, Charles's law, Dalton's law of partial pressures, Gay-Lussac's law, and Henry's law. These are summed up in the ideal gas law. The kinetic theory of gases is based on observations garnered from these laws.

GAY-LUSSAC'S LAW: A statement, derived by French physicist and chemist Joseph Gay-Lussac (1778-1850), which holds that the pressure of a gas is directly related to its absolute temperature. Hence, the ratio of pressure to absolute temperature is a constant.

HENRY'S LAW: A statement, derived

by English chemist William Henry (1774-1836), which holds that the amount of gas dissolved in a liquid is directly proportional to the partial pressure of the gas above the solution. This holds true only for gases, such as hydrogen and oxygen, which do not react chemically to liquids.

IDEAL GAS LAW: A proposition, also known as the combined gas law, which draws on all the gas laws. The ideal gas law can be expressed as the formula $pV = nRT$, where p stands for pressure, V for volume, n for number of moles, and T for temperature. R is known as the universal gas constant, a figure equal to 0.0821 atm • liter/mole • K.

ISOTHERMAL: Referring to a situation in which temperature is kept constant.

ISOBAROMETRIC: Referring to a situation in which pressure is kept constant.

KINETIC ENERGY: The energy that an object possesses by virtue of its motion.

KINETIC THEORY OF GASES: A set of propositions describing a gas as consisting of numerous molecules, relatively far apart in space, which interact by colliding. These collisions are responsible for the production of thermal energy, because when the velocity of the molecules increases—as it does after collision—the temperature increases as well.

MILLIMETER OF MERCURY: Another name for the torr, abbreviated mm Hg.

MOLE: The SI fundamental unit for "amount of substance." The quantity of molecules or atoms in a mole is, generally speaking, the same as Avogadro's number: $6.022137 \cdot 10^{23}$. However, in the more precise SI definition, a mole is equal to the number of carbon atoms in 12.01 g of carbon.

PARTIAL PRESSURE: When two or more gases are present in a container, partial pressure is the pressure that one of them exerts if it alone is in the container. Dalton's law of partial pressure and Henry's law relate to the partial pressure of gases.

PASCAL: The principle SI or metric unit of pressure, abbreviated "Pa" and equal to $1 N/m^2$.

PRESSURE: The ratio of force to surface area, when force is applied in a direction perpendicular to that surface.

THERMAL ENERGY: A form of kinetic energy—commonly called "heat"—that is produced by the movement of atomic or molecular particles. The greater the motion of these particles relative to one another, the greater the thermal energy.

TORR: An SI unit, also known as the millimeter of mercury, that represents the pressure required to raise a column of mercury 1 mm. The torr, equal to 133 Pascals, is named for the Italian physicist Evangelista Torricelli (1608-1647), who invented the barometer.

much, much more slowly. Two seconds after impact, which is an eternity in terms of the processes involved, the pressure inside the bag has returned to 1 atm.

The chemistry of the airbag is particularly interesting. The bag releases inert, or non-reactive, nitrogen gas, which poses no hazard to human life; yet one of the chemical ingredients in the airbag is so lethal that some environmentalist groups have begun to raise concerns over its presence in airbags. This is sodium azide (NaN_3), one of three compounds—along with potassium nitrate (KNO_3) and silicon dioxide (SiO_2)—present in an airbag prior to inflation.

The sodium azide and potassium nitrate react to one another, producing a burst of hot nitrogen gas in two back-to-back reactions. In the fractions of a second during which this occurs, the airbag becomes like a solid-rocket booster, experiencing a relatively slow detonation known as "deflagration."

The first reaction releases nitrogen gas, which fills the bag, while the second reaction leaves behind the by-products potassium oxide (K_2O) and sodium oxide (Na_2O). These combine with the silicon dioxide to produce a safe, stable compound known as alkaline silicate. The latter, similar to the sand used for making glass, is all that remains in the airbag after the nitrogen gas has escaped.

WHERE TO LEARN MORE

"Atmospheric Pressure: The Force Exerted by the Weight of Air" (Web site). <http://kids.earth.nasa.gov/archive/air_pressure/> (April 7, 2001).

"Chemical Sciences Structure: Structure of Matter: Nature of Gases" University of Alberta Chemistry Department (Web site). <http://www.chem.ualberta.ca/~plambeck/che/struct/s0307.htm> (May 12, 2001).

"Chemistry Units: Gas Laws." <http://bio.bio.rpi.edu/MS99/ausemaW/chem/gases.html> (February 21, 2001).

"Homework: Science: Chemistry: Gases" Channelone.com (Web site). <http://www.channelone.com/fasttrack/science/chemistry/gases.html> (May 12, 2001).

"Kinetic Theory of Gases: A Brief Review" University of Virginia Department of Physics (Web site). <http://www.phys.virginia.edu/classes/252/kinetic_theory.html> (April 15, 2001).

Laws of Gases. New York: Arno Press, 1981.

Macaulay, David. The New Way Things Work. Boston: Houghton Mifflin, 1998.

Mebane, Robert C. and Thomas R. Rybolt. Air and Other Gases. Illustrations by Anni Matsick. New York: Twenty-First Century Books, 1995.

"Tutorials—6." Chemistrycoach.com (Web site). <http://www.chemistrycoach.com/tutorials-6.htm> (February 21, 2001).

Zumdahl, Steven S. Introductory Chemistry: A Foundation, 4th ed. Boston: Houghton Mifflin, 2000.

ATOMS
AND MOLECULES

ATOMS

CONCEPT

Our world is made up of atoms, yet the atomic model of the universe is nonetheless considered a "theory." When scientists know beyond all reasonable doubt that a particular principle is the case, then it is dubbed a law. Laws address the fact that certain things happen, as well as how they happen. A theory, on the other hand, attempts to explain why things happen. By definition, an idea that is dubbed a theory has yet to be fully proven, and such is the case with the atomic theory of matter. After all, the atom cannot be seen, even with electron microscopes—yet its behavior can be studied in terms of its effects. Atomic theory explains a great deal about the universe, including the relationship between chemical elements, and therefore (as with Darwin's theory concerning biological evolution), it is generally accepted as fact. The particulars of this theory, including the means by which it evolved over the centuries, are as dramatic as any detective story. Nonetheless, much still remains to be explained about the atom—particularly with regard to the smallest items it contains.

HOW IT WORKS

WHY STUDY ATOMS?

Many accounts of the atom begin with a history of the growth in scientists' understanding of its structure, but here we will take the opposite approach, first discussing the atom in terms of what physicists and chemists today understand. Only then will we examine the many challenges scientists faced in developing the current atomic model: false starts, wrong theories, right roads not taken, incomplete models. In addition, we will explore the many insights added along the way as, piece by piece, the evidence concerning atomic behavior began to accumulate.

People who are not scientifically trained tend to associate studies of the atom with physics, not chemistry. While it is true that physicists study atomic structure, and that much of what scientists know today about atoms comes from the work of physicists, atomic studies are even more integral to chemistry than to physics. At heart, chemistry is about the interaction of different atomic and molecular structures: their properties, their reactions, and the ways in which they bond.

WHAT THE ATOM MEANS TO CHEMISTRY. Just as a writer in English works with the 26 letters of the alphabet, a chemist works with the 100-plus known elements, the fundamental and indivisible substances of all matter. And what differentiates the elements, ultimately, from one another is not their color or texture, or even the phase of matter—solid, gas, or liquid—in which they are normally found. Rather, the defining characteristic of an element is the atom that forms its basic structure.

The number of protons in an atom is the critical factor in differentiating between elements, while the number of neutrons alongside the protons in the nucleus serves to distinguish one isotope from another. However, as important as elements and even isotopes are to the work of a chemist, the components of the atom's nucleus have little direct bearing on the atomic activity that brings about chemical reactions and chemical bonding. All the chemical "work" of an atom is done by particles vastly smaller in mass than

JOHN DALTON.

either the protons or neutrons—fast-moving little bundles of energy called electrons.

Moving rapidly through the space between the nucleus and the edge of the atom, electrons sometimes become dislodged, causing the atom to become a positively charged ion. Conversely, sometimes an atom takes on one or more electrons, thus acquiring a negative charge. Ions are critical to the formation of some kinds of chemical bonds, but the chemical role of the electron is not limited to ionic bonds.

In fact, what defines an atom's ability to bond with another atom, and therefore to form a molecule, is the specific configuration of its electrons. Furthermore, chemical reactions are the result of changes in the arrangement of electrons, not of any activity involving protons or neutrons. So important are electrons to the interactions studied in chemistry that a separate essay has been devoted to them.

WHAT AN ATOM IS

BASIC ATOMIC STRUCTURE. The definitions of atoms and elements seems, at first glance, almost circular: an element is a substance made up of only one kind of atom, and an atom is the smallest particle of an element that retains all the chemical and physical properties of the element. In fact, these two definitions do not form a closed loop, as they would if it were stated that an element is something made up of atoms. Every item of matter that exists, except for the subatomic particles discussed in this essay, is made up of atoms. An element, on the other hand, is—as stated in its definition—made up of only one kind of atom. "Kind of atom" in this context refers to the number of protons in its nucleus.

Protons are one of three basic subatomic particles, the other two being electrons and neutrons. As we shall see, there appear to be particles even smaller than these, but before approaching these "sub-subatomic" particles, it is necessary to address the three most significant components of an atom. These are distinguished from one another in terms of electric charge: protons are positively charged, electrons are negative in charge, and neutrons have no electrical charge. As with the north and south poles of magnets, positive and negative charges attract one another, whereas like charges repel. Atoms have no net charge, meaning that the protons and electrons cancel out one another.

EVOLVING MODELS OF THE ATOM. Scientists originally thought of an atom as a sort of closed sphere with a relatively hard shell, rather like a ball bearing. Nor did they initially understand that atoms themselves are divisible, consisting of the parts named above. Even as awareness of these three parts emerged in the last years of the nineteenth century and the first part of the twentieth, it was not at all clear how they fit together.

At one point, scientists believed that electrons floated in a cloud of positive charges. This was before the discovery of the nucleus, where the protons and neutrons reside at the heart of the atom. It then became clear that electrons were moving around the nucleus, but how? For a time, a planetary model seemed appropriate: in other words, electrons revolved around the nucleus much as planets orbit the Sun. Eventually, however—as is often the case with scientific discovery—this model became unworkable, and had to be replaced by another.

The model of electron behavior accepted today depicts the electrons as forming a cloud around the nucleus—almost exactly the opposite of what physicists believed a century ago. The use of the term "cloud" may perhaps be a bit misleading, implying as it does something that simply hovers. In fact, the electron, under normal circumstances, is constantly moving. The paths of its movement around the nucleus are nothing like that of a planet's orbit, except inasmuch as both models describe a relatively small object moving around a relatively large one.

The furthest edges of the electron's movement define the outer perimeters of the atom. Rather than being a hard-shelled little nugget of matter, an atom—to restate the metaphor mentioned above—is a cloud of electrons surrounding a nucleus. Its perimeters are thus not sharply delineated, just as there is no distinct barrier between Earth's atmosphere and space itself. Just as the air gets thinner the higher one goes, so it is with an atom: the further a point is from the nucleus, the less the likelihood that an electron will pass that point on a given orbital path.

NUCLEONS

MASS NUMBER AND ATOMIC NUMBER. The term nucleon is used generically to describe the relatively heavy particles that make up an atomic nucleus. Just as "sport" can refer to football, basketball, or baseball, or any other item in a similar class, such as soccer or tennis, "nucleon" refers to protons and neutrons. The sum of protons and neutrons is sometimes called the nucleon number, although a more commonly used term is mass number.

Though the electron is the agent of chemical reactions and bonding, it is the number of protons in the nucleus that defines an atom as to its element. Atoms of the same element always have the same number of protons, and since this figure is unique for a given element, each element is assigned an atomic number equal to the number of protons in its nucleus. The atoms are listed in order of atomic number on the periodic table of elements.

ATOMIC MASS AND ISOTOPES. A proton has a mass of $1.673 \cdot 10^{-24}$ g, which is very close to the established figure for measuring atomic mass, the atomic mass unit. At one time, the basic unit of atomic mass was equal to the mass of one hydrogen atom, but hydrogen is so reactive—that is, it tends to combine readily with other atoms to form a molecule, and hence a compound—that it is difficult to isolate. Instead, the atomic mass unit is today defined as 1/12 of

NIELS BOHR.

the mass of a carbon-12 atom. That figure is exactly $1.66053873 \cdot 10^{-24}$ grams.

The mention of carbon-12, a substance found in all living things, brings up the subject of isotopes. The "12" in carbon-12 refers to its mass number, or the sum of protons and neutrons. Two atoms may be of the same element, and thus have the same number of protons, yet differ in their number of neutrons—which means a difference both in mass number and atomic mass. Such differing atoms of the same element are called isotopes. Isotopes are often designated by symbols showing mass number to the upper left of the chemical symbol or element symbol—for instance, ^{12}C for carbon-12.

ELECTRIC CHARGE. Protons have a positive electric charge of 1, designated either as 1+ or +1. Neutrons, on the other hand, have no electric charge. It appears that the 1+ charge of a proton and the 0 charge of a neutron are the products of electric charges on the part of even smaller particles called quarks. A quark may either have a positive electric charge of less than 1+, in which case it is called an "up quark"; or a negative charge of less than 1−, in which case it is called a "down quark."

Research indicates that a proton contains two up quarks, each with a charge of 2/3+, and

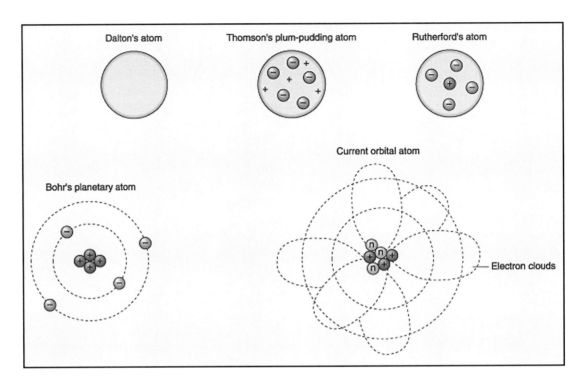

THE EVOLUTION OF ATOMIC THEORY.

one down quark with a charge of 1/3–. This results in a net charge of 1+. On the other hand, a neutron is believed to hold one up quark with a charge of 2/3+, and two down quarks with charges of 1/3– each. Thus, in the neutron, the up and down quarks cancel out one another, and the net charge is zero.

A neutron has about the same mass as a proton, but other than its role in forming isotopes, the neutron's function is not exactly clear. Perhaps, it has been speculated, it binds protons—which, due to their positive charges, tend to repel one another—together at the nucleus.

ELECTRONS

An electron is much smaller than a proton or neutron, and has much less mass; in fact, its mass is equal to 1/1836 that of a proton, and 1/1839 that of a neutron. Yet the area occupied by electrons—the region through which they move—constitutes most of the atom's volume. If the nucleus of an atom were the size of a BB (which, in fact, is billions of times larger than a nucleus), the furthest edge of the atom would be equivalent to the highest ring of seats around an indoor sports arena. Imagine the electrons as incredibly fast-moving insects buzzing constantly through the arena, passing by the BB but then flitting to

the edges or points in between, and you have something approaching an image of the atom's interior.

How fast does an electron move? Speeds vary depending on a number of factors, but it can move nearly as fast as light: 186,000 mi (299,339 km) per second. On the other hand, for an item of matter near absolute zero in temperature, the velocity of the electron is much, much less. In any case, given the fact that an electron has enough negative charge to cancel out that of the proton, it must be highly energized. After all, this would be like an electric generator weighing 1 lb having as much power as a generator that weighed 1 ton.

According to what modern scientists know or hypothesize concerning the inner structure of the atom, electrons are not made up of quarks; rather, they are part of a class of particles called leptons. It appears that leptons, along with quarks and what are called exchange particles, constitute the elementary particles of atoms—particles on a much more fundamental level than that of the proton and neutron.

Electrons are perhaps the most intriguing parts of an atom. Their mass is tiny, even in atomic terms, yet they possess enough charge to counteract a "huge" proton. They are capable, in certain situations, of moving from one atom to another, thus creating ions, and depending on their highly complex configuration and ability to rearrange their configuration, they facilitate or prevent chemical reactions.

REAL-LIFE APPLICATIONS

ANCIENT GREEK THEORIES OF MATTER

The first of the Greek philosophers, and the first individual in Western history who deserves to be called a scientist, was Thales (c. 625-c. 547 B.C.) of Miletus. (Miletus is in Greek Asia Minor, now part of Turkey.) Among his many achievements were the correct prediction of a solar eclipse, and one of the first-ever observations of electricity, when he noted the electrification of amber by friction.

But perhaps the greatest of Thales's legacies was his statement that "Everything is water." This represented the first attempt to characterize the nature of all physical reality. It set off a debate concerning the fundamental nature of matter that consumed Greek philosophers for two centuries. Later, philosophers attempted to characterize matter in terms of fire or air. In time, however, there emerged a school of thought concerned not with identifying matter as one particular thing or another, but with recognizing a structural consistency in all of matter. Among these were the philosophers Leucippus (c. 480-c. 420 B.C.) and his student Democritus (c. 460-370 B.C.)

DEMOCRITUS'S "ATOMS". Leucippus and Democritus proposed a new and highly advanced model for the tiniest point of physical space. Democritus, who actually articulated these ideas (far less is known about Leucippus) began with a "thought experiment," imagining what would happen if an item of matter were subdivided down to its smallest piece. This tiniest fragment, representing an item of matter that could not be cut into smaller pieces, he called by a Greek term meaning "no cut": *atomos.*

Democritus was not necessarily describing matter in a concrete, scientific way: his "atoms" were idealized philosophical constructs rather than purely physical units. Yet, he came amazingly close, and indeed much closer than any thinker for the next 22 centuries, to identifying the fundamental structure of physical reality. Why did it take so long for scientists to come back around to the atomic model? The principal culprit, who advanced an erroneous theory of matter, also happened to be one of the greatest thinkers of all time: Aristotle (384-322 B.C..)

ARISTOTLE'S "ELEMENTS". Aristotle made numerous contributions to science, including his studies in botany and zoology, as well as his explanation of the four causes, a significant attempt to explain events by means other than myth or superstition. In the area of the physical sciences, however, Aristotle's impact was less than beneficial. Most notably, in explaining why objects fall when dropped, he claimed that the ground was their "natural" destination— a fallacy later overturned with the gravitational model developed by Galileo Galilei (1564-1642) and Sir Isaac Newton (1642-1727).

The ideas Aristotle put forward concerning what he called "natural motion" were a product of his equally faulty theories with regard to what today's scientists refer to as chemistry. In ancient times, chemistry, as such, did not exist. Long before Aristotle's time, Egyptian embalmers and metallurgists used chemical processes, but they did so in a practical, applied manner, exerting little effort toward what could be described as scientific theory. Philosophers such as Aristotle, who were some of the first scientists, made little distinction between physical and chemical processes. Thus, whereas physics is understood today as an important background for chemistry, Aristotle's "physics" was actually an outgrowth of his "chemistry."

Rejecting Democritus's atomic model, Aristotle put forward his own view of matter. Like Democritus, he believed that matter was composed of very small components, but these he identified not as atoms, but as "elements": earth, air, fire, and water. He maintained that all objects consisted, in varying degrees, of one or more of these, and based his explanation of gravity on the relative weights of each element. Water sits on top of the earth, he explained, because it is lighter, yet air floats above the water because it is lighter still—and fire, lightest of all, rises highest. Furthermore, he claimed that the planets beyond Earth were made up of a "fifth element," or quintessence, of which little could be known.

In fairness to Aristotle, it should be pointed out that it was not his fault that science all but died out in the Western world during the period from about A.D. 200 to about 1200. Furthermore, he did offer an accurate definition of an element, in a general sense, as "one of those simple bodies into which other bodies can be decomposed, and which itself is not capable of being divided into others." As we shall see, the definition used today is not very different from Aristotle's. However, to define an element scientifically, as modern chemists do, it is necessary to refer to something Aristotle rejected: the atom. So great was his opposition to Democritus's atomic theory, and so enormous was Aristotle's influence on learning for more than 1,500 years following his death, that scientists only began to reconsider atomic theory in the late eighteenth century.

A MATURING CONCEPT OF ELEMENTS

BOYLE'S IDEA OF ELEMENTS. One of the first steps toward an understanding of the chemical elements came with the work of English physicist and chemist Robert Boyle (1627-1691). Building on the usable definition of an element provided by Aristotle, Boyle maintained no substance was an element if it could be broken down into other substances. Thus, air could be eliminated from the list of "elements," because, clearly, it could be separated into more than one elemental substance. (In fact, none of the four "elements" identified by Aristotle even remotely qualifies as an element in modern chemistry.)

Boyle, nonetheless, still clung to aspects of alchemy, a pseudo-science based on the transformation of "base metals," for example, the metamorphosis of iron into gold. Though true chemistry grew out of alchemy, the fundamental proposition of alchemy was faulty: if one metal can be turned into another, then that means that metals are not elements, which, in fact, they are. Nonetheless, Boyle's studies led to the identification of numerous elements—that is, items that really are elements—in the years that followed.

LAVOISIER AND PROUST: CONSTANT COMPOSITION. A few years after Boyle came two French chemists who extended scientific understanding of the elements. Antoine Lavoisier (1743-1794) affirmed the definition of an element as a simple substance that could not be broken down into a simpler substance, and

noted that elements always react with one another in the same proportions.

Joseph-Louis Proust (1754-1826) put forward the law of constant composition, which holds that a given compound always contains the same proportions of mass between elements. Another chemist of the era had claimed that the composition of a compound varies in accordance with the reactants used to produce it. Proust's law of constant composition made it clear that any particular compound will always have the same composition.

EARLY MODERN UNDERSTANDING OF THE ATOM

DALTON AND AVOGADRO: ATOMS AND MOLECULES. The work of Lavoisier and Proust influenced a critical figure in the development of the atomic model: English chemist John Dalton (1766-1844). In *A New System of Chemical Philosophy* (1808), Dalton put forward the idea that nature is composed of tiny particles, and in so doing he adopted Democritus's word "atom" to describe these basic units. This idea, which Dalton had formulated five years earlier, marked the starting-point of modern atomic theory.

Dalton recognized that the structure of atoms in a particular element or compound is uniform, but maintained that compounds are made up of compound atoms: in other words, water, for instance, is a compound of "water atoms." However, water is not an element, and thus, it was necessary to think of its atomic composition in a different way—in terms of molecules rather than atoms. Dalton's contemporary Amedeo Avogadro (1776-1856), an Italian physicist, became the first scientist to clarify the distinction between atoms and molecules.

The later development of the mole, which provided a means whereby equal numbers of molecules could be compared, paid tribute to Avogadro by designating the number of molecules in a mole as "Avogadro's number." Another contemporary, Swedish chemist Jons Berzelius (1779-1848), maintained that equal volumes of gases at the same temperature and pressure contained equal numbers of atoms. Using this idea, he compared the mass of various reacting gases, and developed a system of comparing the mass of various atoms in relation to the lightest one, hydrogen. Berzelius also introduced the system of chemical symbols—H for hydrogen, O for oxygen, and so on—in use today.

BROWNIAN MOTION AND KINETIC THEORY.

Yet another figure whose dates overlapped with those of Dalton, Avogadro, and Berzelius was Scottish botanist Robert Brown (1773-1858). In 1827, Brown noted a phenomenon that later had an enormous impact on the understanding of the atom. While studying pollen grains under a microscope, Brown noticed that the grains underwent a curious zigzagging motion in the water. The pollen assumed the shape of a colloid, a pattern that occurs when particles of one substance are dispersed—but not dissolved—in another substance. At first, Brown assumed that the motion had a biological explanation—that is, it resulted from life processes within the pollen—but later, he discovered that even pollen from long-dead plants behaved in the same way.

Brown never understood what he was witnessing. Nor did a number of other scientists, who began noticing other examples of what came to be known as Brownian motion: the constant but irregular zigzagging of colloidal particles, which can be seen clearly through a microscope. Later, however, Scottish physicist James Clerk Maxwell (1831-1879) and others were able to explain this phenomenon by what came to be known as the kinetic theory of matter.

Kinetic theory is based on the idea that molecules are constantly in motion: hence, the water molecules were moving the pollen grains Brown observed. Pollen grains are many thousands of times as large as water molecules, but since there are so many molecules in even a drop of water, and their motion is so constant but apparently random, they are bound to move a pollen grain once every few thousand collisions.

MENDELEEV AND THE PERIODIC TABLE

In 1869, Russian chemist Dmitri Mendeleev (1834-1907) introduced a highly useful system for organizing the elements, the periodic table. Mendeleev's table is far more than just a handy chart listing elements: at once simple and highly complex, it shows elements in order of increasing atomic mass, and groups together those exhibiting similar forms of chemical behavior and structure.

Reading from right to left and top to bottom, the periodic table, as it is configured today, lists atoms in order of atomic number, generally reflected by a corresponding increase in average atomic mass. As Mendeleev observed, every eighth element on the chart exhibits similar characteristics, and thus the chart is organized in columns representing specific groups of elements.

The patterns Mendeleev observed were so regular that for any "hole" in his table, he predicted that an element would be discovered that would fill that space. For instance, at one point there was a gap between atomic numbers 71 and 73 (lutetium and tantalum, respectively). Mendeleev indicated that an atom would be found for the space, and 15 years after this prediction, the element germanium was isolated.

However, much of what defines an element's place on the chart today relates to subatomic particles—protons, which determine atomic number, and electrons, whose configurations explain certain chemical similarities. Mendeleev was unaware of these particles: from the time he created his table, it was another three decades before the discovery of the first of these particles, the electron. Instead, he listed the elements in an order reflecting outward characteristics now understood to be the result of the quantity and distribution of protons and electrons.

ELECTROMAGNETISM AND RADIATION

The contribution of Mendeleev's contemporary, Maxwell, to the understanding of the atom was not limited to his kinetic theory. Building on the work of British physicist and chemist Michael Faraday (1791-1867) and others, in 1865 he published a paper outlining a theory of a fundamental interaction between electricity and magnetism. The electromagnetic interaction, as it later turned out, explained something that gravitation, the only other form of fundamental interaction known at the time, could not: the force that held together particles in an atom.

The idea of subatomic particles was still a long time in coming, but the model of electromagnetism helped make it possible. In the long run, electromagnetism was understood to encompass a whole spectrum of energy radiation, including radio waves; infrared, visible, and ultraviolet light; x rays; and gamma rays. But this, too, was the product of work on the part of numerous individuals, among whom was English physicist William Crookes (1832-1919).

In the 1870s, Crookes developed an apparatus later termed a Crookes tube, with which he sought to analyze the "rays"—that is, radiation—emitted by metals. The tube consisted of a glass bulb, from which most of the air had been removed, encased between two metal plates or electrodes, referred to as a cathode and an anode. A wire led outside the bulb to an electric source, and when electricity was applied to the electrodes, the cathodes emitted rays. Crookes concluded that the cathode rays were particles with a negative electric charge that came from the metal in the cathode plate.

RADIATION. In 1895, German physicist Wilhelm Röntgen (1845-1923) noticed that photographic plates held near a Crookes tube became fogged, and dubbed the rays that had caused the fogging "x rays." A year after Röntgen's discovery, French physicist Henri Becquerel (1852-1908) left some photographic plates in a drawer with a sample of uranium. Uranium had been discovered more than a century before; however, there were few uses for it until Becquerel discovered that the uranium likewise caused a fogging of the photographic plates.

Thus radioactivity, a type of radiation brought about by atoms that experience radioactive decay was discovered. The term was coined by Polish-French physicist and chemist Marie Curie (1867-1934), who with her husband Pierre (1859-1906), a French physicist, was responsible for the discovery of several radioactive elements.

THE RISE AND FALL OF THE PLUM PUDDING MODEL

Working with a Crookes tube, English physicist J. J. Thomson (1856-1940) hypothesized that the negatively charged particles Crookes had observed were being emitted by atoms, and in 1897, he gave a name to these particles: electrons. The discovery of the electron raised a new question: if Thomson's particles exerted a negative charge, from whence did the counterbalancing positive charge come?

An answer, of sorts, came from William Thomson, not related to the other Thomson and, in any case, better known by his title as Lord Kelvin (1824-1907). Kelvin compared the structure of an atom to an English plum pudding: the electrons were like raisins, floating in a positively charged "pudding"—that is, an undifferentiated cloud of positive charges.

Kelvin's temperature scale contributed greatly to the understanding of molecular motion as encompassed in the kinetic theory of matter. However, his model for the distribution of charges in an atom—charming as it may have been—was incorrect. Nonetheless, for several decades, the "plum pudding model," as it came to be known, remained the most widely accepted depiction of the way that electric charges were distributed in an atom. The overturning of the plum pudding model was the work of English physicist Ernest Rutherford (1871-1937), a student of J. J. Thomson.

RUTHERFORD IDENTIFIES THE NUCLEUS. Rutherford did not set out to disprove the plum pudding model; rather, he was conducting tests to find materials that would block radiation from reaching a photographic plate. The two materials he identified, which were, respectively, positive and negative in electric charge, he dubbed alpha and beta particles. (An alpha particle is a helium nucleus stripped of its electrons, such that it has a positive charge of 2; beta particles are either electrons or positively charged subatomic particles called positrons. The beta particle Rutherford studied was an electron emitted during radioactive decay.)

Using a piece of thin gold foil with photographic plates encircling it, Rutherford bombarded the foil with alpha particles. Most of the alpha particles went straight through the foil—as they should, according to the plum pudding model. However, a few particles were deflected from their course, and some even bounced back. Rutherford later said it was as though he had fired a gun at a piece of tissue paper, only to see the tissue deflect the bullets. Analyzing these results, Rutherford concluded that there was no "pudding" of positive charges: instead, the atom had a positively charged nucleus at its center.

THE NUCLEUS EMERGES

PROTONS AND ISOTOPES. In addition to defining the nucleus, Rutherford also gave a name to the particles that imparted its positive charge: protons. But just as the identification of the electron had raised new questions that, in being answered, led to the discovery of the proton, Rutherford's achievement only brought up new anomalies concerning the behavior of the nucleus.

Together with English chemist Frederick Soddy (1877-1956), Rutherford discovered that when an atom emitted alpha or beta particles, its atomic mass changed. Soddy had a name for atoms that displayed this type of behavior: isotopes. Certain types of isotopes, Soddy and Rutherford went on to conclude, had a tendency to decay, moving toward stabilization, and this decay explained radioactivity.

CLARIFYING THE PERIODIC TABLE. Soddy concluded that atomic mass, as measured by Berzelius, was actually an average of the mass figures for all isotopes within that element. This explained a problem with Mendeleev's periodic table, in which there seemed to be irregularities in the increase of atomic mass from element to element. The answer to these variations in mass, it turned out, related to the number of isotopes associated with a given element: the greater the number of isotopes, the more these affected the overall measure of the element's mass.

By this point, physicists and chemists had come to understand that various levels of energy in matter emitted specific electromagnetic wavelengths. Welsh physicist Henry Moseley (1887-1915) experimented with x rays, bombarding atoms of different elements with high levels of energy and observing the light they gave off as they cooled. In the course of these tests, he uncovered an astounding mathematical relationship: the amount of energy a given element emitted was related to its atomic number.

Furthermore, the atomic number corresponded to the number of positive charges—this was in 1913, before Rutherford had named the proton—in the nucleus. Mendeleev had been able to predict the discovery of new elements, but such predictions had remained problematic. When scientists understood the idea of atomic number, however, it became possible to predict the existence of undiscovered elements with much greater accuracy.

NEUTRONS. Yet again, discoveries—the nucleus, protons, and the relationship between these and atomic number—only created new questions. (This, indeed, is one of the hallmarks of an active scientific theory. Rather than settling questions, science is about raising new ones, and thus improving the quality of the questions that

are asked.) Once Rutherford had identified the proton, and Moseley had established the number of protons, the mystery at the heart of the atom only grew deeper.

Scientists had found that the measured mass of atoms could not be accounted for by the number of protons they contained. Certainly, the electrons had little to do with atomic mass: by then it had been shown that the electron weighed about 0.06% as much as a proton. Yet for all elements other than protium (the first of three hydrogen isotopes), there was a discrepancy between atomic mass and atomic number. Clearly, there had to be something else inside the nucleus.

In 1932, English physicist James Chadwick (1891-1974) identified that "something else." Working with radioactive material, he found that a certain type of subatomic particle could penetrate lead. All other known types of radiation were stopped by the lead, and therefore, Chadwick reasoned that this particle must be neutral in charge. In 1932, he won the Nobel Prize in Physics for his discovery of the neutron.

THE NUCLEAR EXPLOSION

ISOTOPES AND RADIOACTIVITY. Chadwick's discovery clarified another mystery, that of the isotope, which had been raised by Rutherford and Soddy several decades earlier. Obviously, the number of protons in a nucleus did not change, but until the identification of the neutron, it had not been clear what it was that did change. At that point, it was understood that two atoms may have the same atomic number—and hence be of the same element—yet they may differ in number of neutrons, and thus be isotopes.

As the image of what an isotope was became clearer, so too did scientists' comprehension of radioactivity. Radioactivity, it was discovered, was most intense where an isotope was the most unstable—that is, in cases where an isotope had the greatest tendency to experience decay. Uranium had a number of radioactive isotopes, such as ^{235}U, and these found application in the burgeoning realm of nuclear power—both the destructive power of atomic bombs, and later the constructive power of nuclear energy plants.

FISSION VS. FUSION. In nuclear fission, or the splitting of atoms, uranium isotopes (or other radioactive isotopes) are bombarded with neutrons, splitting the uranium nucleus in half and releasing huge amounts of energy. As the nucleus is halved, it emits several extra neutrons, which spin off and split more uranium nuclei, creating still more energy and setting off a chain reaction. This explains the destructive power in an atomic bomb, as well as the constructive power—providing energy to homes and businesses—in a nuclear power plant. Whereas the chain reaction in an atomic bomb becomes an uncontrolled explosion, in a nuclear plant the reaction is slowed and controlled.

Yet nuclear fission is not the most powerful form of atomic reaction. As soon as scientists realized that it was possible to force particles out of a nucleus, they began to wonder if particles could be forced into the nucleus. This type of reaction, known as fusion, puts even nuclear fission, with its awesome capabilities, to shame: nuclear fusion is, after all, the power of the Sun. On the surface of that great star, hydrogen atoms reach incredible temperatures, and their nuclei fuse to create helium. In other words, one element actually transforms into another, releasing enormous amounts of energy in the process.

NUCLEAR ENERGY IN WAR AND PEACE. The atomic bombs dropped by the United States on Japan in 1945 were fission bombs. These were the creation of a group of scientists—legendary figures such as American physicist J. Robert Oppenheimer (1904-1967), American mathematician John von Neumann (1903-1957), American physicist Edward Teller (1908-), and Italian physicist Enrico Fermi (1901-1954)—involved in the Manhattan Project at Las Alamos, New Mexico.

Some of these geniuses, particularly Oppenheimer, were ambivalent about the moral implications of the enormous destructive power they created. However, most military historians believe that far more lives—both Japanese and American—would have been lost if America had been forced to conduct a land invasion of Japan. As it was, the Japanese surrendered shortly after the cities of Hiroshima and Nagasaki suffered the devastating effects of fission-based explosions.

By 1952, U.S. scientists had developed a "hydrogen," or fusion bomb, thus raising the stakes greatly. This was a bomb that possessed far more destructive capability than the ones dropped over Japan. Fortunately, the Hiroshima and Nagasaki bombs were the only ones dropped in wartime, and a ban on atmospheric nuclear testing has greatly reduced the chances of human exposure to nuclear fallout of any kind. With the end of the arms race between the United States and the Soviet Union, the threat of nuclear destruction has receded somewhat—though it will perhaps always be a part of human life.

Nonetheless, fear of nuclear power, spawned as a result of the arms race, continues to cloud the future of nuclear plants that generate electricity—even though these, in fact, emit less radioactive pollution than coal- or gas-burning power plants. At the same time, scientists continue to work on developing a process of power generation by means of nuclear fusion, which, if and when it is achieved, will be one of the great miracles of science.

PARTICLE ACCELERATORS. One of the tools used by scientists researching nuclear fusion is the particle accelerator, which moves streams of charged particles—protons, for instance—faster and faster. These fast particles are then aimed at a thin plate composed of a light element, such as lithium. If the proton manages to be "captured" in the nucleus of a lithium atom, the resulting nucleus is unstable, and breaks into alpha particles.

This method of induced radioactivity is among the most oft- used means of studying nuclear structure and subatomic particles. In 1932, the same year that Chadwick discovered the neutron, English physicist John D. Cockcroft (1897-1967) and Irish physicist Ernest Walton (1903-1995) built the first particle accelerator. Some particle accelerators today race the particles in long straight lines or, to save space, in ringed paths several miles in diameter.

QUANTUM THEORY AND BEYOND

THE CONTRIBUTION OF RELATIVITY. It may seem strange that in this lengthy (though, in fact, quite abbreviated!) overview of developments in understanding of the atom, no mention has been made of the figure most associated with the atom in the popular mind: German-American physicist Albert Einstein (1879-1955). The reasons for this are several. Einstein's relativity theory addresses physical, rather than chemical, processes, and did not directly contribute to enhanced understanding of atomic structure or elements. The heart of relativity theory is the famous formula $E = mc^2$, which means that every item of matter possesses energy proportional to its mass multiplied by the squared speed of light.

The value of mc^2, of course, is an enormous amount of energy, and in order to be released in significant quantities, an article of matter must experience the kinetic energy associated with very, very high speeds—speeds close to that of light. Obviously, the easiest thing to accelerate to such a speed is an atom, and hence, nuclear energy is a result of Einstein's famous equation. Nonetheless, it should be stressed that although Einstein is associated with unlocking the power of the atom, he did little to explain what atoms are.

However, in the course of developing his relativity theory in 1905, Einstein put to rest a question about atoms and molecules that still remained unsettled after more than a century. Einstein's analysis of Brownian motion, combined with the confirmation of his results by French physicist Jean Baptiste Perrin (1870-1942), showed conclusively that yes, atoms and molecules do exist. It may seem amazing that as recently as 1905, this was still in doubt; however, this only serves to illustrate the arduous path scientists must tread in developing a theory that accurately explains the world.

PLANCK'S QUANTUM THEORY. A figure whose name deserves to be as much a household word as Einstein's—though it is not—is German physicist Max Planck (1858-1947). It was Planck who initiated the quantum theory that Einstein developed further, a theory that prevails today in the physical sciences.

At the atomic level, Planck showed, energy is emitted in tiny packets or "quanta." Each of these energy packets is indivisible, and the behavior of quanta redefine the old rules of physics handed down from Newton and Maxwell. Thus, it is Planck's quantum theory, rather than Einstein's relativity, that truly marks the watershed, or "before and after," between classical physics and modern physics.

Quantum theory is important not only to physics, but to chemistry as well. It helps to explain the energy levels of electrons, which are not continuous, as in a spectrum, but jump between certain discrete points. The quantum model is now also applied to the overall behavior

of the electron; but before this could be fully achieved, scientists had to develop a new understanding of the way electrons move around the nucleus.

BOHR'S PLANETARY MODEL OF THE ATOM. As was often the case in the history of the atom, a man otherwise respected as a great scientist put forward a theory of atomic structure that at first seemed convincing, but ultimately turned out to be inaccurate. In this case, it was Danish physicist Niels Bohr (1885-1962), a seminal figure in the development of nuclear fission.

Using the observation, derived from quantum theory, that electrons only occupied specific energy levels, Bohr hypothesized that electrons orbited around a nucleus in the same way that planets orbit the Sun. There is no reason to believe that Bohr formed this hypothesis for any sentimental reasons—though, of course, scientists are just as capable of prejudice as anyone. His work was based on his studies; nonetheless, it is easy to see how this model seemed appealing, showing as it did an order at the subatomic level reflecting an order in the heavens.

ELECTRON CLOUDS. Many people today who are not scientifically trained continue to think that an atom is structured much like the Solar System. This image is reinforced by symbolism, inherited from the 1950s, that represents "nuclear power" by showing a dot (the nucleus) surrounded by ovals at angles to one another, representing the orbital paths of electrons. However, by the 1950s, this model of the atom had already been overturned.

In 1923, French physicist Louis de Broglie (1892-1987) introduced the particle-wave hypothesis, which indicated that electrons could sometimes have the properties of waves—an eventuality not encompassed in the Bohr model. It became clear that though Bohr was correct in maintaining that electrons occupy specific energy levels, his planetary model was inadequate for explaining the behavior of electrons.

Two years later, in 1925, German physicist Werner Heisenberg (1901-1976) introduced what came to be known as the Heisenberg Uncertainty Principle, showing that the precise position and speed of an electron cannot be known at the same time. Austrian physicist Erwin Schrödinger (1887-1961) developed an

KEY TERMS

ATOM: The smallest particle of an element that retains all the chemical and physical properties of the element. An atom can exist either alone or in combination with other atoms in a molecule. Atoms are made up of protons, neutrons, and electrons.

ATOMIC MASS UNIT: An SI unit (abbreviated amu), equal to $1.66 \cdot 10^{-24}$ g, for measuring the mass of atoms.

ATOMIC NUMBER: The number of protons in the nucleus of an atom. Since this number is different for each element, elements are listed on the periodic table of elements in order of atomic number.

AVERAGE ATOMIC MASS: A figure used by chemists to specify the mass—in atomic mass units—of the average atom in a large sample.

CHEMICAL SYMBOL: A one- or two-letter abbreviation for the name of an element.

COMPOUND: A substance made up of atoms of more than one element. These atoms are usually joined in molecules.

ELECTRON: Negatively charged particles in an atom. Electrons, which spin around the protons and neutrons that make up the atom's nucleus, constitute a very small portion of the atom's mass. The number of electrons and protons is the same, thus canceling out one another; on the other hand, if an atom loses or gains electrons, it becomes an ion.

ELEMENT SYMBOL: Another term for chemical symbol.

ION: An atom or atoms that has lost or gained one or more electrons, and thus has a net electric charge.

ISOTOPES: Atoms that have an equal number of protons, and hence are of the same element, but differ in their number of neutrons.

MASS NUMBER: The sum of protons and neutrons in an atom's nucleus.

MOLECULE: A group of atoms, usually (but not always) representing more than one element, joined in a structure. Compounds are typically made up of molecules.

equation for calculating how an electron with a certain energy moves, identifying regions in an atom where an electron possessing a certain energy level is likely to be. Schrödinger's equation cannot, however, identify the location exactly.

Rather than being called orbits, which suggest the orderly pattern of Bohr's model, Schrödinger's regions of probability are called orbitals. Moving within these orbitals, electrons describe the shape of a cloud, as discussed much earlier in this essay; as a result, the "electron cloud" theory prevails today. This theory incorporates aspects of Bohr's model, inasmuch as electrons move from one orbital to another by absorbing or emitting a quantum of energy.

WHERE TO LEARN MORE

"The Atom." Thinkquest (Web site). <http://library.thinkquest.org/17940/texts/atom/atom.html> (May 18, 2001).

"Elements" (Web site). <http://home.school.net.hk/~chem/main/F5notes/atom/element.html> (May 18, 2001).

"Explore the Atom" CERN—European Organization for Nuclear Research (Web site). <http://public.web.cern.ch/Public/SCIENCE/Welcome.html> (May 18, 2001).

KEY TERMS CONTINUED

NEUTRON: A subatomic particle that has no electric charge. Neutrons are found at the nucleus of an atom, alongside protons.

NUCLEON: A generic term for the heavy particles—protons and neutrons—that make up the nucleus of an atom.

NUCLEON NUMBER: Another term for mass number.

NUCLEUS: The center of an atom, a region where protons and neutrons are located, and around which electrons spin.

PERIODIC TABLE OF ELEMENTS: A chart that shows the elements arranged in order of atomic number, along with chemical symbol and the average atomic mass (in atomic mass units) for that particular element. Vertical columns within the periodic table indicate groups or "families" of elements with similar chemical characteristics.

PROTON: A positively charged particle in an atom. Protons and neutrons, which together form the nucleus around which electrons spin, have approximately the same mass—a mass that is many times greater than that of an electron.

QUARK: A particle believed to be a component of protons and neutrons. A quark may either have a positive electric charge of less than 1+, in which case it is called an "up quark"; or a negative charge of less than 1-, in which case it is called a "down quark."

RADIATION: In a general sense, radiation can refer to anything that travels in a stream, whether that stream be composed of subatomic particles or electromagnetic waves. In a more specific sense, the term relates to the radiation from radioactive materials, which can be harmful to human beings.

RADIOACTIVITY: A term describing a phenomenon whereby certain isotopes are subject to a form of decay brought about by the emission of high-energy particles or radiation, such as alpha particles, beta particles, or gamma rays.

Gallant, Roy A. *The Ever-Changing Atom.* New York: Benchmark Books, 1999.

Goldstein, Natalie. *The Nature of the Atom.* New York: Rosen Publishing Group, 2001.

"A Look Inside the Atom" (Web site). <http://www.aip.org/history/electron/jjhome.htm> (May 18, 2001).

"Portrait of the Atom" (Web site). <http://www.inetarena.com/~pdx4d/snelson/Portrait.html> (May 18, 2001).

"A Science Odyssey: You Try It: Atom Builder." PBS—Public Broadcasting System (Web site). <http://www.pbs.org/wgbh/aso/tryit/atom/> (May 18, 2001).

Spangenburg, Ray and Diane K. Moser. *The History of Science in the Nineteenth Century.* New York: Facts on File, 1994.

Zumdahl, Steven S. *Introductory Chemistry: A Foundation,* 4th ed. Boston: Houghton Mifflin, 2000.

ATOMIC MASS

CONCEPT

Every known item of matter in the universe has some amount of mass, even if it is very small. But what about something so insignificant in mass that comparing it to a gram is like comparing a millimeter to the distance between Earth and the nearest galaxy? Obviously, special units are needed for such measurements; then again, one might ask why it is necessary to weigh atoms at all. One answer is that everything is made of atoms. More specifically, the work of a chemist requires the use of accurate atomic proportions in forming the molecules that make up a compound. The measurement of atomic mass was thus a historic challenge that had to be overcome, and the story of the ways that scientists met this challenge is an intriguing one.

HOW IT WORKS

WHY MASS AND NOT WEIGHT?

Some textbooks and other sources use the term atomic weight instead of atomic mass. The first of these is not as accurate as the second, which explains why atomic mass was chosen as the subject of this essay. Indeed, the use of "atomic weight" today merely reflects the fact that scientists in the past used that expression and spoke of "weighing" atoms. Though "weigh" is used as a verb in this essay, this is only because it is less cumbersome than "measure the mass of." (In addition, "atomic weight" may be mentioned when discussing studies by scientists of the nineteenth century, who applied that term rather than atomic mass.)

One might ask why such pains have been taken to make the distinction. Mass is, after all, basically the same as weight, is it not? In fact it is not, though people are accustomed to thinking in those terms since most weight scales provide measurements in both pounds and kilograms. However, the pound is a unit of weight in the English system, whereas a kilogram is a unit of mass in the metric and SI systems. Though the two are relatively convertible on Earth, they are actually quite different. (Of course it would make no more sense to measure atoms in pounds or kilograms than to measure the width of a hair in light-years; but pounds and kilograms are the most familiar units of weight and mass respectively.)

Weight is a measure of force affected by Earth's gravitational pull. Therefore a person's weight varies according to gravity, and would be different if measured on the Moon, whereas mass is the same throughout the universe. Its invariability makes mass preferable to weight as a parameter of scientific measure.

PUTTING AN ATOM'S SIZE AND MASS IN CONTEXT

Mass does not necessarily relate to size, though there is enough of a loose correlation that more often than not, we can say that an item of very small size will have very small mass. And atoms are very, very small—so much so that, until the early twentieth century, chemists and physicists had no accurate means of isolating them to determine their mass.

The diameter of an atom is about 10^{-8} cm. This is equal to about 0.000000003937 in—or to put it another way, an inch is about as long as 250 million atoms lined up side by side. Obviously, special units are required for describing the size of atoms. Usually, measurements are provided in terms of the angstrom, equal to 10^{-10} m. (In other words, there are 10 million angstroms in a millimeter.)

Measuring the spatial dimensions of an atom, however, is not nearly as important for chemists' laboratory work as measuring its mass. The mass of an atom is almost inconceivably small. It takes about $5.0 \cdot 10^{23}$ carbon atoms to equal just one gram in mass. At first, 10^{23} does not seem like such a huge number, until one considers that 10^6 is already a million, meaning that 10^{23} is a million times a million times a million times 100,000. If $5.0 \cdot 10^{23}$ angstrom lengths—angstroms, not meters or even millimeters—were laid end to end, they would stretch from Earth to the Sun and back 107,765 times!

ATOMIC MASS UNITS

It is obvious, then, that an entirely different unit is needed for measuring the mass of an atom, and for this purpose, chemists and other scientists use an atom mass unit (abbreviated amu), which is equal to $1.66 \cdot 10^{-24}$ g.

Though the abbreviation amu is used in this book, atomic mass units are sometimes designated simply by a u. On the other hand, they may be presented as numbers without any unit of measure—as for instance on the periodic table of elements.

Within the context of biochemistry and microbiology, often the term dalton (abbreviated Da or D) is used. This is useful for describing the mass of large organic molecules, typically rendered in kilodaltons (kDa). The Latin prefix kilo- indicates 1,000 of something, and "kilodalton" is much less of a tongue-twister than "kilo-amu". The term "dalton" honors English chemist John Dalton (1766-1844), who, as we shall see, introduced the concept of the atom to science.

AVERAGE ATOMIC MASS. Since 1960, when its value was standardized, the atomic mass unit has been officially known as the "unified atomic mass unit." The addition of the word "unified" reflects the fact that atoms are not weighed individually—a labor that would be problematic at the very least. In any case, to do so would be to reinvent the wheel, as it were,

ERNEST RUTHERFORD.

because average atomic mass figures have been established for each element.

Average atomic mass figures range from 1.008 amu for hydrogen, the first element listed on the periodic table of elements, to over 250 amu for elements of very high atomic number. Figures for average atomic mass can be used to determine the average mass of a molecule as well, since a molecule is just a group of atoms joined in a structure. The mass of a molecule can be determined simply by adding together average atomic mass figures for each atom the molecule contains. A water molecule, for instance, consists of two hydrogen atoms and one oxygen atom; therefore, its mass is equal to the average atomic mass of hydrogen multiplied by two, and added to the average atomic mass of oxygen.

AVOGADRO'S NUMBER AND THE MOLE

Atomic mass units and average atomic mass are not the only components necessary for obtaining accurate mass figures where atoms are concerned. Obviously, as suggested several times already, it would be fruitless to determine the mass of individual atoms or molecules. Nor would it do to measure the mass of a few hundred, or even a few million, of these particles.

FREDERICK SODDY.

After all, as we have seen, it takes about 500,000 trillion million carbon atoms to equal just one gram—and a gram, after all, is still rather small in mass compared to most objects encountered in daily life. (There are 1,000 g in a kilogram, and a pound is equal to about 454 g.)

In addition, scientists need some means for comparing atoms or molecules of different substances in such a way that they know they are analyzing equal numbers of particles. This cannot be done in terms of mass, because the number of atoms in each sample varies: a gram of hydrogen, for instance, contains about 12 times as many atoms as a gram of carbon, which has an average atomic mass of 12.01 amu. What is needed, instead, is a way to designate a certain number of atoms or molecules, such that accurate comparisons are possible.

In order to do this, chemists make use of a figure known as Avogadro's number. Named after Italian physicist Amedeo Avogadro (1776-1856), it is equal to 6.022137×10^{23}. Avogadro's number, which is 6,022,137 followed by 17 zeroes, designates the quantity of molecules in a mole (abbreviated mol). The mole, a fundamental SI unit for "amount of substance," is defined precisely as the number of carbon atoms in 12.01 g of carbon. It is here that the value of Avogadro's number

becomes clear: as noted, carbon has an average atomic mass of 12.01 amu, and multiplication of the average atomic mass by Avogadro's number yields a figure in grams equal to the value of the average atomic mass in atomic mass units.

REAL-LIFE APPLICATIONS

EARLY IDEAS OF ATOMIC MASS

Dalton was not the first to put forth the idea of the atom: that concept, originated by the ancient Greeks, had been around for more than 2,000 years. However, atomic theory had never taken hold in the world of science—or, at least, what passed for science prior to the seventeenth century revolution in thinking brought about by Galileo Galilei (1564-1642) and others.

Influenced by several distinguished predecessors, Dalton in 1803 formulated the theory that nature is formed of tiny particles, an idea he presented in *A New System of Chemical Philosophy* (1808). Dalton was the first to treat atoms as fully physical constructs; by contrast, ancient proponents of atomism conceived these fundamental particles in ideal or spiritual terms. Dalton described atoms as hard, solid, indivisible particles with no inner spaces—a definition that did not endure, as later scientific inquiry revealed the complexities of the atom. Yet he was correct in identifying atoms as having weight—or, as scientists say today, mass.

THE FIRST TABLE OF ATOMIC WEIGHTS. The question was, how could anyone determine the weight of something as small as an atom? A year after the publication of Dalton's book, a discovery by French chemist and physicist Joseph Gay-Lussac (1778-1850) and German naturalist Alexander von Humboldt (1769-1859) offered a clue. Humboldt and Gay-Lussac—famous for his gas law associating pressure and temperature—found that gases combine to form compounds in simple proportions by volume.

For instance, as Humboldt and Gay-Lussac discovered, water is composed of only two elements: hydrogen and oxygen, and these two combine in a whole-number ratio of 8:1. By separating water into its components, they found that for every part of oxygen, there were eight parts of hydrogen. Today we know that water molecules are formed by two hydrogen atoms, with an average atomic mass of 1.008 amu each, and one oxygen atom. The ratio between the average atomic mass of oxygen (16.00 amu) and that of the two hydrogen atoms is indeed very nearly 8:1.

In the early nineteenth century, however, chemists had no concept of molecular structure, or any knowledge of the atomic masses of elements. They could only go on guesswork: hence Dalton, in preparing the world's first "Table of Atomic Weights," had to make some assumptions based on Humboldt's and Gay-Lussac's findings. Presumably, Dalton reasoned, only one atom of hydrogen combines with one atom of oxygen to form a "water atom." He assigned to hydrogen a weight of 1, and according to this, calculated the weight of oxygen as 8.

AVOGADRO AND BERZELIUS IMPROVE ON DALTON'S WORK. The implications of Gay-Lussac's discovery that substances combined in whole-number ratios were astounding. (Gay-Lussac, who studied gases for much of his career, is usually given more credit than Humboldt, an explorer and botanist who had his hand in many things.) On the one hand, the more scientists learned about nature, the more complex it seemed; yet here was something amazingly simple. Instead of combining in proportions of, say, 8.3907 to 1.4723, oxygen and hydrogen molecules formed a nice, clean, ratio of 8 to 1. This served to illustrate the fact that, as Dalton had stated, the fundamental particles of matter must be incredibly tiny; otherwise, it would be impossible for every possible quantity of hydrogen and oxygen in water to have the same ratio.

Intrigued by the work of Gay-Lussac, Avogadro in 1811 proposed that equal volumes of gases have the same number of particles if measured at the same temperature and pressure. He also went on to address a problem raised by Dalton's work. If atoms were indivisible, as Dalton had indicated, how could oxygen exist both as its own atom and also as part of a water "atom"? Water, as Avogadro correctly hypothesized, is not composed of atoms but of molecules, which are themselves formed by the joining of two hydrogen atoms with one oxygen atom.

Avogadro's molecular theory opened the way to the clarification of atomic mass and the development of the mole, which, as we have seen, makes it possible to determine mass for large quantities of molecules. However, his ideas did not immediately gain acceptance. Only in 1860, four years after Avogadro's death, did Italian chemist Stanislao Cannizzaro (1826-1910) resurrect the concept of the molecule as a way of addressing disagreements among scientists regarding the determination of atomic mass.

In the meantime, Swedish chemist Jons Berzelius (1779-1848) had adopted Dalton's method of comparing all "atomic weights" to that of hydrogen. In 1828, Berzelius published a table of atomic weights, listing 54 elements along with their weight relative to that of hydrogen. Thus carbon, in Berzelius's system, had a weight of 12. Unlike Dalton's figures, Berzelius's are very close to those used by scientists today. By the time Russian chemist Dmitri Ivanovitch Mendeleev (1834-1907) created his periodic table in 1869, there were 63 known elements. That first table retained the system of measuring atomic mass in comparison to hydrogen.

THE DISCOVERY OF SUBATOMIC STRUCTURES

Until scientists began to discover the existence of subatomic structures, measurements of atomic mass could not really progress. Then in 1897, English physicist J. J. Thomson (1856-1940) identified the electron. A particle possessing negative charge, the electron contributes little to an atom's mass, but it pointed the way to the existence of other particles within an atom. First of all, there had to be a positive charge to offset that of the electron, and secondly, the item or items providing this positive charge had to account for the majority of the atom's mass.

Early in the twentieth century, Thomson's student Ernest Rutherford (1871-1937) discovered that the atom has a nucleus, a center around which electrons move, and that the nucleus contains positively charged particles called protons. Protons have a mass 1,836 times as great as that of an electron, and thus seemed to account for the total atomic mass. Later, however, Rutherford and English chemist Frederick Soddy (1877-1956) discovered that when an atom emitted certain types of particles, its atomic mass changed.

ISOTOPES AND ATOMIC MASS. Rutherford and Soddy named these atoms of differing mass isotopes, though at that point—because the neutron had yet to be discovered—

they did not know exactly what had caused the change in mass. Certain types of isotopes, Soddy and Rutherford concluded, had a tendency to decay, moving (sometimes over a great period of time) toward stabilization. Such isotopes were radioactive.

Soddy concluded that atomic mass, as measured by Berzelius, was actually an average of the mass figures for all isotopes within that element. This explained a problem with Mendeleev's periodic table, in which there seemed to be irregularities in the increase of atomic mass from element to element. The answer to these variations in mass, it turned out, related to the number of isotopes associated with a given element: the greater the number of isotopes, the more these affected the overall measure of the element's mass.

A NEW DEFINITION OF ATOMIC NUMBER. Up to this point, the term "atomic number" had a different, much less precise, meaning than it does today. As we have seen, the early twentieth century periodic table listed elements in order of their atomic mass in relation to hydrogen, and thus atomic number referred simply to an element's position in this ordering. Then, just a few years after Rutherford and Soddy discovered isotopes, Welsh physicist Henry Moseley (1887-1915) determined that every element has a unique number of protons in its nucleus.

Today, the number of protons in the nucleus, rather than the mass of the atom, determines the atomic number of an element. Carbon, for instance, has an atomic number of 6, not because there are five elements lighter—though this is also true—but because it has six protons in its nucleus. The ordering by atomic number happens to correspond to the ordering by atomic mass, but atomic number provides a much more precise means of distinguishing elements. For one thing, atomic number is always a whole integer—1 for hydrogen, for instance, or 17 for chlorine, or 92 for uranium. Figures for mass, on the other hand, are almost always rendered with decimal fractions (for example, 1.008 for hydrogen).

NEUTRONS COMPLETE THE PICTURE. As with many other discoveries along the way to uncovering the structure of the atom, Moseley's identification of atomic number with the proton raised still more questions. In particular, if the unique number of protons identified an element, what was it that made isotopes

of the same element different from one another? Hydrogen, as it turned out, indeed had a mass very nearly equal to that of one proton—thus justifying its designation as the basic unit of atomic mass. Were it not for the isotope known as deuterium, which has a mass nearly twice as great as that of hydrogen, the element would have an atomic mass of exactly 1 amu.

A discovery by English physicist James Chadwick (1891-1974) in 1932 finally explained what made an isotope an isotope. It was Chadwick who identified the neutron, a particle with no electric charge, which resides in the nucleus alongside the protons. In deuterium, which has one proton, one neutron, and one electron, the electron accounts for only 0.0272% of the total mass—a negligible figure. The proton, on the other hand, makes up 49.9392% of the mass. Until the discovery of the neutron, there had been no explanation of the other 50.0336% of the mass in an atom with just one proton and one electron.

AVERAGE ATOMIC MASS TODAY

Thanks to Chadwick's discovery of the neutron, it became clear why deuterium weighs almost twice as much as ordinary hydrogen. This in turn is the reason why a large sample of hydrogen, containing as it does a few molecules of deuterium here and there, does not have the same average atomic mass as a proton. Today scientists know that there are literally thousands of isotopes—many of them stable, but many more of them unstable or radioactive—for the 100-plus elements on the periodic table. Each isotope, of course, has a slightly different atomic mass. This realization has led to clarification of atomic mass figures.

One might ask how figures of atomic mass are determined. In the past, as we have seen, it was largely a matter of guesswork, but today chemists and physicist use a highly sophisticated instrument called a mass spectrometer. First, atoms are vaporized, then changed to positively charged ions, or cations, by "knocking off" electrons. The cations are then passed through a magnetic field, and this causes them to be deflected by specific amounts, depending on the size of the charge and its atomic mass. The particles eventually wind up on a deflector plate, where the amount of deflection can be measured and compared with the charge. Since 1 amu has been calculated to equal approximately 931.494 MeV, or mega electron-volts, very accurate figures can be determined.

CALIBRATION OF THE ATOMIC MASS UNIT. When 1 is divided by Avogadro's number, the result is $1.66 \cdot 10^{-24}$—the value, in grams, of 1 amu. However, in accordance with a 1960 agreement among members of the international scientific community, measurements of atomic mass take as their reference point the mass of carbon-12. Not only is the carbon-12 isotope found in all living things, but hydrogen is a problematic standard because it bonds so readily with other elements. According to the 1960 agreement, 1 amu is officially 1/12 the mass of a carbon-12 atom, whose exact value (retested in 1998), is $1.6653873 \cdot 10^{-24}$ g.

Carbon-12, sometimes represented as $^{12}_{6}C$, contains six protons and six neutrons. (As explained in the essay on Isotopes, where an isotope is indicated, the number to the upper left of the chemical symbol indicates the total number of protons and neutrons. Sometimes this is the only number shown; but if a number is included on the lower left, this indicates only the number of protons, which remains the same for each element.) The value of 1 amu thus obtained is, in effect, an average of the mass for a proton and neutron—a usable figure, given the fact that a neutron weighs only 0.163% more than a proton.

Of all the carbon found in nature (as opposed to radioactive isotopes created in laboratories), 98.89% of it is carbon-12. The remainder is mostly carbon-13, with traces of carbon-14, an unstable isotope produced in nature. By definition, carbon-12 has an atomic mass of exactly 12 amu; that of carbon-13 (about 1.11% of all carbon) is 13 amu. Thus the atomic mass of carbon, listed on the periodic table as 12.01 amu, is obtained by taking 98.89% of the mass of carbon-12, combined with 1.11% of the mass of carbon-13.

ATOMIC MASS UNITS AND THE PERIODIC TABLE. The periodic table as it is used today includes figures, in atomic mass units, for the average mass of each atom. As it turns out, Berzelius was not so far off in his use of hydrogen as a standard, since its mass is almost exactly 1 amu—but not quite, because (as noted above) deuterium increases the average mass somewhat. Figures increase from there along the

periodic table, though not by a regular pattern. Sometimes the increase from one element to the next is by just over 1 amu, and in other cases, the increase is by more than 3 amu. This only serves to prove that atomic number, rather than atomic mass, is a more straightforward means of ordering the elements.

Mass figures for many elements that tend to appear in the form of radioactive isotopes are usually shown in parentheses. This is particularly true for elements with atomic numbers above 92, because samples of these elements do not stay around long enough to be measured. Some have a half-life—the period in which half the isotopes decay to a stable form—of just a few minutes, and for others, the half-life is but a fraction of a second. Therefore, atomic mass figures represent the mass of the longest-lived isotope.

USES OF ATOMIC MASS IN CHEMISTRY

MOLAR MASS. Just as the value of atomic mass units has been calibrated to the mass of carbon-12, the mole is no longer officially defined in terms of Avogadro's number, though in general its value has not changed. By international scientific agreement, the mole equals the number of carbon atoms in 12.01 g of carbon. Note that, as stated earlier, carbon has an average atomic mass of 12.01 amu.

This is no coincidence, of course: multiplication of the average atomic mass by Avogadro's number yields a figure in grams equal to the value of the average atomic mass in amu. A mole of helium, with an average atomic mass of 4.003, is 4.003 g. Iron, on the other hand, has an average atomic mass of 55.85, so a mole of iron is 55.85 g. These figures represent the molar mass—the mass of 1 mole—for each of the elements mentioned.

THE NEED FOR EXACT PROPORTIONS. When chemists discover new substances in nature or create new ones in the laboratory, the first thing they need to determine is the chemical formula—in other words, the exact quantities and proportions of elements in each molecule. By chemical means, they separate the compound into its constituent elements, then determine how much of each element is present.

KEY TERMS

ATOM: The smallest particle of an element that retains all the chemical and physical properties of that element.

ATOMIC MASS UNIT: An SI unit (abbreviated amu), equal to $1.66 \cdot 10^{-24}$ g, for measuring the mass of atoms.

ATOMIC NUMBER: The number of protons in the nucleus of an atom. Since this number is different for each element, elements are listed on the periodic table of elements in order of atomic number.

ATOMIC WEIGHT: An old term for atomic mass. Since weight varies depending on gravitational field, whereas mass is the same throughout the universe, scientists typically use the term "atomic mass" instead.

AVERAGE ATOMIC MASS: A figure used by chemists to specify the mass—in atomic mass units—of the average atom in a large sample.

AVOGADRO'S NUMBER: A figure, named after Italian physicist Amedeo Avogadro (1776-1856), equal to 6.022137×10^{23}. Avogadro's number indicates the number of atoms or molecules in a mole.

COMPOUND: A substance made up of atoms of more than one element. These atoms are usually joined in molecules.

DALTON: An alternate term for atomic mass units, used in biochemistry and microbiology for describing the mass of large organic molecules. The dalton (abbreviated Da or D) is named after English chemist John Dalton (1766-1844), who introduced the concept of the atom to science.

ELEMENT: A substance made up of only one kind of atom, which cannot be chemically broken into other substances.

HALF-LIFE: The length of time it takes a substance to diminish to one-half its initial amount.

Since they are using samples in relatively large quantities, molar mass figures for each element make it possible to determine the chemical composition. To use a very simple example, suppose a quantity of water is separated, and the result is 2.016 g of hydrogen and 16 g of oxygen. The latter is the molar mass of oxygen, and the former is the molar mass of hydrogen multiplied by two. Thus we know that there are two moles of hydrogen and one mole of oxygen, which combine to make one mole of water.

Of course the calculations used by chemists working in the research laboratories of universities, government institutions, and corporations are much, much more complex than the example we have given. In any case, it is critical that a chemist be exact in making these determinations,

so as to know the amount of reactants needed to produce a given amount of product, or the amount of product that can be produced from a given amount of reactant.

When a company produces millions or billions of a single item in a given year, a savings of very small quantities in materials—thanks to proper chemical measurement—can result in a savings of billions of dollars on the bottom line. Proper chemical measurement can also save lives. Again, to use a very simple example, if a mole of compounds weighs 44.01 g and is found to contain two moles of oxygen and one of carbon, then it is merely carbon dioxide—a compound essential to plant life. But if it weighs 28.01 g and has one mole of oxygen with one mole of carbon, it is poisonous carbon monoxide.

KEY TERMS CONTINUED

ION: An atom or group of atoms that has lost or gained one or more electrons, and thus has a net electric charge.

ISOTOPES: Atoms of the same element (that is, they have the same number of protons) that differ in terms of mass. Isotopes may be either stable or unstable. The latter type, known as radioisotopes, are radioactive.

MASS: The amount of matter an object contains.

MOLAR MASS: The mass, in grams, of 1 mole of a given substance. The value in grams of molar mass is always equal to the value, in atomic mass units, of the average atomic mass of that substance. Thus carbon has a molar mass of 12.01 g, and an average atomic mass of 12.01 amu.

MOLE: The SI fundamental unit for "amount of substance." A mole is, generally speaking, Avogadro's number of atoms or molecules; however, in the more precise SI definition, a mole is equal to the number of carbon atoms in 12.01 g of carbon.

MOLECULE: A group of atoms, usually, but not always, representing more than one element, joined in a structure. Compounds are typically made of up molecules.

PERIODIC TABLE OF ELEMENTS: A chart that shows the elements arranged in order of atomic number, along with chemical symbol and the average atomic mass (in atomic mass units) for that particular element.

RADIOACTIVITY: A term describing a phenomenon whereby certain isotopes known as radioisotopes are subject to a form of decay brought about by the emission of high-energy particles. "Decay" does not mean that the isotope "rots"; rather, it decays to form another isotope, until eventually (though this may take a long time), it becomes stable.

WHERE TO LEARN MORE

"Atomic Weight" (Web site). <http://www.colorado.edu/physics/2000/periodic_table/atomic_weight.html> (May 23, 2001).

"An Experiment with 'Atomic Mass'" (Web site). <http://www.carlton.paschools.pa.sk.ca/chemical/molemass/moles3a.htm> (May 23, 2001).

Knapp, Brian J. and David Woodroffe. *The Periodic Table.* Danbury, CT: Grolier Educational, 1998.

Oxlade, Chris. *Elements and Compounds.* Chicago: Heinemann Library, 2001.

"Periodic Table: Atomic Mass." *ChemicalElements.com* (Web site). <http://www.chemicalelements.com/show/mass.html> (May 23, 2001).

"Relative Atomic Mass" (Web site). <http://www.chemsoc.org/viselements/pages/mass.html> (May 23, 2001).

"What Are Atomic Number and Atomic Weight?" (Web site). <http://tis.eh.doe.gov/ohre/roadmap/achre/intro_9_3.html> (May 23, 2001).

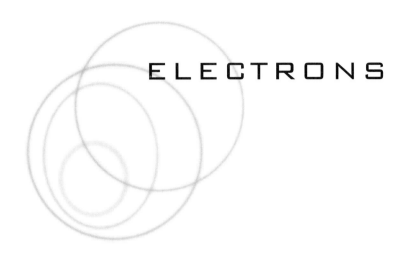

ELECTRONS

CONCEPT

No one can see an electron. Even an electron microscope, used for imaging the activities of these subatomic particles, does not offer a glimpse of an electron as one can look at an amoeba; instead, the microscope detects the patterns of electron deflection. In any case, a single-cell organism is gargantuan in comparison to an electron. Even when compared to a proton or a neutron, particles at the center of an atom, electrons are minuscule, being slightly more than 1/2000 the size of either. Yet the electron is the key to understanding the chemical process of bonding, and electron configurations clarify a number of aspects of the periodic table that may, at first, seem confusing.

HOW IT WORKS

THE ELECTRON'S PLACE IN THE ATOM

The atom is discussed in detail elsewhere in this book; here, the particulars of atomic structure will be presented in an abbreviated form, so that the discussion of electrons may proceed. At the center of an atom, the smallest particle of an element, there is a nucleus, which contains protons and neutrons. Protons are positive in charge, while neutrons exert no charge.

Moving around the nucleus are electrons, negatively charged particles whose mass is very small in comparison to the proton or neutron: 1/1836 and 1/1839 the mass of the proton and neutron respectively. The mass of an electron is $9.109389 \cdot 10^{-33}$ g. Compare the number 9 to the number 1,000,000,000,000,000,000,000,000,000,000,000, and this gives some idea of the ratio between an electron's mass and a gram. The large number— 1 followed by 33 zeroes—is many trillions of times longer than the age of the universe in seconds.

IONS AND CHEMICAL BONDS. Electrons, though very small, are exceedingly powerful, and they are critical to both physical and chemical processes. The negative charge of an electron, designated by the symbol 1- or -1, is enough to counteract the positive charge (1+ or +1) of a proton, even though the proton has much greater mass. As a result, atoms (because they possess equal numbers of protons and electrons) have an electric charge of zero.

Sometimes an atom will release an electron, or a released electron will work its way into the structure of another atom. In either case, an atom that formerly had no electric charge acquires one, becoming an ion. In the first of the instances described, the atom that has lost an electron or electrons becomes a positively charged ion, or cation. In the second instance, an atom that gains an electron or electrons becomes a negatively charged ion, or anion.

One of the first forms of chemical bonding discovered was ionic bonding, in which electrons clearly play a role. In fact, electrons are critical to virtually all forms of chemical bonding, which relates to the interactions between electrons of different atoms.

MOVEMENT OF THE ELECTRON ABOUT THE NUCLEUS. If the size of the nucleus were compared to a grape, the edge of the atom itself would form a radius of about a mile. Because it is much smaller than the nucle-

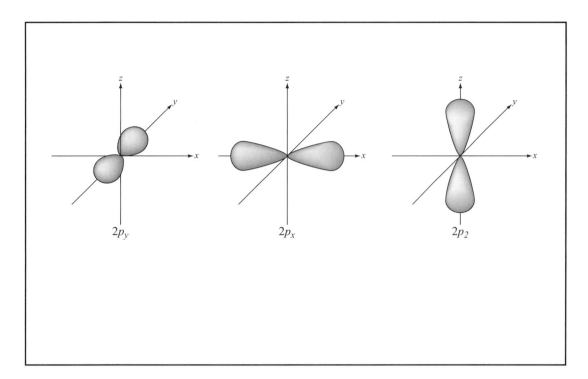

TWO-LOBED ORBITALS.

us, an electron might be depicted in this scenario as a speck of dust—but a speck of dust with incredible energy, which crosses the mile radius of this enlarged atom at amazing speeds. In an item of matter that has been frozen to absolute zero, an electron moves relatively little. On the other hand, it can attain speeds comparable to that of the speed of light: 186,000 mi (299,339 km) per second.

The electron does not move around the nucleus as a planet orbits the Sun—a model of electron behavior that was once accepted, but which has since been overturned. On the other hand, as we shall see, the electron does not simply "go where it pleases": it acts in accordance with complex patterns described by the quantum theory of physics. Indeed, to some extent the behavior of the electron is so apparently erratic that the word "pattern" seems hardly to describe it. However, to understand the electron clearly, one has to set aside all ideas about how objects behave in the physical world.

EMERGING MODELS OF ELECTRON BEHAVIOR

The idea that matter is composed of atoms originated in ancient Greece, but did not take hold until early in the nineteenth century, with the atomic theory of English chemist John Dalton (1766-1844). In the years that followed, numerous figures—among them Russian chemist Dmitri Mendeleev (1834-1907), father of the periodic table of elements—contributed to the emerging understanding of the atomic model.

Yet Mendeleev, despite his awareness that the mass of an atom differentiated one element from another, had no concept of subatomic particles. No one did: until very late in the nineteenth century, the atom might as well have been a hard-shelled ball of matter, for all that scientists understood about its internal structure. The electron was the first subatomic particle discovered, in 1897—nearly a century after the scientific beginning of atomic theory. Yet the first hints regarding its existence had begun to appear some 60 years before.

DISCOVERY OF THE ELEC-TRON. In 1838, British physicist and chemist Michael Faraday (1791-1867) was working with a set of electrodes—metal plates used to emit or collect electric charge—which he had placed at either end of an evacuated glass tube. (In other words, most of the air and other matter had been removed from the tube.) He applied a charge of several thousand volts between the electrodes, and discovered that an electric current flowed between them.

A COMPUTER-GENERATED MODEL OF A NEON ATOM. THE NUCLEUS, AT CENTER, IS TOO SMALL TO BE SEEN AT THIS SCALE AND IS REPRESENTED BY THE FLASH OF LIGHT. SURROUNDING THE NUCLEUS ARE THE ATOM'S ELECTRON ORBITALS: 1S (SMALL SPHERE), 2S (LARGE SPHERE), AND 2P (LOBED). *(Photograph by Kenneth Eward/BioGrafx. National Audubon Society Collection/Photo Researchers, Inc. Reproduced by permission.)*

This seemed to suggest the existence of particles carrying an electric charge, and four decades later, English physicist William Crookes (1832-1919) expanded on these findings with his experiments using an apparatus that came to be known as a Crookes tube. As with Faraday's device, the Crookes tube used an evacuated glass tube encased between two electrodes—a cathode at the negatively charged end, and an anode at the positively charged end. A wire led outside the bulb to an electric source, and when electricity was applied to the electrodes, the cathodes emitted rays. Crookes concluded that the cathode rays were particles with a negative electric charge that came from the metal in the cathode plate.

English physicist J. J. Thomson (1856-1940) hypothesized that the negatively charged particles Crookes had observed were being emitted by atoms, and in 1897 he gave a name to these particles: electrons. The discovery of the electron raised questions concerning its place in the atom: obviously, there had to be a counterbalancing positive charge, and if so, from whence did it come?

FROM PLUM PUDDINGS TO PLANETS. Around the beginning of the twentieth century, the prevailing explanation of atomic structure was the "plum pudding model," which depicted electrons as floating like raisins in a "pudding" of positive charges. This was overturned by the discovery of the nucleus, and, subsequently, of the proton and neutron it contained.

In 1911, the great Danish physicist Niels Bohr (1885-1962) studied hydrogen atoms, and concluded that electrons move around the nucleus in much the same way that planets move around the Sun. This worked well when describing the behavior of hydrogen, which, in its simplest form—the isotope protium—has only one electron and one proton, without any neutrons. The model did not work as well when applied to other elements, however, and within less than two decades Bohr's planetary model was overturned.

As it turns out, the paths of an electron's movement around the nucleus are nothing like that of a planet's orbit—except inasmuch as both models describe a relatively small object moving around a relatively large one. The reality is much more complex, and to comprehend the secret of the electron's apparently random behavior is truly a mind-expanding experience. Yet Bohr is still considered among the greatest scientists of the twentieth century: it was he, after all, who first explained the quantum behavior of electrons, examined below.

REAL-LIFE APPLICATIONS

QUANTUM THEORY AND THE ATOM

Much of what scientists understand today about the atom in general, and the electron in particular, comes from the quantum theory introduced by German physicist Max Planck (1858-1947). Planck showed that, at the atomic level, energy is emitted in tiny packets, or "quanta." Applying this idea to the electron, Bohr developed an idea of the levels at which an electron moves around the nucleus. Though his conclusions led him to the erroneous planetary model, Bohr's explanation of energy levels still prevails.

As has been suggested, the interaction between electrons and protons is electromagnetic, and electromagnetic energy is emitted in the form of radiation, or a stream of waves and particles. The Sun, for instance, emits electromagnetic radiation along a broad spectrum that includes radio waves, infrared light, visible light, ultraviolet light, x rays, and gamma rays. These are listed in ascending order of their energy levels, and the energy emitted can be analyzed in terms of wavelength and frequency: the shorter the wavelength, the greater the frequency and the greater the energy level.

When an atom is at its ordinary energy level, it is said to be in a ground state, but when it acquires excess energy, it is referred to as being in an excited state. It may release some of that energy in the form of a photon, a particle of electromagnetic radiation. The amount of energy involved can be analyzed in terms of the wavelengths of light the atom emits in the form of photons, and such analysis reveals some surprising things about the energy levels of atoms.

If one studies the photons emitted by an atom as it moves between a ground state and an excited state, one discovers that it emits only certain kinds of photons. From this, Bohr concluded that the energy levels of an atom do not exist on a continuum; rather, there are only certain energy levels possible for an atom of a given element. The energy levels are therefore said to be quantized.

THE WAVE MECHANICAL MODEL

In everyday terms, quantization can be compared to the way that a person moves up a set of stairs: by discrete steps. If one step directly follows another, there is no step in between, nor is there any gradual way of moving from step to step, as one would move up a ramp. The movement of electrons from one energy level to another is not a steady progression, like the movement of a person up a ramp; rather, it is a series of quantum steps, like those a person makes when climbing a set of stairs.

The idea of quantization was ultimately applied to describing the paths that an electron makes around the nucleus, but this required some clarification along the way. It had been believed that an electron could move through any point between the nucleus and the edge of the atom (again, like a ramp), but it later became clear that the electrons could only move along specific energy levels. As we have seen, Bohr believed that these corresponded to the orbits of planets around the Sun; but this explanation would be discarded in light of new ideas that emerged in the 1920s.

During the early part of that decade, French physicist Louis de Broglie (1892-1987) and Austrian physicist Erwin Schrödinger (1887-1961) introduced what came to be known as the wave mechanical model, also known as the particle-wave hypothesis. Because light appeared to have the properties of both particles and waves, they reasoned, electrons (possessing electromagnetic energy as they did) might behave in the same fashion. In other words, electrons were not just particles: in some sense, they were waves as well.

UNDERSTANDING ORBITALS.
The wave mechanical model depicted the movement of electrons, not as smooth orbits, but as orbitals—regions in which there is the highest probability that an electron will be found. An orbital is nothing like the shape of a solar system, but, perhaps ironically, it can be compared to the photographs astronomers have taken of galaxies. In most of these photographs, one sees an area of

intense light emitted by the stars in the center. Further from this high-energy region, the distribution of stars (and hence of light) becomes increasingly less dense as one moves from the center of the galaxy to the edges.

Replace the center of the galaxy with the nucleus of the atom, and the stars with electrons, and this is an approximation of an orbital. Just as a galaxy looks like a cloud of stars, scientists use the term electron cloud to describe the pattern formed by orbitals. The positions of electrons cannot be predicted; rather, it is only possible to assign probabilities as to where they will be. Naturally, they are most drawn to the positive charges in the nucleus, and hence an orbital depicts a high-density region of probabilities at the center—much like the very bright center of a galaxy.

The further away from the nucleus, the less the probability that an electron will be in that position. Hence in models of an orbital, the dots are concentrated at the center, and become less dense the further away from the nucleus they are. As befits the comparison to a cloud, the edge of an orbital is fuzzy. Contrary to the earlier belief that an atom was a clearly defined little pellet of matter, there is no certainty regarding the exact edge of a given atom; rather, scientists define the sphere of the atom as the region encompassing 90% of the total electron probability.

THE MYSTERIOUS ELECTRON. As complex as this description of electron behavior may seem, one can rest assured that the reality is infinitely more complex than this simplified explanation suggests. Among the great mysteries of the universe is the question of why an electron moves as it does, or even exactly how it does so. Nor does a probability model give us any way of knowing when an electron will occupy a particular position.

In fact, as German physicist Werner Heisenberg (1901-1976) showed with what came to be known as the Heisenberg Uncertainty Principle, it is impossible to know both the speed of an electron and its precise position at the same time. This, of course, goes against every law of physics that prevailed until about 1920, and in fact quantum theory offers an entirely different model of reality than the one accepted during the seventeenth, eighteenth, and nineteenth centuries.

ORBITALS

Given the challenges involved in understanding electron behavior, it is amazing just how much scientists do know about electrons—particularly where energy levels are concerned. This, in turn, makes possible an understanding of the periodic table that would astound Mendeleev.

Every element has a specific configuration of energy levels that becomes increasingly complex as one moves along the periodic table. In the present context, these configurations will be explained as simply as possible, but the reader is encouraged to consult a reliable chemistry textbook for a more detailed explanation.

PRINCIPAL ENERGY LEVELS AND SUBLEVELS. The principal energy level of an atom indicates a distance that an electron may move away from the nucleus. This is designated by a whole-number integer, beginning with 1 and moving upward: the higher the number, the further the electron is from the nucleus, and hence the greater the energy in the atom. Each principal energy level is divided into sublevels corresponding to the number n of the principal energy level: thus principal energy level 1 has one sublevel, principal energy level 2 has two, and so on.

The simplest imaginable atom, a hydrogen atom in a ground state, has an orbital designated as $1s^1$. The s indicates that an electron at energy level 1 can be located in a region described by a sphere. As for the significance of the superscript 1, this will be explained shortly.

Suppose the hydrogen atom is excited enough to be elevated to principal energy level 2. Now there are two sublevels, $2s$ (for now we will dispense with the superscript 1) and $2p$. A p orbital is rather like the shape of a figure eight, with its center of gravity located on the nucleus, and thus unlike the s sublevel, p orbitals can have a specific directional orientation. Depending on whether it is oriented along an x-, y-, or z-axis, orbitals in sublevel p are designated as $2p_x$, $2p_y$, or $2p_z$.

If the hydrogen atom is further excited, and therefore raised to principal energy level 3, it now has three possible sublevels, designated as s, p, and d. Some of the d orbitals can be imagined as two figure eights at right angles to one another, once again with their centers of gravity along the nucleus of the atom. Because of their more complex shape, there are five possible spatial orientations for orbitals at the *d* sublevel.

Even more complex is the model of an atom at principal level 4, with four sublevels—*s*,*p*, *d*, and *f*, which has a total of seven spatial orientations. Obviously, things get very, very complex at increased energy levels. The greater the energy level, the further the electron can move from the nucleus, and hence the greater the possible number of orbitals and corresponding shapes.

ELECTRON SPIN AND THE PAULI EXCLUSION PRINCIPLE. Every electron spins in one of two directions, and these are indicated by the symbols ↑ and ↓. According to the Pauli exclusion principle, named after the Austrian-Swiss physicist Wolfgang Pauli (1900-1958), no more than two electrons can occupy the same orbital, and those two electrons must have spins opposite one another.

This explains the use of the superscript 1, which indicates the number of electrons in a given orbital. This number is never greater than two: hence, the electron configuration of helium is written as $1s^2$. It is understood that these two electrons must be spinning in opposite directions, but sometimes this is indicated by an orbital diagram showing both an upward- and downward-pointing arrow in an orbital that has been filled, or only an upward-pointing arrow in an orbital possessing just one electron.

ELECTRON CONFIGURATION AND THE PERIODIC TABLE

THE FIRST 18 ELEMENTS. As one moves up the periodic table from atomic number 1 (hydrogen) to 18 (argon), a regular pattern emerges. The orbitals are filled in a neat progression: from helium (atomic number 2) onward, all of principal level 1 is filled; beginning with beryllium (atomic number 4), sublevel $2s$ is filled; from neon (atomic number 10), sublevel $2p$—and hence principal level 2 as a whole—is filled, and so on. The electron configuration for neon, thus, is written as $1s^22s^22p^6$.

Note that if one adds together all the superscript numbers, one obtains the atomic number of neon. This is appropriate, of course, since atomic number is defined by the number of protons, and an atom in a non-ionized state has an equal number of protons and electrons. In noticing the electron configurations of an element, pay close attention to the last or highest principal energy level represented. These are the valence electrons, the ones involved in chemical bonding. By contrast, the core electrons, or the ones that are at lower energy levels, play no role in the bonding of atoms.

SHIFTS IN ELECTRON CONFIGURATION PATTERNS. After argon, however, as one moves to the element occupying the nineteenth position on the periodic table—potassium—the rules change. Argon has an electron configuration of $1s^22s^22p^63s^23p^6$, and by the pattern established with the first 18 elements, potassium should begin filling principal level $3d$. Instead, it "skips" $3d$ and moves on to $4s$. The element following argon, calcium, adds a second electron to the $4s$ level.

After calcium, the pattern again changes. Scandium (atomic number 21) is the first of the transition metals, a group of elements on the periodic table in which the $3d$ orbitals are filled. This explains why the transition metals are indicated by a shading separating them from the rest of the elements on the periodic table.

But what about the two rows at the very bottom of the chart, representing groups of elements that are completely set apart from the periodic table? These are the lanthanide and actinide series, which are the only elements that involve *f* sublevels. In the lanthanide series, the seven $4f$ orbitals are filled, while the actinide series reflects the filling of the seven $5f$ orbitals.

As noted earlier, the patterns involved in the *f* sublevel are ultra-complex. Thus it is not surprising than the members of the lanthanide series, with their intricately configured valence electrons, were very difficult to extract from one another, and from other elements: hence their old designation as the "rare earth metals." However, there are a number of other factors—relating to electrons, if not necessarily electron configuration—that explain why one element bonds as it does to another.

CHANGES IN ATOMIC SIZE. With the tools provided by the basic discussion of electrons presented in this essay, the reader is encouraged to consult the essays on Chemical Bonding, as well as The Periodic Table of the Elements, both of which explore the consequences of electron arrangement in chemistry. Not only are electrons the key to chemical bonding, understanding their configurations is critical to an understanding of the periodic table.

KEY TERMS

ANION: The negative ion that results when an atom gains one or more electrons.

ANODE: An electrode at the positively charged end of a supply of electric current.

ATOM: The smallest particle of an element.

ATOMIC NUMBER: The number of protons in the nucleus of an atom. Since this number is different for each element, elements are listed on the periodic table of elements in order of atomic number.

CATHODE: An electrode at the negatively charged end of a supply of electric current.

CATION: The positive ion that results when an atom loses one or more electrons.

ELECTRODE: A structure, often a metal plate or grid, that conducts electricity, and which is used to emit or collect electric charge.

ELECTRON: A negatively charged particle in an atom.

ELECTRON CLOUD: A term used to describe the pattern formed by orbitals.

EXCITED STATE: A term describing the characteristics of an atom that has acquired excess energy.

GROUND STATE: A term describing the state of an atom at its ordinary energy level.

ION: An atom or atoms that has lost or gained one or more electrons, and thus has a net electric charge.

ISOTOPES: Atoms that have an equal number of protons, and hence are of the same element, but differ in their number of neutrons.

NEUTRON: A subatomic particle that has no electric charge. Neutrons are found in the nucleus of an atom, alongside protons.

NUCLEUS: The center of an atom, a region where protons and neutrons are located, and around which electrons spin.

One of the curious things about the periodic table, for instance, is the fact that the sizes of atoms decrease as one moves from left to right across a row or period, even though the sizes increase as one moves from top to bottom along a column or group. The latter fact—the increase of atomic size in a group, as a function of increasing atomic number—is easy enough to explain: the higher the atomic number, the higher the principal energy level, and the greater the distance from the nucleus to the furthest probability range.

On the other hand, the decrease in size across a period (row) is a bit more challenging to comprehend. However, all the elements in a period have their outermost electrons at a particular principal energy level corresponding to the number of the period. For instance, the elements on period 5 all have principal energy levels 1 through 5. Yet as one moves along a period from left to right, there is a corresponding increase in the number of protons within the nucleus. This means a stronger positive charge pulling the electrons inward; therefore, the "electron cloud" is drawn ever closer toward the increasingly powerful charge at the center of the atom.

WHERE TO LEARN MORE

"Chemical Bond" (Web site). <http://www.science.uwaterloo.ca/~cchieh/cact/c120/chembond.html> (May 18, 2001).

"The Discovery of the Electron" (Web site). <http://www.aip.org/history/electron/> (May 18, 2001).

KEY TERMS CONTINUED

ORBITAL: A pattern of probabilities regarding the regions that an electron can occupy within an atom in a particular energy state. The orbital, complex and imprecise as it may seem, is a much more accurate depiction of electron behavior than the model once used, which depicted electrons moving in precisely defined orbits around the nucleus, rather as planets move around the Sun.

PERIODIC TABLE OF ELEMENTS: A chart that shows the elements arranged in order of atomic number. Vertical columns within the periodic table indicate groups or "families" of elements with similar chemical characteristics.

PHOTON: A particle of electromagnetic radiation.

PRINCIPAL ENERGY LEVEL: A value indicating the distance that an electron may move away from the nucleus of an atom. This is designated by a whole-num-ber integer, beginning with 1 and moving upward. The higher the number, the further the electron is from the nucleus, and hence the greater the energy in the atom.

PROTON: A positively charged particle in an atom.

QUANTIZATION: A term describing any property that has only certain discrete values, as opposed to values distributed along a continuum. The quantization of an atom means that it does not have a continuous range of energy levels; rather, it can exist only at certain levels of energy from the ground state through various excited states.

RADIATION: In a general sense, radiation can refer to anything that travels in a stream, whether that stream be composed of subatomic particles or electromagnetic waves.

Ebbing, Darrell D.; R. A. D. Wentworth; and James P. Birk. *Introductory Chemistry.* Boston: Houghton Mifflin, 1995.

Gallant, Roy A. *The Ever-Changing Atom.* New York: Benchmark Books, 1999.

Goldstein, Natalie. *The Nature of the Atom.* New York: Rosen Publishing Group, 2001.

"*Life, the Universe, and the Electron*" (Web site). <http://www.iop.org/Physics/Electron/Exhibition/ (May 18, 2001).

"*A Look Inside the Atom*" (Web site). <http://www.aip.org/history/electron/jjhome.htm> (May 18, 2001).

"*Valence Shell Electron Pair Repulsion (VSEPR).*" <http://www.shef.ac.uk/~chem/vsepr/chime/vsepr.html> (May 18, 2001).

"*What Are Electron Microscopes?*" (Web site). <http://www.unl.edu/CMRAcfem/em.htm> (May 18, 2001).

Zumdahl, Steven S. *Introductory Chemistry: A Foundation,* 4th ed. Boston: Houghton Mifflin, 2000.

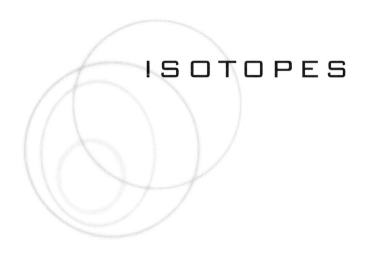

ISOTOPES

CONCEPT

Isotopes are atoms of the same element that have different masses due to differences in the number of neutrons they contain. Many isotopes are stable, meaning that they are not subject to radioactive decay, but many more are radioactive. The latter, also known as radioisotopes, play a significant role in modern life. Carbon-14, for instance, is used for estimating the age of objects within a relatively recent span of time—up to about 5,000 years—whereas geologists and other scientists use uranium-238 to date minerals of an age on a scale with that of the Earth. Concerns over nuclear power and nuclear weapons testing in the atmosphere have heightened awareness of the dangers posed by certain kinds of radioactive isotopes, which can indeed be hazardous to human life. However, the reality is that people are subjected to considerably more radiation from nonnuclear sources.

HOW IT WORKS

ATOMS AND ELEMENTS

The elements are substances that cannot be broken down into other matter by chemical means, and an atom is the fundamental particle in an element. As of 2001, there were 112 known elements, 88 of which occur in nature; the rest were created in laboratories. Due to their high levels of radioactivity, they exist only for extremely short periods of time. Whatever the number of elements—and obviously that number will increase over time, as new elements are synthesized—the same number of basic atomic structures exists in the universe.

What distinguishes one element from another is the number of protons, subatomic particles with a positive electric charge, in the nucleus, or center, of the atom. The number of protons, whatever it may be, is unique to an element. Thus if an atom has one proton, it is an atom of hydrogen, because hydrogen has an atomic number of 1, as shown on the periodic table of elements. If an atom has 109 protons, on the other hand, it is meitnerium. (Meitnerium, synthesized at a German laboratory in 1982, is the last element on the periodic table to have been assigned a name as of 2001.)

THE NUCLEUS AND ELECTRONS. Together with protons in the nucleus are neutrons, which exert no charge. The discovery of these particles, integral to the formation of isotopes, is discussed below. The nucleus, with a diameter about 1/10,000 that of the atom itself, makes up only a tiny portion of the atom's volume, but the vast majority of its mass. Thus a change in the mass of a nucleus, as occurs when an isotope is formed, is reflected by a noticeable change in the mass of the atom itself.

Far from the nucleus (in relative terms, of course), at the perimeter of the atom, are the electrons, which have a negative electric charge. Whereas the protons and neutrons have about the same mass, the mass of an electron is less than 0.06% of either a proton or neutron. Nonetheless, electrons play a highly significant role in chemical reactions and chemical bonding. Just as isotopes are the result of changes in the number of neutrons, ions—atoms that are either positive or negative in electric charge—are the result of changes in the number of electrons.

Unless it loses or gains an electron, thus becoming an ion, an atom is neutral in charge, and it maintains this electric-charge neutrality by having an equal number of protons and electrons. There is, however, no law of the universe stating that an atom must have the same number of neutrons as it does protons and electrons: some do, but this is far from universal, as we shall see.

NEUTRONS

The number of neutrons is variable within an element precisely because they exert no charge, and thus while their addition or removal changes the mass, it does not affect the electric charge of the atom. Therefore, whereas the importance of the proton and the electron is very clear to anyone who studies atomic behavior, neutrons, on the other hand, might seem at first glance as though they are only "along for the ride." Yet they are all-important to the formation of isotopes.

Not surprisingly, given their lack of electric charge, neutrons were the last of the three major subatomic particles to be discovered. English physicist J. J. Thomson (1856-1940) identified the electron in 1897, and another English physicist, Ernest Rutherford (1871-1937), discovered the proton in 1914. Rutherford's discovery overturned the old "plum pudding" model, whereby atoms were depicted as consisting of electrons floating in a positively charged cloud, rather like raisins in an English plum pudding. As Rutherford showed, the atom must have a nucleus—yet protons alone could not account for the mass of the nucleus.

There must be something else at the heart of the atom, and in 1932, yet another English physicist, James Chadwick (1891-1974), identified what it was. Working with radioactive material, he found that a certain type of subatomic particle could penetrate lead. All types of radiation known at the time were stopped by the lead, and therefore Chadwick reasoned that this particle must be neutral in charge. In 1932, he won the Nobel Prize in physics for his discovery of the neutron.

NEUTRONS AND NUCLEAR FUSION. Neutrons played a critical role in the development of the atomic bomb during the 1940s. In nuclear fission, atoms of uranium are bombarded with neutrons. The result is that the uranium nucleus splits in half, releasing huge

A MUSHROOM CLOUD RISES AFTER THE DETONATION OF A HYDROGEN BOMB BY FRANCE IN A 1968 TEST. DEUTERIUM IS A CRUCIAL PART OF THE DETONATING DEVICE FOR HYDROGEN BOMBS. *(Bettmann/Corbis. Reproduced by permission.)*

amounts of energy. As it does so, it emits several extra neutrons, which split more uranium nuclei, creating still more energy and setting off a chain reaction.

This explains the destructive power in an atomic bomb, as well as the constructive power—providing energy to homes and businesses—in a nuclear power plant. Whereas the chain reaction in an atomic bomb becomes an uncontrolled explosion, in a nuclear plant, the reaction is slowed and controlled. One of the means used to do this is by the application of "heavy water," which, as we shall see, is water made with a hydrogen isotope.

ISOTOPES: THE BASICS

Two atoms may have the same number of protons, and thus be of the same element, yet differ in their number of neutrons. Such atoms are called isotopes, atoms of the same element having different masses. The name comes from the Greek phrase *isos topos*, meaning "same place": because they have the same atomic number,

ENRICO FERMI.

isotopes of the same element occupy the same position on the periodic table.

Also called nuclides, isotopes are represented symbolically as follows: $\frac{m}{a}S$, where S is the symbol of the element, a is the atomic number, and m is the mass number—the sum of protons and neutrons in the atom's nucleus. For the stable silver isotope designated as $\frac{93}{47}Ag$, for instance, Ag is the element symbol; 47 its atomic number; and 93 the mass number. From this, it is easy to discern that this particular stable isotope has 46 neutrons in its nucleus.

Because the atomic number of any element is established, sometimes isotopes are represented simply with the mass number, thus: ^{93}Ag. They may also be designated with a subscript notation indicating the number of neutrons, so that this information can be obtained at a glance without having to do the arithmetic. For the silver isotope shown here, this is written as $\frac{93}{47}Ag_{46}$. Isotopes can also be indicated by simple nomenclature: for instance, carbon-12 or carbon-13.

STABLE AND UNSTABLE ISOTOPES

Radioactivity is a term describing a phenomenon whereby certain materials are subject to a form of decay brought about by the emission of high-energy particles or radiation. Forms of particles or energy emitted in radiation include alpha particles (positively charged helium nuclei); beta particles (either electrons or subatomic particles called positrons); or gamma rays, which occupy the highest energy level in the electromagnetic radiation emitted by the Sun. Radioactivity will be discussed below, but for the present, the principal concern is with radioactive properties as a distinguishing factor between the two varieties of isotope.

Isotopes are either stable or unstable. The unstable variety, known as radioisotopes, are subject to radioactive decay, but in this context, "decay" does not mean what it usually does. A radioisotope does not "rot"; it decays by turning into another isotope of the same element—or even into another element entirely. (For example, uranium-238 decays by emitting alpha particles, ultimately becoming lead-206.) A stable isotope, on the other hand, has already become what it is going to be, and will not experience further decay.

Most elements have between two and six stable isotopes. On the other hand, a few elements—for example, technetium—have no stable isotopes. Twenty elements, among them gold, fluorine, sodium, aluminum, and phosphorus, have only one stable isotope each. The element with the most stable isotopes is easy to remember because its name is almost the same as its number of stable isotopes: tin, with 10.

As for unstable isotopes, there are over 1,000, some of which exist in nature, but most of which have been created synthetically in laboratories. This number is not fixed; in any case, it is not necessarily important, because many of these highly radioactive isotopes last only for fractions of a second before decaying to form a stable isotope. Yet radioisotopes in general have so many uses, in comparison to stable isotopes, that they are often referred to simply as "isotopes."

UNDERSTANDING ISOTOPES

Before proceeding with a discussion of isotopes and their uses, it is necessary to address a point raised earlier, when it was stated that some atoms do have the same numbers of neutrons and protons, but that this is far from universal. In fact, nuclear stability is in part a function of neutron-to-proton ratio.

Stable nuclei with low atomic numbers (up to about 20) have approximately the same number of neutrons and protons. For example, the most stable and abundant form of carbon is carbon-12, with six protons and six neutrons. Beyond atomic number 20 or so, however, the number of neutrons begins to grow: in other words, the lowest mass number is increasingly high in comparison to the atomic number.

For example, uranium has an atomic number of 92, but the lowest mass number for a uranium isotope is not 184, or 92 multiplied by two; it is 218. The ratio of neutrons to protons necessary for a stable isotope creeps upward along the periodic table: tin, with an atomic number of 50, has a stable isotope with a mass number of 120, indicating a 1.4 to 1 ratio of neutrons to protons. For mercury-200, the ratio is 1.5 to 1.

The higher the atomic number, by definition, the greater the number of protons in the nucleus. This means that more neutrons are required to "bind" the nucleus together. In fact, all nuclei with 84 protons or more (i.e., starting at polonium and moving along the periodic table) are radioactive, for the simple reason that it is increasingly difficult for the neutrons to withstand the strain of keeping so many protons in place.

One can predict the mode of radioactive decay by noting whether the nucleus is neutron-rich or neutron-poor. Neutron-rich nuclei undergo beta emission, which decreases the numbers of protons in the nucleus. Neutron-poor nuclei typically undergo positron emission or electron capture, the first of these being more prevalent among the lighter nuclei. Elements with atomic numbers of 84 or greater generally undergo alpha emission, which decreases the numbers of protons and neutrons by two each.

REAL-LIFE APPLICATIONS

DEUTERIUM AND TRITIUM

Only three isotopes are considered significant enough to have names of their own, as opposed to being named after a parent atom (for example, carbon-12, uranium-238). These are protium, deuterium, and tritium, all three isotopes of hydrogen. Protium, or ^1H, is hydrogen in its most basic form—one proton, no neutrons—and the name "protium" is only applied when necessary to distinguish it from the other two isotopes. Therefore we will focus primarily on the two others.

Deuterium, designated as ^2H, is a stable isotope, whereas tritium—^3H—is radioactive. Both, in fact, have chemical symbols (D and T respectively), just as though they were elements on the periodic table. What makes these two so special? They are, as it were, "the products of a good home"—in other words, their parent atom is the most basic and plentiful element in the universe. Indeed, the vast majority of the universe is hydrogen, along with helium, which is formed by the fusion of hydrogen atoms. If all atoms were numbers, then hydrogen would be 1; but of course, this is more than a metaphor, since its atomic number is indeed 1.

Ordinary hydrogen or protium, as noted, consists of a single proton and a single electron, the simplest possible atomic form possible. Its simplicity has made it a model for understanding the atom, and therefore when physicists discovered the existence of two hydrogen atoms that were just a bit more complex, they were intrigued.

Just as hydrogen represented the standard against which atoms could be measured, scientists reasoned, deuterium and tritium could offer valuable information regarding stable and unstable isotopes respectively. Furthermore, the pronounced tendency of hydrogen to bond with other substances—it almost never appears by itself on Earth—presented endless opportunities for study regarding hydrogen isotopes in association with other elements.

ISOLATION OF DEUTERIUM.
Deuterium is sometimes called "heavy hydrogen," and its nucleus—with one proton and one neutron—is called a deuteron. It was first isolated in 1931 by American chemist Harold Clayton Urey (1893-1981), who was awarded the 1934 Nobel Prize in Chemistry for his discovery.

Serving at that time as a professor of chemistry at Columbia University in New York City, Urey started with the assumption that any hydrogen isotopes other than protium must exist in very minute quantities. This assumption, in turn, followed from an awareness that hydrogen's average atomic mass—measured in atomic mass units—was only slightly higher than 1. There must be, as Urey correctly reasoned, a very small quantity of "heavy hydrogen" on Earth.

To separate deuterium, Urey collected a relatively large sample of liquid hydrogen: 4.2 quarts (4 l). Then he allowed the liquid to evaporate very slowly, predicting that the more abundant protium would evaporate more quickly than the

isotope whose existence he had hypothesized. After all but 0.034 oz (1 ml) of the sample had evaporated, he submitted the remainder to a form of analysis called spectroscopy, adding a burst of energy to the atoms and then analyzing the light spectrum they emitted for evidence of differing varieties of atom.

CHARACTERISTICS AND USES OF DEUTERIUM.
Deuterium, with an atomic mass of 2.014102 amu, is almost exactly twice as heavy as protium, which has an atomic mass of 1.007825. Its melting point, or the temperature at which it changes from a solid to a liquid -426°F (-254°C), is much higher than for protium, which melts at -434°F (-259°C). The same relationship holds for its boiling point, or the temperature at which it changes from a liquid to its normal state on Earth, as a gas: -417°F (-249°C), as compared to -423°F (-253°C) for protium. Deuterium is also much, much less plentiful than protium: protium represents 99.985% of all the hydrogen that occurs naturally, meaning that deuterium accounts for just 0.015%.

Often, deuterium is applied as a tracer, an atom or group of atoms whose participation in a chemical, physical, or biological reaction can be easily observed. Radioisotopes are most often used as tracers, precisely because of their radioactive emissions; deuterium, on the other hand, is effective due to its almost 2:1 mass ratio in comparison to protium. In addition, it bonds with other atoms in a fashion slightly different from that of protium, and this contrast makes its presence easier to trace.

Its higher boiling and melting points mean that when deuterium is combined with oxygen to form "heavy water" (D_2O), the water likewise has higher boiling and melting points than ordinary water. Heavy water is often used in nuclear fission reactors to slow down the fission process, or the splitting of atoms.

DEUTERIUM IN NUCLEAR FUSION.
Deuterium is also applied in a type of nuclear reaction much more powerful that fission: fusion, or the joining of atomic nuclei. The Sun produces energy by fusion, a thermonuclear reaction that takes places at temperatures of many millions of degrees Celsius. In solar fusion, it appears that two protium nuclei join to form a single deuteron.

During the period shortly after World War II, physicists developed a means of duplicating the thermonuclear fusion process. The result was the hydrogen bomb—more properly called a fusion bomb—whose detonating device was a compound of lithium and deuterium called lithium deuteride. Vastly more powerful than the "atomic" (that is, fission) bombs dropped by the United States over Japan in 1945, the hydrogen bomb greatly increased the threat of worldwide nuclear annihilation in the postwar years.

Yet the power that could destroy the world also has the potential to provide safe, abundant fusion energy from power plants—a dream that as yet remains unrealized. Among the approaches being attempted by physicists studying nuclear fusion is a process in which two deuterons are fused. The result is a triton, the nucleus of tritium, along with a single proton. The triton and deuteron would then be fused to create a helium nucleus, with a resulting release of vast amounts of energy.

TRITIUM. Whereas deuterium has a single neutron, tritium—as its mass number of 3 indicates—has two. And just as deuterium has approximately twice the mass of protium, tritium has about three times the mass, 3.016 amu. As is expected, the thermal properties of tritium are different from those of protium. Again, the melting and boiling points are higher: thus tritium heavy water (T_2O) melts at 40°F (4.5°C), as compared with 32°F (0°C) for H_2O.

Because it is radioactive, tritium is often described in terms of half-life, the length of time it takes for a substance to diminish to one-half its initial amount. The half-life of tritium is 12.26 years. As it decays, its nucleus emits a low-energy beta particle, and this results in the creation of the helium-3 isotope. Due to the low energy levels involved, the radioactive decay of tritium poses little danger to humans.

Like deuterium, tritium is applied in nuclear fusion, though due to its scarcity, it is usually combined with deuterium. Furthermore, tritium decay requires that hydrogen bombs containing the radioisotope be recharged periodically. Also, like deuterium, tritium is an effective tracer. Sometimes it is released in small quantities into groundwater as a means of monitoring subterranean water flow. It is also used as a tracer in biochemical processes.

SEPARATING ISOTOPES

As noted in the discussion of deuterium, tritium can only be separated from protium due to the differences in mass. The chemical properties of isotopes with the same parent element make them otherwise indistinguishable, and hence purely chemical means cannot be used to separate them.

Physicists working on the Manhattan Project, the U.S. effort to develop atomic weaponry during World War II, were faced with the need to separate ^{235}U from ^{238}U. Uranium-238 is far more abundant, but what they wanted was the uranium-235, highly fissionable and thus useful in the processes they were attempting.

Their solution was to allow a gaseous uranium compound to diffuse, or separate, the uranium through porous barriers. Because uranium-238 was heavier, it tended to move more slowly through the barriers, much like grains of rice getting caught in a sifter. Another means of separating isotopes is by mass spectrometry.

RADIOACTIVITY

One of the scientists working on the Manhattan Project was Italian physicist Enrico Fermi (1901-1954), who used radium and beryllium powder to construct a neutron source for making new radioactive materials. Fermi and his associates succeeded in producing radioisotopes of sodium, iron, copper, gold, and numerous other elements. As a result of Fermi's work, for which he won the 1938 Nobel Prize in Physics, scientists have been able to develop radioactive versions of virtually all elements.

Interestingly, the ideas of radioactivity, fission reactions, and fusion reactions collectively represent the realization of a goal sought by the medieval alchemists: the transformation of one element into another. The alchemists, forerunners of chemists, believed they could transform ordinary metals into gold by using various potions—an impossible dream. Yet as noted in the preceding paragraph, among the radioisotopes generated by Fermi's neutron source was gold. The "catch," of course, is that this gold was unstable; furthermore, the amount of energy and human mental effort required to generate it far outweighed the monetary value of the gold itself.

Radioactivity is, in the modern imagination, typically associated with fallout from nuclear war, or with hazards resulting from nuclear power—hazards that, as it turns out, have been greatly exaggerated. Nor is radioactivity always harmful to humans. For instance, with its applications in medicine—as a means of diagnosing and treating thyroid problems, or as a treatment for cancer patients—it can actually save lives.

HAZARDS ASSOCIATED WITH RADIOACTIVITY. It is a good thing that radiation, even the harmful variety known as ionizing radiation, is not fatal in small doses, because every person on Earth is exposed to small quantities of radiation every year. About 82% of this comes from natural sources, and 18% from manmade sources. Of course, some people are at much greater risk of radiation exposure than others: coal miners are exposed to higher levels of the radon-222 isotope present underground, while cigarette smokers ingest much higher levels of radiation than ordinary people, due to the polonium-210, lead-210, and radon-222 isotopes present in the nitrogen fertilizers used to grow tobacco.

Nuclear weapons, as most people know, produce a great deal of radioactive pollution. However, atmospheric testing of nuclear armaments has long been banned, and though the isotopes released in such tests are expected to remain in the atmosphere for about a century, they do not constitute a significant health hazard to most Americans. (It should be noted that nations not inclined to abide by international protocols might still conduct atmospheric tests in defiance of the test bans.) Nuclear power plants, despite the great deal of attention they have received from the media and environmentalist groups, do not pose the hazard that has often been claimed: in fact, coal- and oil-burning power plants are responsible for far more radioactive pollution in the United States.

This is not to say that nuclear energy poses no dangers, as the disaster at Chernobyl in the former Soviet Union has shown. In April 1986, an accident at a nuclear reactor in what is now the Ukraine killed 31 workers immediately, and ultimately led to the deaths of some 10,000 people. The fact that the radiation was allowed to spread had much to do with the secretive tactics of the Communist government, which attempted to cover up the problem rather than evacuate the area.

Another danger associated with nuclear power plants is radioactive waste. Spent fuel rods and other waste products from these plants have to be dumped somewhere, but it cannot simply be buried in the ground because it will create a continuing health hazard through the water supply. No fully fail-safe storage system has been developed, and the problem of radioactive waste poses a continuing threat due to the extremely long half-lives of some of the isotopes involved.

DATING TECHNIQUES

In addition to their uses in applications related to nuclear energy, isotopes play a significant role in dating techniques. The latter may sound like a subject that has something to do with romance, but it does not: dating techniques involve the use of materials, including isotopes, to estimate the age of both organic and inorganic materials.

Uranium-238, for instance, has a half-life of $4.47 \cdot 10^9$ years, which is nearly the age of Earth; in fact, uranium-dating techniques have been used to determine the planet's age, which is estimated at about 4.7 billion years. As noted elsewhere in this volume, potassium-argon dating, which involves the isotopes potassium-40 and argon-40, has been used to date volcanic layers in east Africa. Because the half-life of potassium-40 is 1.3 billion years, this method is useful for dating activities that are distant in the human scale of time, but fairly recent in geological terms.

Another dating technique is radiocarbon dating, used for estimating the age of things that were once alive. All living things contain carbon, both in the form of the stable isotope carbon-12 and the radioisotope carbon-14. While a plant or animal is living, there is a certain proportion between the amounts of these two isotopes in the organism's body, with carbon-12 being far more abundant. When the organism dies, however, it ceases to acquire new carbon, and the carbon-14 present in the body begins to decay into nitrogen-14. The amount of nitrogen-14 that has been formed is thus an indication of the amount of time that has passed since the organism was alive.

Because it has a half-life of 5,730 years, carbon-14 is useful for dating activities within the span of human history, though it is not without controversy. Some scientists contend, for instance, that samples may be contaminated by carbon from the surrounding soils, thus affecting ratios and leading to inaccurate dates.

KEY TERMS

ATOM: The smallest particle of an element. Atoms are made up of protons, neutrons, and electrons. Atoms that have the same number of protons—that is, are of the same element—but differ in number of neutrons are known as isotopes.

ATOMIC MASS UNIT: An SI unit (abbreviated amu), equal to $1.66 \cdot 10^{-24}$ g, for measuring the mass of atoms.

ATOMIC NUMBER: The number of protons in the nucleus of an atom. Since this number is different for each element, elements are listed on the periodic table of elements in order of atomic number.

AVERAGE ATOMIC MASS: A figure used by chemists to specify the mass—in atomic mass units—of the average atom in a large sample.

ELEMENT: A substance made up of only one kind of atom. Hence an element cannot be chemically broken into other substances.

ELECTRON: Negatively charged particles in an atom, which spin around the protons and neutrons that make up the atom's nucleus.

HALF-LIFE: The length of time it takes a substance to diminish to one-half its initial amount.

ION: An atom or atoms that has lost or gained one or more electrons, and thus has a net electric charge.

ISOTOPES: Atoms that have an equal number of protons, and hence are of the same element, but differ in their number of neutrons. This results in a difference of mass. Isotopes may be either stable or unstable. The latter type is known as a radioisotope.

NEUTRON: A subatomic particle that has no electric charge. Neutrons are found at the nucleus of an atom, alongside protons.

NUCLEUS: The center of an atom, a region where protons and neutrons are located, and around which electrons spin. The plural of "nucleus" is nuclei.

NUCLIDES: Another name for isotopes.

PERIODIC TABLE OF ELEMENTS: A chart that shows the elements arranged in order of atomic number.

PROTON: A positively charged particle in an atom. Protons and neutrons, which together form the nucleus around which electrons spin, have approximately the same mass—a mass that is many times greater than that of an electron. The number of protons in the nucleus of an atom is the atomic number of an element.

KEY TERMS continued

RADIATION: In a general sense, radiation can refer to anything that travels in a stream, whether that stream be composed of subatomic particles or electromagnetic waves. In a more specific sense, the term relates to the radiation from radioactive materials, which can be harmful to human beings.

RADIOACTIVITY: A term describing a phenomenon whereby certain materials are subject to a form of decay brought about by the emission of high-energy particles or radiation, including alpha particles, beta particles, or gamma rays.

RADIOISOTOPE: An isotope subject to the decay associated with radioactivity. A radioisotope is thus an unstable isotope.

TRACER: An atom or group of atoms whose participation in a chemical, physical, or biological reaction can be easily observed. Radioisotopes are often used as tracers.

WHERE TO LEARN MORE

"*Carbon 14 Dating Calculator*" (Web site). <http://www.museum.mq.edu.au/eegypt2/carbdate.html> (May 15, 2001).

Ebbing, Darrell D.; R. A. D. Wentworth; and James P. Birk. *Introductory Chemistry.* Boston: Houghton Mifflin, 1995.

"*Exploring the Table of Isotopes*" (Web site). <http://ie.lbl.gov/education/isotopes.htm> (May 15, 2001).

Goldstein, Natalie. *The Nature of the Atom.* New York: Rosen Publishing Group, 2001.

"*The Isotopes*" (Web site). <http://chemlab.pc.maricopa.edu/periodic/isotopes.html> (May 15, 2001).

"*Isotopes*" *University of Colorado Department of Physics* (Web site). <http://www.colorado.edu/physics/2000/isotopes/index.html> (May 15, 2001).

Milne, Lorus Johnson and Margery Milne. *Understanding Radioactivity.* Illustrated by Bill Hiscock. New York: Atheneum, 1989.

Smith, Norman F. *Millions and Billions of Years Ago: Dating Our Earth and Its Life.* New York: F. Watts, 1993.

"*Stable Isotope Group.*" *Martek Biosciences* (Web site). <http://www.martekbio.com/frmain.htm> (May 15, 2001).

"*Tracking with Isotopes*" (Web site). <http://whyfiles.org/083isotope/2.html> (May 15, 2001).

IONS AND IONIZATION

CONCEPT

Atoms have no electric charge; if they acquire one, they are called ions. Ions are involved in a form of chemical bonding that produces extremely strong bonds between metals, or between a metal and a nonmetal. These substances, of which table salt is an example, are called ionic compounds. Ionization is the process whereby electrons are removed from an atom or molecule, as well as the process whereby an ionic substance, such as salt, is dissociated into its component ions in a solution such as water. There are several varieties of ionization, including field ionization, which almost everyone has experienced in the form of static electricity. Ion exchange, or the replacement of one ion by another, is used in applications such as water purification, while chemists and physicists use ions in mass spectrometry, to discover mass and structural information concerning atoms and molecules. Another example of ions at work (and a particularly frightening example at that) is ionizing radiation, associated with the radioactive decay following a nuclear explosion.

HOW IT WORKS

IONS: POSITIVE AND NEGATIVE

Atoms have no electric charge, because they maintain an equal number of protons (positively charged subatomic particles) and electrons, subatomic particles with a negative charge. In certain situations, however, the atom may lose or gain one or more electrons and acquire a net charge, becoming an ion.

Aluminum, for instance, has an atomic number of 13, which tells us that an aluminum atom will have 13 protons. Given the fact that every proton has a positive charge, and that most atoms tend to be neutral in charge, this means that there are usually 13 electrons, with a negative charge, present in an atom of aluminum. Yet like all metals, aluminum is capable of forming an ion by losing electrons—in this case, three.

CATIONS. Initially, the aluminum atom had a charge of $+13 + (-13) = 0$; in other words, its charge was neutral due to the equal numbers of protons and electrons. When it becomes an ion, it loses 3 electrons, leaving behind only 10. Now the charge is $+13 + (-10) = +3$. Thus the remaining aluminum ion is said to have a net positive charge of 3, represented as +3 or 3+. Chemists differ as to whether they represent the plus sign (or the minus sign, in the case of a negatively charged ion) before or after the number. Because both systems of notation are used, these will be applied interchangeably throughout the course of this essay.

When a neutral atom loses one or more electrons, the result is a positively charged ion, or cation (pronounced KAT-ie-un). Cations are usually represented by a superscript number and plus sign: Al^{+3} or Al^{3+}, for instance, represents the aluminum cation described above. A cation is named after the element of which it is an ion: thus the ion we have described is either called the aluminum ion, or the aluminum cation.

ANIONS. When a neutrally charged atom gains electrons, acquiring a negative charge as a result, this type of ion is known as an anion (AN-ie-un). Anions can be represented symbolically in much the same way as cations: Cl^-, for

A COMMON FORM OF FIELD IONIZATION IS STATIC ELECTRICITY. HERE, A GIRL PLACES HER HAND ON A STATIC ELECTRICITY GENERATOR. *(Paul A. Souders/Corbis. Reproduced by permission.)*

instance, is an anion of chlorine that forms when it acquires an electron, thus assuming a net charge of -1. Note that the 1 is not represented in the superscript notation, much as people do not write 10^1. In both cases, the 1 is assumed, but any number higher than 1 is shown.

The anion described here is never called a chlorine anion; rather, anions have a special nomenclature. If the anion represents, as was the case here, a single element, it is named by adding the suffix -ide to the name of the original element name: chloride. Such is the case, for instance, with a deadly mixture of carbon and nitrogen (CN^-), better known as cyanide.

Most often the -ide suffix is used, but in the case of most anions involving more than one element (polyatomic anions), as well as with oxyanions (anions containing oxygen), the rules can get fairly complicated. The general principles for naming anions are as follows:

- -ide: A single element with a negative charge. Note, however, that both hydroxide (OH^-) and cyanide (CN^-) also receive the -ide suffix, even though they involve more than one element.
- -ate: An oxyanion with the normal number of oxygen atoms, a number that depends on

the nature of the compound. Examples include oxalate ($C_2O_4^{-2}$) or chlorate (ClO_3^-).

- -ite: An oxyanion containing 1 less oxygen than normal. Examples include chlorite (ClO_2^-).
- hypo____ite: An oxyanion with 2 less oxygens than normal, but with the normal charge. An example is hypochlorite, or ClO^-.
- per____ate: An oxyanion with 1 more oxygen than normal, but with the normal charge. Perchlorate, or ClO_4^-, is an example.
- thio-: An anion in which sulfur has replaced an oxygen. Thus, SO_4^{-2} is called sulfate, whereas $S_2O_3^{-2}$ is called thiosulfate.

ELEMENTS AND ION CHARGES

As one might expect, given the many differences among families of elements on the periodic table, different elements form ions in different ways. Yet precisely because many of these can be grouped into families, primarily according to the column or group they occupy on the periodic table, it is possible to predict the ways in which they will form ions. The table below provides a few rules of thumb. (All group numbers refer to the North American version of the periodic table;

THE DAMAGED REACTOR AT THE CHERNOBYL NUCLEAR PLANT IN THE FORMER SOVIET UNION. THE 1986 ACCIDENT
AT THE PLANT RELEASED IONIZING RADIATION INTO THE ATMOSPHERE. *(AP/Wide World Photos. Reproduced by permission.)*

see Periodic Table of Elements essay for an explanation of the differences between this and the IUPAC version.)

- Alkali metals (Group 1) form 1+ cations. For example, the ion of lithium (Li) is always Li^+.
- Alkaline earth metals (Group 2) form 2+ cations. Thus, beryllium (Be), for instance, forms a Be^{2+} ion.
- Most Group 3 metals (aluminum, gallium, and indium) form 3+ cations. The cation of aluminum, thus, is designated as Al_{3+}.
- Group 6 nonmetals and metalloids (oxygen, sulfur, selenium, and tellurium) form 2– anions. Oxygen, in its normal ionized state, is shown as O^{2-}.
- Halogens (Group 7) form 1– anions. Fluorine's anion would therefore be designated as Fl^-.

The metals always form positive ions, or cations; indeed, one of the defining characteristics of a metal is that it tends to lose electrons. However, the many elements of the transition metals family form cations with a variety of different charges; for this reason, there is no easy way to classify the ways in which these elements form cations.

Likewise, it should be evident from the above table that nonmetals, such as oxygen or fluorine, gain electrons to form anions. This, too, is a defining characteristic of this broad grouping of elements. The reasons why these elements—both metals and nonmetals—behave as they do are complex, involving the numbers of valence electrons (the electrons involved in chemical bonding) for each group on the periodic table, as well as the octet rule of chemical bonding, whereby elements typically bond so that each atom has eight valence electrons.

REAL-LIFE APPLICATIONS

IONIC BONDS, COMPOUNDS, AND SOLIDS

German chemist Richard Abegg (1869-1910), whose work with noble gases led to the discovery of the octet rule, hypothesized that atoms combine with one another because they exchange electrons in such a way that both end up with eight valence electrons. This was an early model of ionic bonding, which results from attractions between ions with opposite electric charges:

when they bond, these ions "complete" one another.

As noted, ionic bonds occur when a metal bonds with a nonmetal, and these bonds are extremely strong. Salt, for instance, is formed by an ionic bond between the metal sodium (a +1 cation) and the nonmetal chlorine, a -1 anion. Thus Na^+ joins with Cl^- to form NaCl, or table salt. The strength of the bond in salt is reflected by its melting point of 1,472°F (800°C), which is much higher than that of water at 32°F (0°C).

In ionic bonding, two ions start out with different charges and end up forming a bond in which both have eight valence electrons. In a covalent bond, as is formed between nonmetals, two atoms start out as most atoms do, with a net charge of zero. Each ends up possessing eight valence electrons, but neither atom "owns" them; rather, they share electrons. Today, chemists understand that most bonds are neither purely ionic nor purely covalent; rather, there is a wide range of hybrids between the two extremes. The degree to which elements attract one another is a function of electronegativity, or the relative ability of an atom to attract valence electrons.

IONIC COMPOUNDS AND SOLIDS. Not only is salt formed by an ionic bond, but it is an ionic compound—that is, a compound containing at least one metal and nonmetal, in which ions are present. Salt is also an example of an ionic solid, or a crystalline solid that contains ions. A crystalline solidis a type of solid in which the constituent parts are arranged in a simple, definite geometric pattern that is repeated in all directions. There are three types of crystalline solid: a molecular solid (for example, sucrose or table sugar), in which the molecules have a neutral electric charge; an atomic solid (a diamond, for instance, which is pure carbon); and an ionic solid.

Salt is not formed of ordinary molecules, in the way that water or carbon dioxide are. (Both of these are molecular compounds, though neither is a solid at room temperature.) The internal structure of salt can be depicted as a repeating series of chloride anions and sodium cations packed closely together like oranges in a crate. Actually, it makes more sense to picture them as cantaloupes and oranges packed together, with the larger chloride anions as the cantaloupes and the smaller sodium cations as the oranges.

If a grocer had to pack these two fruits together in a crate, he would probably put down a layer of cantaloupes, then follow this with a layer of oranges in the spaces between the cantaloupes. This pattern would be repeated as the crate was packed, with the smaller oranges filling the spaces. Much the same is true of the way that chlorine "cantaloupes" and sodium "oranges" are packed together in salt.

This close packing of positive and negative charges helps to form a tight bond, and therefore salt must be heated to a high temperature before it will melt. Solid salt does not conduct electricity, but when melted, it makes for an extremely good conductor. When it is solid, the ions are tightly packed, and thus there is no freedom of movement to carry the electric charge along; but when the structure is disturbed by melting, movement of ions is therefore possible.

IONIZATION AND IONIZATION ENERGY

Water is not a good conductor, although it will certainly allow an electric current to flow through it, which is why it is dangerous to operate an electrical appliance near water. We have already seen that salt becomes a good conductor when melted, but this can also be achieved by dissolving it in water. This is one type of ionization, which can be defined as the process in which one or more electrons are removed from an atom or molecule to create an ion, or the process in which an ionic solid, such as salt, dissociates into its component ions upon being dissolved in a solution.

The amount of energy required to achieve ionization is called ionization energy or ionization potential. When an atom is at its normal energy level, it is said to be in a ground state. At that point, electrons occupy their normal orbital patterns. There is always a high degree of attraction between the electron and the positively charged nucleus, where the protons reside. The energy required to move an electron to a higher orbital pattern increases the overall energy of the atom, which is then said to be in an excited state.

The excited state of the atom is simply a step along the way toward ionizing it by removing the electron. "Step" is an appropriate metaphor, because electrons do not simply drift along a continuum from one energy level to the next, the way a person walks gently up a ramp. Rather, they make discrete steps, like a person climbing a flight of stairs or a ladder. This is one of the key principles of quantum mechanics, a cutting-edge area of physics that also has numerous applications to chemistry. Just as we speak of a sudden change as a "quantum leap," electrons make quantum jumps from one energy level to another.

Due to the high attraction between the electron and the nucleus, the first electron to be removed is on the outermost orbital—that is, one of the valence electrons. This amount of energy is called the first ionization energy. To remove a second electron will be considerably more difficult, because now the atom is a cation, and the positive charge of the protons in the nucleus is greater than the negative charge of the electrons. Hence the second ionization energy, required to remove a second electron, is much higher than the first ionization energy.

IONIZATION ENERGIES OF ELEMENTS AND COMPOUNDS.

First and second ionization energy levels for given elements have been established, though it should be noted that hydrogen, because it has only one electron, has only a first ionization energy. In general, the figures for ionization energy increase from left to right along a period or row on the periodic table, and decrease from top to bottom along a column or group.

The reason ionization energy increases along a period is that nonmetals, on the right side of the table, have higher ionization energies than metals, which are on the left side. Ionization energies decrease along a group because elements lower on the table have higher atomic numbers, which means more protons and thus more electrons. It is therefore easier for them to give up one of their electrons than it is for an element with a lower atomic number—just as it would be easier for a millionaire to lose a dollar than it would be for a person earning minimum wage.

For molecules in compounds, the ionization energies are generally related to those of the elements whose atoms make up the molecule. Just as elements with fewer electrons are typically less inclined to give up one, so are molecules with only a few atoms. Thus the ionization energy of carbon dioxide (CO_2), with just three atoms, is relatively high. Conversely, in larger molecules as with larger atoms, there are more electrons to

give up, and therefore it is easier to separate one of these from the molecules.

IONIZATION PROCESSES.

A number of methods are used to produce ions for mass spectrometry (discussed below) or other applications. The most common of these methods is electron impact, produced by bombarding a sample of gas with a stream of fast-moving electrons. Though easier than some other methods, this one is not particularly efficient, because it supplies more energy than is needed to remove the electron. An electron gun, usually a heated tungsten wire, produces huge amounts of electrons, which are then fired at the gas. Because electrons are so small, this is rather like using a rapid-fire machine gun to kill mosquitoes: it is almost inevitable that some of the mosquitoes will get hit, but plenty of the rounds will fire into the air without hitting a single insect.

Another ionization process is field ionization, in which ionization is produced by subjecting a molecule to a very intense electric field. Field ionization occurs in daily life, when static electricity builds up on a dry day: the small spark that jumps from the tip of your finger when you touch a doorknob is actually a stream of electrons. In field ionization as applied in laboratories, fine, sharpened wires are used. This process is much more efficient than electron impact ionization, employing far less energy in relation to that required to remove the electrons. Because it deposits less energy into the ion source, the parent ion, it is often used when it is necessary not to damage the ionized specimen.

Chemical ionization employs a method similar to that of electron impact ionization, except that instead of electrons, a beam of positively charged molecular ions is used to bombard and ionize the sample. The ions used in this bombardment are typically small molecules, such as those in methane, propane, or ammonia. Nonetheless, the molecular ion is much larger than an electron, and these collisions are highly reactive, yet they tend to be much more efficient than electron impact ionization. Many mass spectrometers use a source capable of both electron impact and chemical ionization.

Ionization can also be supplied by electromagnetic radiation, with wavelengths shorter than those of visible light—that is, ultraviolet light, x rays, or gamma rays. This process, called photoionization, can ionize small molecules, such as those of oxygen (O_2). Photoionization occurs in the upper atmosphere, where ultraviolet radiation from the Sun causes ionization of oxygen and nitrogen (N_2) in their molecular forms.

In addition to these other ionization processes, ionization can be produced in a fairly simple way by subjecting atoms or molecules to the heat from a flame. The temperatures, however, must be several thousand degrees to be effective, and therefore specialized flames, such as an electrical arc, spark, or plasma, are typically used.

MASS SPECTROMETRY. As not-ed, a number of the ionization methods described above are used in mass spectrometry, a means of obtaining structure and mass information concerning atoms or molecules. In mass spectrometry, ionized particles are accelerated in a curved path through an electromagnetic field. The field will tend to deflect lighter particles from the curve more easily than heavier ones. By the time the particles reach the detector, which measures the ratio between mass and charge, the ions will have been separated into groups according to their respective mass-to-charge ratios.

When molecules are subjected to mass spectrometry, fragmentation occurs. Each molecule breaks apart in a characteristic fashion, and this makes it possible for a skilled observer to interpret the mass spectrum of the particles generated. Mass spectrometry is used to establish values for ionization energy, as well as to ascertain the mass of substances when that mass is not known. It can also be used to determine the chemical makeup of a substance. Mass spectrometry is applied by chemists, not only in pure research, but in applications within the environmental, pharmaceutical, and forensic (crime-solving) fields. Chemists for petroleum companies use it to analyze hydrocarbons, as do scientists working in areas that require flavor and fragrance analysis.

ION EXCHANGE

The process of replacing one ion with another of a similar charge is called ion exchange. When an ionic solution—for example, salt dissolved in water—is placed in contact with a solid containing ions that are only weakly bonded within its crystalline structure, ion exchange is possible. Throughout the exchange, electric neutrality is maintained: in other words, the total number of positive charges in the solid and solution equals the total number of negative charges. The only thing that changes is the type of ion in the solid and in the solution.

There are natural ion exchangers, such as zeolites, a class of minerals that contain aluminum, silicon, oxygen, and a loosely held cation of an alkali metal or alkaline earth metal (Group 1 and Group 2 of the periodic table, respectively). When the zeolite is placed in an ionic solution, exchange occurs between the loosely held zeolite cation and the dissolved cation. In synthetic ion exchangers, used in laboratories, a charged group is attached to a rigid structural framework. One end of the charged group is permanently fixed to the frame, while a positively or negatively charged portion kept loose at the other end attracts other ions in the solution.

RESINS AND SEMIPERMEABLE MEMBRANES. Anion resins and cation resins are both solid materials, but in an anion resin, the positive ions are tightly bonded, while the negative ions are loosely bonded. The negative ions will exchange with negative ions in the solution. The reverse is true for a cation resin, in which the negative ions are tightly bonded, while the loosely bonded positive ions exchange with positive ions in the solution. These resins have applications in scientific studies, where they are used to isolate and collect various types of ionic substances. They are also used to purify water, by removing all ions.

Semipermeable membranes are natural or synthetic materials with the ability to selectively permit or retard the passage of charged and uncharged molecules through their surfaces. In the cells of living things, semipermeable membranes regulate the balance between sodium and potassium cations. Kidney dialysis, which removes harmful wastes from the urine of patients with kidneys that do not function properly, is an example of the use of semipermeable membranes to purify large ionic molecules. When seawater is purified by removal of salt, or otherwise unsafe water is freed of its impurities, the water is passed through a semipermeable membrane in a process known as reverse osmosis.

IONIZING RADIATION

We have observed a number of ways in which ionization is useful, and indeed part of nature. But ionizing radiation, which causes ionization

KEY TERMS

ANION: The negatively charged ion that results when an atom gains one or more electrons. An anion (pronounced "AN-ie-un") of a single element is named by adding the suffix -ide to the name of the original element—hence, "chloride." Other rules apply for more complex anions.

CATION: The positively charged ion that results when an atom loses one or more electrons. A cation (pronounced KAT-ie-un) is named after the element of which it is an ion and, thus, is called, for instance, the aluminum ion or the aluminum cation.

CRYSTALLINE SOLID: A type of solid in which the constituent parts have a simple and definite geometric arrangement that is repeated in all directions. Types of crystalline solids include atomic solids, ionic solids, and molecular solids.

ELECTRON: Negatively charged particles in an atom. The number of electrons and protons in an atom is the same, thus canceling out one another. When an atom loses or gains one or more electrons, however—thus becoming an ion—it acquires a net electric charge.

EXCITED STATE: A term describing the characteristics of an atom that has acquired excess energy.

GROUND STATE: A term describing the characteristics of an atom that is at its ordinary energy level.

ION: An atom or group of atoms that has lost or gained one or more electrons, and thus has a net electric charge. There are two types of ions: anions and cations.

ION EXCHANGE: The process of replacing one ion with another of similar charge.

IONIC BONDING: A form of chemical bonding that results from attractions between ions with opposite electric charges.

IONIC COMPOUND: A compound (two or more elements chemically bonded to one another), in which ions are present. Ionic compounds contain at least one metal and nonmetal, joined by an ionic bond.

IONIC SOLID: A form of crystalline solid that contains ions. When mixed with a solvent such as water, ions from table salt—an example of an ionic solid—move freely throughout the solution, making it possible to conduct an electric current.

IONIZATION: A term that refers to two different processes. In one kind of ionization, one or more electrons are removed from an atom or molecule to create an ion. A second kind of ionization occurs when an ionic solid, such as salt dissociates into its component ions upon being dissolved in a solution.

IONIZATION ENERGY: The amount of energy required to achieve ionization in which one or more electrons are removed from an atom or molecule. Another term for this is ionization potential.

NUCLEUS: The center of an atom, a region where protons and neutrons are located, and around which electrons spin.

ORBITAL: A pattern of probabilities regarding the position of an electron for an atom in a particular energy state.

OXYANION: An anion containing oxygen.

POLYATOMIC ANION: An anion involving more than one element.

PROTON: A positively charged particle in an atom.

in the substance through which it passes, is extremely harmful. It should be noted that not all radiation is unhealthy: after all, Earth receives heat and light from the Sun by means of radiation. Ionizing radiation, on the other hand, is the kind of radiation associated with radioactive fall-out from nuclear warfare, and with nuclear disasters such as the one at Chernobyl in the former Soviet Union in 1986.

Whereas thermal radiation from the Sun can be harmful if one is exposed to it for too long, ionizing radiation is much more so, and involves much greater quantities of energy deposited per area per second. In ionizing radiation, an ionizing particle knocks an electron off an atom or molecule in a living system (that is, human, animal, or plant), freeing an electron. The molecule left behind becomes a free radical, a highly reactive group of atoms with unpaired electrons. These can spawn other free radicals, inducing chemical changes that can cause cancer and genetic damage.

WHERE TO LEARN MORE

"The Atom." Thinkquest (Web site). <http://library. thinkquest.org/17940/texts/atom/atom.html> (May 18, 2001).

"Explore the Atom" CERN—European Organization for Nuclear Research (Web site). <http://public.web. cern.ch/Public/SCIENCE/Welcome.html> (May 18, 2001).

Gallant, Roy A. The Ever-Changing Atom. New York: Benchmark Books, 1999.

"Ion Exchange Chromatography" (Web site). <http://ntri. tamuk.edu/fplc/ion.html> (June 1, 2001).

"Just After Big Bang, Colliding Gold Ions Exploded." Uni-Sci: Daily University Science News (Web site). <http://unisci.com/stories/20012/0501012.htm> (June 1, 2001).

"A Look Inside the Atom" (Web site). <http://www.aip. org/history/electron/jjhome.htm> (May 18, 2001).

"Portrait of the Atom" (Web site). <http://www. inetarena.com/~pdx4d/snelson/Portrait.html> (May 18, 2001).

"A Science Odyssey: You Try It: Atom Builder." PBS—Public Broadcasting System (Web site). <http://www.pbs. org/wgbh/aso/tryit/atom/> (May 18, 2001).

"Table of Inorganic Ions and Rules for Naming Inorganic Compounds." Augustana College (Web site.) <http://inst.augie.edu/~dew/242mat.htm> (June 1, 2001).

"Virtual Chemistry Lab." University of Oxford Department of Chemistry (Web site). <http://www.chem.ox. ac.uk/vrchemistry/complex/menubar.html> (June 1, 2001).

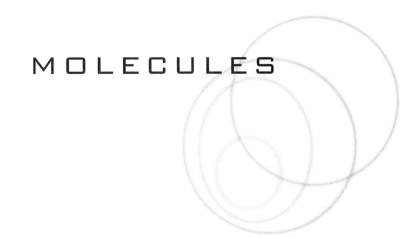

MOLECULES

CONCEPT

Prior to the nineteenth century, chemists pursued science simply by taking measurements, before and after a chemical reaction, of the substances involved. This was an external approach, rather like a person reaching into a box and feeling of the contents without actually being able to see them. With the evolution of atomic theory, chemistry took on much greater definition: for the first time, chemists understood that the materials with which they worked were interacting on a level much too small to see. The effects, of course, could be witnessed, but the activities themselves involved the interactions of atoms in molecules. Just as an atom is the most basic particle of an element, a molecule is the basic particle of a compound. Whereas there are only about 90 elements that occur in nature, many millions of compounds are formed naturally or artificially. Hence the study of the molecule is at least as important to the pursuit of modern chemistry as the study of the atom. Among the most important subjects in chemistry are the ways in which atoms join to form molecules—not just the numbers and types of atoms involved, but the shape that they form together in the molecular structure.

HOW IT WORKS

INTRODUCTION TO THE MOLECULE

Sucrose or common table sugar, of course, is grainy and sweet, yet it is made of three elements that share none of those characteristics. The formula for sugar is $C_{12}H_{22}O_{11}$, meaning that each molecule is formed by the joining of 12 carbon atoms, 22 hydrogens, and 11 atoms of oxygen. Coal is nothing like sugar—for one thing, it is as black as sugar is white, yet it is almost pure carbon. Carbon, at least, is a solid at room temperature, like sugar. The other two components of sugar, on the other hand, are gases, and highly flammable ones at that.

The question of how elements react to one another, producing compounds that are altogether unlike the constituent parts, is one of the most fascinating aspects of chemistry and, indeed, of science in general. Combined in other ways and in other proportions, the elements in sugar could become water (H_2O), carbon dioxide (CO_2), or even petroleum, which is formed by the joining of carbon and hydrogen.

Two different compounds of hydrogen and oxygen serve to further illustrate the curiosities involved in the study of molecules. As noted, hydrogen and oxygen are both flammable, yet when they form a molecule of water, they can be used to extinguish most fires. On the other hand, when two hydrogens join with two oxygens to form a molecule of hydrogen peroxide (H_2O_2), the resulting compound is quite different from water. In relatively high concentrations, hydrogen peroxide can burn the skin, and in still higher concentrations, it is used as rocket fuel. And whereas water is essential to life, pure hydrogen peroxide is highly toxic.

THE QUESTION OF MOLECULAR STRUCTURE. It is not enough, however, to know that a certain combination of atoms forms a certain molecule, because molecules may have identical formulas and yet be quite different substances. In English, for

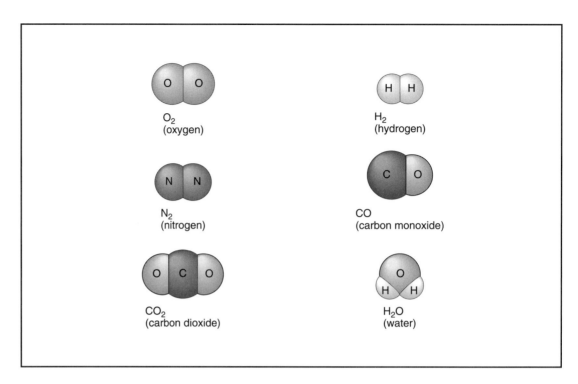

MODELS OF VARIOUS ELEMENTS.

instance, there is the word "rose." Simply seeing the word, however, does not tell us whether it is a noun, referring to a flower, or a verb, as in "she rose through the ranks." Similarly, the formula of a compound does not necessarily tell what it is, and this can be crucial.

For instance, the formula C_2H_6O identifies two very different substances. One of these is ethyl alcohol, the type of alcohol found in beer and wine. Note that the elements involved are the same as those in sugar, though the proportions are different: in fact, some aspects of the body's reaction to ethyl alcohol are not so different from its response to sugar, since both lead to unhealthy weight gain. In reasonable small quantities, of course, ethyl alcohol is not toxic, or at least only mildly so; yet methyl ether—which has an identical formula—is a toxin.

But the distinction is not simply an external one, as simple as the difference between beer and a substance such as methyl ether, sometimes used as a refrigerant. To put it another way, the external difference reflects an internal disparity: though the formulas for ethyl alcohol and methyl ether are the same, the arrangements of the atoms within the molecules of each are not. The substances are therefore said to be isomers.

In fact C_2H_6O is just one of three types of formula for a compound: an empirical formula, or one that shows the smallest possible whole-number ratio of the atoms involved. By contrast, a molecular formula—a formula that indicates the types and numbers of atoms involved— shows the actual proportions of atoms. If the formula for glucose, a type of sugar ($C_6H_{12}O_6$), were rendered in empirical form, it would be CH_2O, which would reveal less about its actual structure. Most revealing of all, however, is a structural formula—a diagram that shows how the atoms are bonded together, complete with lines representing covalent bonds. (Structural formulas such as those that apply the Couper or Lewis systems are discussed in the Chemical Bonding essay, which also examines the subject of covalent bonds.)

Chemists involved in the area of stereo-chemistry, discussed below, attempt to develop three-dimensional models to show how atoms are arranged in a molecule. Such models for ethyl alcohol and methyl ether, for instance, would reveal that they are quite different, much as the two definitions of rose mentioned above illustrate the two distinctly different meanings. Because stereochemistry is a highly involved and complex subject, it can only be touched upon very briefly in this essay; nonetheless, an under-

TWO HIGH SCHOOL STUDENTS POSE WITH A MODEL OF A DNA MOLECULE. *(James A. Sugar/Corbis. Reproduced by permission.)*

standing of a molecule's actual shape is critical to the work of a professional chemist.

MOLECULES AND COMPOUNDS.

A molecule can be most properly defined as a group of atoms joined in a specific structure. A compound, on the other hand, is a substance made up of more than one type of atom—in other words, more than one type of element. Not all compounds are composed of discrete molecules, however. For instance, table salt (NaCl) is an ionic compound formed by endlessly repeating clusters of sodium and chlorine that are not, in the strictest sense of the word, molecules.

Salt is an example of a crystalline solid, or a solid in which the constituent parts are arranged in a simple, definite geometric pattern repeated in all directions. There are three kinds of crystalline solids, only one of which has a truly molecular structure. In an ionic solid such as table salt, ions (atoms, or groups of atoms, with an electric charge) bond a metal to a nonmetal—in this case, the metal sodium and the nonmetal chlorine. Another type of crystalline solid, an atomic solid, is formed by atoms of one element bonding to one another. A diamond, made of pure carbon, is an example. Only the third type of crystalline solid is truly molecular in structure:

a molecular solid—sugar, for example—is one in which the molecules have a neutral electric charge.

Not all solids are crystalline; nor, of course, are all compounds solids: water, obviously, is a liquid at room temperature, while carbon dioxide is a gas. Nor is every molecule composed of more than one element. Oxygen, for instance, is ordinarily diatomic, meaning that even in its elemental form, it is composed of two atoms that join in an O_2 molecule. It is obvious, then, that the defining of molecules is more complex than it seems. One can safely say, however, that the vast majority of compounds are made up of molecules in which atoms are arranged in a definite structure.

In the essay that follows, we will discuss the ways atoms join to form molecules, a subject explored in more depth within the Chemical Bonding essay. (In addition, compounds themselves are examined in somewhat more detail within the Compounds essay.) We will also briefly examine how molecules bond to other molecules in the formation of solids and liquids. First, however, a little history is in order: as noted in the introduction to this essay, chemists did not always possess a clear understanding of the nature of a molecule.

A Brief History of the Molecule

In ancient and medieval times, early chemists—some of whom subscribed to an unscientific system known as alchemy—believed that one element could be transformed into another. Thus many an alchemist devoted an entire career to the vain pursuit of turning lead into gold. The alchemists were at least partially right, however: though one element cannot be transformed into another (except by nuclear fusion), it is possible to change the nature of a compound by altering the relations of the elements within it.

Modern understanding of the elements began to emerge in the seventeenth century, but the true turning point came late in the eighteenth century. It was then that French chemist Antoine Lavoisier (1743-1794) defined an element as a simple substance that could not be separated into simpler substances by chemical means. Around the same time, another French chemist, Joseph-Louis Proust (1754-1826) stated that a given compound always contained the same proportions of mass between elements. The ideas of Lavoisier and Proust were revolutionary at the time, and these concepts pointed to a substructure, invisible to the naked eye, underlying all matter.

In 1803, English chemist John Dalton (1766-1844) defined that substructure by introducing the idea that the material world is composed of tiny particles called atoms. Despite the enormous leap forward that his work afforded to chemists, Dalton failed to recognize that matter is not made simply of atoms. Water, for instance, is not just a collection of "water atoms": clearly, there is some sort of intermediary structure in which atoms are combined. This is the molecule, a concept introduced by Italian physicist Amedeo Avogadro (1776-1856).

AVOGADRO AND THE IDEA OF THE MOLECULE. French chemist and physicist Joseph Gay-Lussac (1778-1850) had announced in 1809 that gases combine to form compounds in simple proportions by volume. As Gay-Lussac explained, the ratio, by weight, between hydrogen and oxygen in water is eight to one. The fact that this ratio was so "clean," involving whole numbers rather than decimals, intrigued Avogadro, who in 1811 proposed that equal volumes of gases have the same number of particles if measured at the same temperature and pressure. This, in turn, led him to the hypothesis that water is not composed simply of atoms, but of molecules in which hydrogen and oxygen combine.

For several decades, however, chemists largely ignored Avogadro's idea of the molecule. Only in 1860, four years after his death, was the concept resurrected by Italian chemist Stanislao Cannizzaro (1826-1910). Of course, the understanding of the molecule has progressed enormously in the years since then, and much of this progress is an outcome of advances in the study of subatomic structure. Only in the early twentieth century did physicists finally identify the electron, the negatively charged subatomic particle critical to the bonding of atoms.

REAL-LIFE APPLICATIONS

Molecular Mass

Just as the atoms of elements have a definite mass, so do molecules—a mass equal to that of the combined atoms in the molecule. The figures for the atomic mass of all elements are established, and can be found on the periodic table; therefore, when one knows the mass of a hydrogen atom and an oxygen atom, as well as the fact that there are two hydrogens and one oxygen in a molecule of water, it is easy to calculate the mass of a water molecule.

Individual molecules cannot easily be studied; therefore, the mass of molecules is compared by use of a unit known as the mole. The mole contains 6.022137×10^{23} molecules, a figure known as Avogadro's number, in honor of the man who introduced the concept of the molecule. When necessary, it is possible today to study individual molecules, or even atoms and subatomic particles, using techniques such as mass spectrometry.

Bonding Within Molecules

Note that the mass of an atom in a molecule does not change; nor, indeed, do the identities of the individual atoms. An oxygen atom in water is the same oxygen atom in sugar, or in any number of other compounds. With regard to compounds, it should be noted that these are not the same thing as a mixture, or a solution. Sugar or salt can be dissolved in water at the appropriate temperatures, but the resulting solution is not a compound; the substances are joined physically, but they are not chemically bonded.

Chemical bonding is the joining, through electromagnetic force, of atoms representing different elements. Each atom possesses a certain valency, which determines its ability to bond with atoms of other elements. Valency, in turn, is governed by the configuration of valence electrons at the highest energy level (the shell) of the atom.

While studying noble gases, noted for their tendency not to bond, German chemist Richard Abegg (1869-1910) discovered that these gases always have eight valence electrons. This led to the formation of the octet rule: most elements (with the exception of hydrogen and a few others) are inclined to bond in such a way that they end up with eight valence electrons.

When a metal bonds to a nonmetal, this is known as ionic bonding, which results from attractions between ions with opposite electric charges. In ionic bonding, two ions start out with different charges and form a bond in which both have eight valence electrons. Nonmetals, however, tend to form covalent bonds. In a covalent bond, two atoms start out as most atoms do, with a net charge of zero. Each ends up possessing eight valence electrons, but neither atom "owns" them; rather, they share electrons.

ELECTRONEGATIVITY. Not all elements bond covalently in the same way. Each has a certain value of electronegativity—the relative ability of an atom to attract valence electrons. Elements capable of bonding are assigned an electronegativity value ranging from a minimum of 0.7 for cesium to a maximum of 4.0 for fluorine. The greater the electronegativity value, the greater the tendency of an element to attract valence electrons.

When substances of differing electronegativity values form a covalent bond, this is described as polar covalent bonding. Water is an example of a molecule with a polar covalent bond. Because oxygen has a much higher electronegativity (3.5) than hydrogen (2.1), the electrons tend to gravitate toward the oxygen atom. By contrast, molecules of petroleum, a combination of carbon and hydrogen, tend to be nonpolar, because carbon (2.5) and hydrogen have very similar electronegativity values.

A knowledge of electronegativity values can be used to make predictions concerning bond polarities. Bonds that involve atoms whose electronegativities differ by more than 2 units are substantially ionic, whereas bonds between atoms whose electronegativities differ by less than 2 units are polar covalent. If the atoms have the same or similar electronegativity values, the bond is covalent.

ATTRACTIONS BETWEEN MOLECULES

The energy required to pull apart a molecule is known as bond energy. Covalent bonds that involve hydrogen are among the weakest bonds between atoms, and hence it is relatively easy to separate water into its constituent parts, hydrogen and oxygen. (This is sometimes done by electrolysis, which involves the use of an electric current to disperse atoms.) Double and triple covalent bonds are stronger, but strongest of all is an ionic bond. The strength of the bond energy in salt, for instance, is reflected by its melting point of 1,472°F (800°C), much higher than that of water, at 32°F (0°C).

Bond energy relates to the attraction between atoms in a molecule, but in considering various substances, it is also important to recognize the varieties of bonds between molecules—that is, intermolecular bonding. For example, the polar quality of a water molecule gives it a great attraction for ions, and thus ionic substances such as salt and any number of minerals dissolve easily in water. On the other hand, we have seen that petroleum is essentially nonpolar, and therefore, an oil molecule offers no electric charge to bond it with a water molecule. For this reason, oil and water do not mix.

The bonding between water molecules is known as a dipole-dipole attraction. This type of intermolecular bond can be fairly strong in the liquid or solid state, though it is only about 1% as strong as a covalent bond within a molecule. When a substance containing molecules joined by dipole-dipole attraction is heated to become a gas, the molecules spread far apart, and these bonds become very weak. On the other hand, when hydrogen bonds to an atom with a high value of electronegativity (fluorine, for example), the dipole-dipole attraction between these molecules is particularly strong. This is known as hydrogen bonding.

KEY TERMS

AVOGADRO'S NUMBER: A figure, named after Italian physicist Amedeo Avogadro (1776-1856), equal to 6.022137×10^{23}. Avogadro's number indicates the number of molecules in a mole.

BOND ENERGY: The energy required to pull apart the atoms in a chemical bond.

CHEMICAL BONDING: The joining, through electromagnetic force, of atoms representing different elements.

COMPOUND: A substance made up of atoms of more than one element. These atoms are usually joined in molecules.

COVALENT BONDING: A type of chemical bonding in which two atoms share valence electrons.

DIATOMIC: A term describing an element that exists as molecules composed of two atoms.

DIPOLE-DIPOLE ATTRACTION: A form of intermolecular bonding between molecules formed by a polar covalent bond.

ELECTRON: A negatively charged particle in an atom.

ELECTRONEGATIVITY: The relative ability of an atom to attract valence electrons.

EMPIRICAL FORMULA: A chemical formula that shows the smallest possible whole-number ratio of the atoms involved. Compare with molecular formula and structural formula.

HYDROGEN BONDING: A kind of dipole-dipole attraction between molecules formed of hydrogen along with an element having a high electronegativity.

INTERMOLECULAR BONDING: The bonding that exists between molecules. This is not to be confused with chemical bonding, the bonding of atoms within a molecule.

ION: An atom or group of atoms that has lost or gained one or more electrons, and thus has a net electric charge.

IONIC BONDING: A form of chemical bonding resulting from attractions between ions with opposite electric charges.

ISOMERS: Substances having the same chemical formula, but which are chemically dissimilar due to differences in the arrangement of atoms.

LONDON DISPERSION FORCES: A term describing the weak intermolecular bond between molecules that are not formed by a polar covalent bond.

Even a nonpolar molecule, however, must have some attraction to other nonpolar molecules. The same is true of helium and the other noble gases, which are highly nonattractive but can be turned into liquids or even solids at extremely low temperatures. The type of intermolecular attraction that exists in such a situation is described by the term London dispersion forces. The name has nothing to do with the capital of England: it is a reference to German-

American physicist Fritz Wolfgang London (1900-1954), who in the 1920s studied the molecule from the standpoint of quantum mechanics.

Because electrons are not uniformly distributed around the nucleus of an atom at every possible moment, instantaneous dipoles are formed when most of the electrons happen to be on one side of an atom. Of course, this only happens for an infinitesimal fraction of time, but it serves to create a weak attraction. Only at very low tem-

KEY TERMS CONTINUED

MOLE: The SI fundamental unit for "amount of substance." A mole is, generally speaking, Avogadro's number of molecules; however, in the more precise SI definition, a mole is equal to the number of carbon atoms in 12.01 g of carbon.

MOLECULAR FORMULA: A chemical formula that indicates the types and numbers of atoms involved, showing the actual proportions of atoms in a molecule. Compare with empirical formula and structural formula.

MOLECULAR SOLID: A form of crystalline solid—a solid in which the constituent parts have a simple and definite geometric arrangement repeated in all directions—in which the molecules have a neutral electric charge. Table sugar (sucrose) is an example.

MOLECULE: A group of atoms, usually but not always representing more than one element, joined in a structure. Compounds are typically made up of molecules.

OCTET RULE: A term describing the distribution of valence electrons that takes place in chemical bonding for most elements, which end up with eight valence electrons.

POLAR COVALENT BONDING: The type of chemical bonding between atoms that have differing values of electronegativity. Water molecules are an example of a polar covalent bond.

SHELL: The orbital pattern of the valence electrons at the outside of an atom.

STEREOCHEMISTRY: The area of chemistry devoted to the three-dimensional arrangement of atoms in a molecule.

STRUCTURAL FORMULA: A diagram that shows how the atoms are bonded together, complete with lines representing covalent bonds. Compare with empirical formula and molecular formula.

VALENCE ELECTRONS: Electrons that occupy the highest energy levels in an atom. These are the only electrons involved in chemical bonding.

VALENCY: The property of the atom of one element that determines its ability to bond with atoms of other elements.

VSEPR (VALENCE SHELL ELECTRON PAIR REPULSION) MODEL: A means of representing the three-dimensional structure of atoms in a molecule.

---

peratures do London dispersion forces become strong enough to result in the formation of a solid. (Thus, for instance, oil and rubbing alcohol freeze only at low temperatures.)

MOLECULAR STRUCTURE

The Couper system and Lewis structures, discussed in the Chemical Bonding essay, provide a means of representing the atoms that make up a molecule. Though Lewis structures show the dis-

tribution of valence electrons, they do not represent the three-dimensional structure of the molecule. As noted earlier in this essay, the structure is highly important, because two compounds may be isomers, meaning that they have the same proportions of the same elements, yet are different substances.

Stereochemistry is the realm of chemistry devoted to the three-dimensional arrangement of atoms in a molecule. One of the most important methods used is known as the VSEPR model (valence shell electron pair repulsion). In bonding, elements always share at least one pair of electrons, and the VSEPR model begins with the assumption that the electron pairs must be as far apart as possible to minimize their repulsion, since like charges repel.

VSEPR structures can be very complex, and the rules governing them will not be discussed here, but a few examples can be given. If there are just two electron pairs in a bond between three atoms, the structure of a VSEPR model is like that of a stick speared through a ball, with two other balls attached at each end. The "ball" is an atom, and the "stick" represents the electron pairs. In water, there are four electron pairs, but still only three atoms and two bonds. In order to keep the electron pairs as far apart as possible, the angle between the two hydrogen atoms attached to the oxygen is 109.5°.

WHERE TO LEARN MORE

Basmajian, Ronald; Thomas Rodella; and Allen E. Breed. *Through the Molecular Maze: A Helpful Guide to the Elements of Chemistry for Beginning Life Science Students.* Merced, CA: Bioventure Associates, 1990.

Burnie, David. *Microlife.* New York: DK Publishing, 1997.

"*Common Molecules*" (Web site). <http://www.recipnet.indiana.edu/common/common.html> (June 2, 2001).

Cooper, Christopher. *Matter.* New York: DK Publishing, 2000.

Mebane, Robert C. and Thomas R. Rybolt. *Adventures with Atoms and Molecules, Book V: Chemistry Experiments for Young People.* Springfield, NJ: Enslow Publishers, 1995.

"*The Molecules of Life*" (Web site). <http://biop.ox.ac.uk/www/mol_of_life/Molecules_of_Life.html> (June 2, 2001).

"*Molecules of the Month.*" University of Oxford (Web site). <http://www.chem.ox.ac.uk/mom/> (June 2, 2001).

"*Molecules with Silly or Unusual Names*" (Web site). <http://www.bris.ac.uk/Depts/Chemistry/MOTM/silly/sillymols.htm> (June 2, 2001).

"*Theory of Atoms in Molecules*" (Web site). <http://www.chemistry.mcmaster.ca/faculty/bader/aim/> (June 2, 2001).

Zumdahl, Steven S. *Introductory Chemistry: A Foundation,* 4th ed. Boston: Houghton Mifflin, 2000.

ELEMENTS

ELEMENTS

PERIODIC TABLE OF ELEMENTS

FAMILIES OF ELEMENTS

ELEMENTS

CONCEPT

The elements are at the heart of chemistry, and indeed they are at the heart of life as well. Every physical substance encountered in daily life is either an element, or more likely a compound containing more than one element. There are millions of compounds, but only about 100 elements, of which only 88 occur naturally on Earth. How can such vast complexity be created from such a small amount of elements? Consider the English alphabet, with just 26 letters, with which an almost infinite number of things can be said or written. Even more is possible with the elements, which greatly outnumber the letters of the alphabet. Despite the relatively great quantity of elements, however, just two make up almost the entire mass of the universe—and these two are far from abundant on Earth. A very small number of elements, in fact, are essential to life on this planet, and to the existence of human beings.

HOW IT WORKS

THE FOCAL POINT OF CHEMISTRY

Like physics, chemistry is concerned with basic, underlying processes that explain how the universe works. Indeed, these two sciences, along with astronomy and a few specialized fields, are the only ones that address phenomena both on the Earth and in the universe as a whole. By contrast, unless or until life on another planet is discovered, biology has little concern with existence beyond Earth's atmosphere, except inasmuch as the processes and properties in outer space affect astronauts.

While physics and chemistry address many of the same fundamental issues, they do so in very different ways. To make a gross generalization—subject to numerous exceptions, but nonetheless useful in clarifying the basic difference between the two sciences—physicists are concerned with external phenomena, and chemists with internal ones.

For instance, when physicists and chemists study the interactions between atoms, unless physicists focus on some specialized area of atomic research, they tend to treat all atoms as more or less the same. Chemists, on the other hand, can never treat atoms as though they are just undifferentiated particles colliding in space. The difference in structure between one kind of atom and another, in fact, is the starting-point of chemical study.

STRUCTURE OF THE ATOM

An atom is the fundamental particle in a chemical element, or a substance that cannot be broken down into another substance by chemical means. Clustered at the center, or nucleus, of the atom are protons, which have a positive electric charge, and neutrons, which possess no charge.

Spinning around the nucleus are electrons, which are negatively charged. The vast majority of the atom's mass is made up by the protons and neutrons, which have approximately the same mass, whereas that of the electron is much smaller. The mass of an electron is about 1/1836 that of proton, and 1/1839 that of a neutron.

It should be noted that the nucleus, though it constitutes most of the atom's mass, is only a tiny portion of the atom's volume. If the nucleus were the size of a grape, in fact, the electrons would, on average, be located about a mile away.

| Element | Symbol | Percent of all atoms* | | | | Characteristics under ordinary room conditions |
		In the universe	In the earth's crust	In sea water	In the human body	
Aluminum	Al	—	6.3	—	—	A lightweight, silvery metal
Calcium	Ca	—	2.1	—	0.2	Common in minerals, seashells, and bones
Carbon	C	—	—	—	10.7	Basic in all living things
Chlorine	Cl	—	—	0.3	—	A toxic gas
Copper	Cu	—	—	—	—	The only red metal
Gold	Au	—	—	—	—	The only yellow metal
Helium	He	7.1	—	—	—	A very light gas
Hydrogen	H	92.8	2.9	66.2	60.6	The lightest of all elements; a gas
Iodine	I	—	—	—	—	A nonmetal; used as antiseptic
Iron	Fe	—	2.1	—	—	A magnetic metal; used in steel
Lead	Pb	—	—	—	—	A soft, heavy metal
Magnesium	Mg	—	2.0	—	—	A very light metal
Mercury	Hg	—	—	—	—	A liquid metal; one of the two liquid elements
Nickel	Ni	—	—	—	—	A noncorroding metal; used in coins
Nitrogen	N	—	—	—	2.4	A gas; the major component of air
Oxygen	O	—	60.1	33.1	25.7	A gas; the second major component of air
Phosphorus	P	—	—	—	0.1	A nonmetal; essential to plants
Potassium	K	—	1.1	—	—	A metal; essential to plants; commonly called "potash"
Silicon	Si	—	20.8	—	—	A semiconductor; used in electronics
Silver	Ag	—	—	—	—	A very shiny, valuable metal
Sodium	Na	—	2.2	0.3	—	A soft metal; reacts readily with water, air
Sulfur	S	—	—	—	0.1	A yellow nonmetal; flammable
Titanium	Ti	—	0.3	—	—	A light, strong, noncorroding metal used in space vehicles
Uranium	U	—	—	—	—	A very heavy metal; fuel for nuclear power

*If no number is entered, the element constitutes less than 0.1 percent.

THE MOST COMMON AND/OR IMPORTANT CHEMICAL ELEMENTS.

ISOTOPES

Atoms of the same element always have the same number of protons, and since this figure is unique for a given element, each element is assigned an atomic number equal to the number of protons in its nucleus. Two atoms may have the same number of protons, and thus be of the same element, yet differ in their number of neutrons. Such atoms are called isotopes.

Isotopes are generally represented as follows: $^{m}_{a}S$, where S is the symbol of the element, a is the atomic number, and m is the mass number—the sum of protons and neutrons in the atom's nucleus. For the stable silver isotope designated as $^{93}_{47}Ag$, for instance, Ag is the element symbol (discussed below); 47 its atomic number; and 93 the mass number. From this, it is easy to discern that this particular stable isotope has 46 neutrons in its nucleus.

Because the atomic number of any element is established, sometimes isotopes are represented simply with the mass number thus: ^{93}Ag. They may also be designated with a subscript notation indicating the number of neutrons, so that one can obtain this information at a glance without having to do the arithmetic. For the silver isotope shown here, this is written as $^{93}_{47}Ag_{46}$. Isotopes are sometimes indicated by simple nomenclature as well: for instance, carbon-12 or carbon-13.

IONS

The number of electrons in an atom is usually the same as the number of protons, and thus most atoms have a neutral charge. In certain situations, however, the atom may lose or gain one or more electrons and acquire a net charge, becoming an ion. The electrons are not "lost" when an atom becomes an ion: they simply go elsewhere.

Aluminum (Al), for instance, has an atomic number of 13, which tells us that an aluminum atom will have 13 protons. Given the fact that every proton has a positive charge, and that most atoms tend to be neutral in charge, this means that there are usually 13 electrons, with a negative charge, present in an atom of aluminum. Aluminum may, however, form an ion by losing three electrons.

CATIONS. After its three electrons have departed, the remaining aluminum ion has a net positive charge of 3, represented as +3. How do we know this? Initially the atom had a charge of $+13 + (-13) = 0$. With the exit of the 3 electrons, leaving behind only 10, the picture changes: now the charge is $+13 + (-10) = +3$.

When a neutral atom loses one or more electrons, the result is a positively charged ion, or cation (pronounced KAT-ie-un). Cations are represented by a superscript number and plus sign after the element symbol: Al^{3+}, for instance,

represents the aluminum cation described above. (Some chemists represent this with the plus sign before the number—for example, Al^{+3}.) A cation is named after the element of which it is an ion: thus the ion we have described is called either the aluminum ion, or the aluminum cation.

ANIONS. When a neutrally charged atom gains electrons, and as a result acquires a negative charge, this type of ion is known as an anion (AN-ie-un). Anions can be represented symbolically in much the same way of cations: Cl^{-}, for instance, is an anion of chlorine that forms when it acquires an electron, thus assuming a net charge of -1. Note that the 1 is not represented in the superscript notation, much as people do not write 10^{1}. In both cases, the 1 is assumed, whereas any number higher than 1 is shown.

The anion described here is never called a chlorine anion; rather, anions have a special nomenclature. If the anion represents, as is the case here, a single element, it is named by adding the suffix -ide to the name of the original element name: hence it would be called chloride. Other anions involve more than one element, and in these cases other rules apply for designating names. A few two-element anions use the -ide ending; such is the case, for instance, with a deadly mixture of carbon and nitrogen (CN^{-}), better known as cyanide.

For anions involving oxygen, there may be different prefixes and suffixes, depending on the relative number of oxygen atoms in the anion.

REAL-LIFE APPLICATIONS

THE PERIODIC TABLE

Note that an ion is never formed by a change in the number of protons: that number, as noted earlier, is a defining characteristic of an element. If all we know about a particular atom is that it has one proton, we can be certain that it is an atom of hydrogen. Likewise, if an atom has 79 protons, it is gold.

Knowing these quantities is not a matter of memorization: rather, one can learn this and much more by consulting the periodic table of elements. The periodic table is a chart, present in virtually every chemistry classroom in the world, showing the elements arranged in order of atomic number. Elements are represented by boxes containing the atomic number, element symbol, and average atomic mass, in atomic mass units, for that particular element. Vertical columns within the periodic table indicate groups or "families" of elements with similar chemical characteristics.

These groups include alkali metals, alkaline earth metals, halogens, and noble gases. In the middle of the periodic table is a wide range of vertical columns representing the transition metals, and at the bottom of the table, separated from it, are two other rows for the lanthanides and actinides.

AN OVERVIEW OF THE ELEMENTS

As of 2001, there 112 elements, of which 88 occur naturally on Earth. (Some sources show 92 naturally occurring elements; however, a few of the elements with atomic numbers below 92 have not actually been found in nature.) The others were created synthetically, usually in a laboratory, and because these are highly radioactive, they exist only for fractions of a second. The number of elements thus continues to grow, but these "new" elements have little to do with the daily lives of ordinary people. Indeed, this is true even for some of the naturally occurring elements: few people who are not chemically trained, for instance, are able to identify thulium, which has an atomic number of 69.

Though an element can theoretically exist as a gas, liquid, or a solid, in fact the vast majority of elements are solids. Only 11 elements—the six noble gases, along with hydrogen, nitrogen, oxygen, fluorine, and chlorine—exist in the gaseous state at a normal temperature of about 77°F (25°C). Just two are liquids at normal temperature: mercury, a metal, and the non-metal bromine. (The metal gallium becomes liquid at just 85.6°F, or 29.76°C.) The rest are all solids.

The noble gases are monatomic, meaning that they exist purely as single atoms. So too are the "noble metals," such as gold, silver, and platinum. "Noble" in this context means "set apart": noble gases and noble metals are known for their tendency not to react to, and hence not to bond with, other elements. On the other hand, a number of other elements are described as diatomic, meaning that two atoms join to form a molecule.

NAMES OF THE ELEMENTS

ELEMENT SYMBOL. As noted above, the periodic table includes the element symbol or chemical symbol—a one-or two-letter abbreviation for the name of the element. Many of these are simple one-letter designations: O for oxygen, or C for carbon. Others are two-letter abbreviations, such as Ne for neon or Si for silicon. Note that the first letter is always capitalized, and the second is always lowercase.

In many cases, the two-letter symbols indicate the first and second letters of the element's name, but this is far from universal. Cadmium, for example, is abbreviated Cd, while platinum is Pt. In other cases, the symbol seems to have nothing to do with the element name as it is normally used: for instance, Au for gold or Fe for iron.

Many of the one-letter symbols indicate elements discovered early in history. For instance, carbon is represented by C, and later "C" elements took two-letter designations: Ce for cerium, Cr for chromium, and so on. But many of those elements with apparently strange symbols were among the first discovered, and this is precisely why the symbols make little sense to a person who does not recognize the historical origins of the name.

HISTORICAL BACKGROUND ON SOME ELEMENT NAMES. For many years, Latin was the language of communication between scientists from different nations; hence the use of Latin names such as *aurum* ("shining dawn") for gold, or *ferrum*, the Latin word for iron. Likewise, lead (Pb) and sodium (Na) are designated by their Latin names, *plumbum* and *natrium*, respectively.

Some chemical elements are named for Greek or German words describing properties of the element—for example, bromine (Br), which comes from a Greek word meaning "stench." The name of cobalt comes from a German term meaning "underground gnome," because miners considered the metal a troublemaker. The names of several elements with high atomic numbers reflect the places where they were originally discovered or created: francium, germanium, americium, californium.

Americium and californium, with atomic numbers of 95 and 98 respectively, are among those elements that do not occur naturally, but were created artificially. The same is true of several elements named after scientists—among them einsteinium, after Albert Einstein (1879-1955), and nobelium after Alfred Nobel (1833-1896), the Swedish inventor of dynamite who established the Nobel Prize.

ABUNDANCE OF ELEMENTS

IN THE UNIVERSE. The first two elements on the periodic table, hydrogen and helium, represent 99.9% of the matter in the entire universe. This may seem astounding, but Earth is a tiny dot within the vastness of space, and hydrogen and helium are the principal elements in stars.

All elements, except for those created artificially, exist both on Earth and throughout the universe. Yet the distribution of elements on Earth is very, very different from that in other places—as well it should be, given the fact that Earth is the only planet known to support life. Hydrogen, for instance, constitutes only about 0.87%, by mass, of the elements found in the planet's crust, waters, and atmosphere. As for helium, it is not even among the 18 most abundant elements on Earth.

ON EARTH. That great element essential to animal life, oxygen, is by far the most plentiful on Earth, representing nearly half—49.2%—of the total mass of atoms found on this planet. (Here the term "mass" refers to the known elemental mass of the planet's atmosphere, waters, and crust; below the crust, scientists can only speculate, though it is likely that much of Earth's interior consists of iron.) Together with silicon (25.7%), oxygen accounts for almost exactly three-quarters of the elemental mass of Earth. Add in aluminum (7.5%), iron (4.71%), calcium (3.39%), sodium (2.63%), potassium (2.4%), and magnesium (1.93%), and these eight elements make up about 97.46% of Earth's material.

In addition to hydrogen, whose distribution is given above, nine other elements account for a total of 2% of Earth's composition: titanium (0.58%), chlorine (0.19%), phosphorus (0.11%), manganese (0.09%), carbon (0.08%), sulfur (0.06%), barium (0.04%), nitrogen (0.03%), and fluorine (0.03%). The remaining 0.49% is made up of various elements.

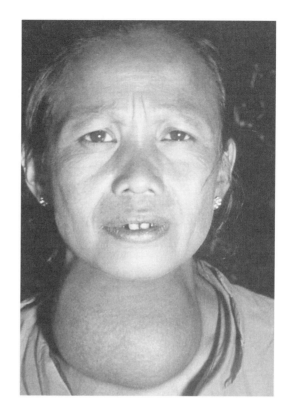

THIS WOMAN'S GOITER IS PROBABLY THE RESULT OF A LACK OF IODINE IN HER DIET. *(Lester V. Bergman/Corbis. Reproduced by permission.)*

ELEMENTS IN THE HUMAN BODY

Fans of science-fiction are familiar with the phrase "carbon-based life form," which is used, for instance, by aliens in sci-fi movies to describe humans. In fact, the term is a virtual redundancy on Earth, since all living things contain carbon.

Essential though it is to life, carbon as a component of the human body takes second place to oxygen, which is an even larger proportion of the body's mass—65.0%—than it is of the Earth. Carbon accounts for 18%, and hydrogen for 10%, meaning that these three elements make up 93% of the body's mass. Most of the remainder is taken up by 10 other elements: nitrogen (3%), calcium (1.4%), phosphorus (1.0%), magnesium (0.50%), potassium (0.34%), sulfur (0.26%), sodium (0.14%), chlorine (0.14%), iron (0.004%), and zinc (0.003%).

TRACE ELEMENTS. As small as the amount of zinc is in the human body, there are still other elements found in even smaller quantities. These are known as trace elements, because only traces of them are present in the body. Yet they are essential to human well-being: without enough iodine, for instance, a person can devel-

KEY TERMS

ANION: The negative ion that results when an atom gains one or more electrons. An anion (pronounced "AN-ie-un") of an element is never called, for instance, the chlorine anion. Rather, for an anion involving a single element, it is named by adding the suffix -ide to the name of the original element—hence, "chloride." Other rules apply for more complex anions.

ATOM: The smallest particle of an element. An atom can exist either alone or in combination with other atoms in a molecule. Atoms are made up of protons, neutrons, and electrons. An atom that loses or gains one or more electrons, and thus has a net charge, is an ion. Atoms that have the same number of protons—that is, are of the same element—but differ in number of neutrons are known as isotopes.

ATOMIC MASS UNIT: An SI unit (abbreviated amu), equal to $1.66 \cdot 10^{-24}$ g, for measuring the mass of atoms.

ATOMIC NUMBER: The number of protons in the nucleus of an atom. Since this number is different for each element, elements are listed on the periodic table in order of atomic number.

AVERAGE ATOMIC MASS: A figure used by chemists to specify the mass—in atomic mass units—of the average atom in a large sample. If a substance is a compound, the average atomic mass of all atoms in a molecule of that substance must be added together to yield the average molecular mass of that substance.

CATION: The positive ion that results when an atom loses one or more electrons. A cation (pronounced KAT-ie-un) is named after the element of which it is an ion and thus is called, for instance, the aluminum ion or the aluminum cation.

CHEMICAL SYMBOL: Another term for element symbol.

COMPOUND: A substance made up of atoms of more than one element. These atoms are usually joined in molecules.

DIATOMIC: A term describing an element that exists as molecules composed of two atoms. This is in contrast to monatomic elements.

ELECTRON: Negatively charged particles in an atom. Electrons, which spin around the protons and neutrons that make up the atom's nucleus, constitute a very small portion of the atom's mass. The number of electrons and protons is the same, thus canceling out one another. But when an atom loses or gains one or more electrons, however—thus becoming an ion—it acquires a net electric charge.

op a goiter, a large swelling in the neck area. Chromium helps the body metabolize sugars, which is why people concerned with losing weight and/or toning their bodies through exercise may take a chromium supplement.

Even arsenic, lethal in large quantities, is a trace element in the human body, and medicines for treating illnesses such as the infection known as "sleeping sickness" contain tiny amounts of arsenic. Other trace elements include cobalt, copper, fluorine, manganese, molybdenum, nickel, selenium, silicon, and vanadium.

MAINTAINING AND IMPROVING HEALTH. Though these elements are present in trace quantities within the human body, that does not mean that exposure to large amounts of them is healthy. Arsenic, of course, is a good example; so too is aluminum. Aluminum is present in unexpected places: in baked goods, for instance, where a compound containing alu-

KEY TERMS continued

ELEMENT: A substance made up of only one kind of atom. Unlike compounds, elements cannot be chemically broken into other substances.

ELEMENT SYMBOL: A one- or two-letter abbreviation for the name of an element. These may be a single capitalized letter (O for oxygen), or a capitalized letter followed by a lowercase one (Ne for neon). Sometimes the second letter is not the second letter of the element name (for example, Pt for platinum). In addition, the symbol may refer to an original Greek or Latin name, rather than the name used now, is the case with Au (aurum) for gold.

ION: An atom that has lost or gained one or more electrons, and thus has a net electric charge.

ISOTOPES: Atoms that have an equal number of protons, and hence are of the same element, but differ in their number of neutrons.

MASS NUMBER: The sum of protons and neutrons in an atom's nucleus. Where an isotope is represented, the mass number is placed above the atomic number to the left of the element symbol.

MOLECULE: A group of atoms, usu-ally but not always representing more than one element, joined in a structure.

Compounds are typically made of up molecules.

MONATOMIC: A term describing an element that exists as single atoms. This in contrast to diatomic elements.

NEUTRON: A subatomic particle that has no electric charge. Neutrons are found at the nucleus of an atom, alongside protons.

NUCLEUS: The center of an atom, a region where protons and neutrons are located, and around which electrons spin.

PERIODIC TABLE OF ELEMENTS: A chart that shows the elements arranged in order of atomic number, along with element symbol and the average atomic mass (in atomic mass units) for that particular element. Vertical columns within the periodic table indicate groups or "families" of elements with similar chemical characteristics.

PROTON: A positively charged particle in an atom. Protons and neutrons, which together form the nucleus around which electrons spin, have approximately the same mass—a mass that is many times greater than that of an electron. The number of protons in the nucleus of an atom is the atomic number of an element.

minum (baking powder) is sometimes used in the leavening process; or even in cheeses, as an aid to melting when heated. The relatively high concentrations of aluminum in these products, as well as fluorine in drinking water, has raised concerns among some scientists.

Generally speaking, an element is healthy for the human body in proportion to its presence in the body. With trace elements and others that are found in smaller quantities, however, it is some-

times possible and even advisable to increase the presence of those elements by taking dietary supplements. Hence a typical multivitamin contains calcium, iron, iodine, magnesium, zinc, selenium, copper, chromium, manganese, molybdenum, boron, and vanadium.

For most of these, recommended daily allowances (RDA) have been established by the federal government. Usually, people do not take sodium as a supplement, though—Americans already get more than their RDA of sodium through salt, which is overly abundant in the American diet.

WHERE TO LEARN MORE

Challoner, Jack. *The Visual Dictionary of Chemistry.* New York: DK Publishing, 1996.

"*Classification of Elements and Compounds*" (Web site). <http://dl.clackamas.cc.or.us/ch104-04/classifi.htm> (May 14, 2001).

"*Elements and Compounds*" (Web site). <http://wine1.sb.fsu.edu/chm1045/notes/Intro/Elements/Intro02.htm> (May 14, 2001).

"*Elements, Compounds, and Mixtures.*" Purdue University Department of Chemistry. (Web site). <http://chemed.chem.purdue.edu/genchem/topicreview/bp/ch2/mix.html> (May 14, 2001).

"*Elements, Compounds, and Mixtures*" (Web site). <http://www.juniorcert.net/serve/cont.php3?pg=SC2CEC0339> (May 14, 2001).

Knapp, Brian J. *Elements.* Illustrated by David Woodroffe and David Hardy. Danbury, CT: Grolier Educational, 1996.

"*Matter—Elements, Compounds, and Mixtures*" (Web site). <http://chem.sci.gu.edu.au/help_desk/Matter.htm> (May 14, 2001).

Oxlade, Chris. *Elements and Compounds.* Chicago, IL: Heinemann Library, 2001.

Stwertka, Albert. *A Guide to the Elements.* New York: Oxford University Press, 1998.

Zumdahl, Steven S. *Introductory Chemistry: A Foundation,* 4th ed. Boston: Houghton Mifflin, 2000.

PERIODIC TABLE OF ELEMENTS

CONCEPT

In virtually every chemistry classroom on the planet, there is a chart known as the periodic table of elements. At first glance, it looks like a mere series of boxes, with letters and numbers in them, arranged according to some kind of code not immediately clear to the observer. The boxes would form a rectangle, 18 across and 7 deep, but there are gaps in the rectangle, particularly along the top. To further complicate matters, two rows of boxes are shown along the bottom, separated from one another and from the rest of the table. Even when one begins to appreciate all the information contained in these boxes, the periodic table might appear to be a mere chart, rather than what it really is: one of the most sophisticated and usable means ever designed for representing complex interactions between the building blocks of matter.

HOW IT WORKS

INTRODUCTION TO THE PERIODIC TABLE

As a testament to its durability, the periodic table—created in 1869—is still in use today. Along the way, it has incorporated modifications involving subatomic properties unknown to the man who designed it, Russian chemist Dmitri Ivanovitch Mendeleev (1834-1907). Yet Mendeleev's original model, which we will discuss shortly, was essentially sound, inasmuch as it was based on the knowledge available to chemists at the time.

In 1869, the electromagnetic force fundamental to chemical interactions had only recent-ly been identified; the modern idea of the atom was less than 70 years old; and another three decades were to elapse before scientists began uncovering the substructure of atoms that causes them to behave as they do. Despite these limitations in the knowledge available to Mendeleev, his original table was sound enough that it has never had to be discarded, but merely clarified and modified, in the years since he developed it.

The rows of the periodic table of elements are called periods, and the columns are known as groups. Each box in the table represents an element by its chemical symbol, along with its atomic number and its average atomic mass in atomic mass units. Already a great deal has been said, and a number of terms need to be explained. These explanations will require the length of this essay, beginning with a little historical background, because chemists' understanding of the periodic table—and of the elements and atoms it represents—has evolved considerably since 1869.

ELEMENTS AND ATOMS

An element is a substance that cannot be broken down chemically into another substance. An atom is the smallest particle of an element that retains all the chemical and physical properties of the element, and elements contain only one kind of atom. The scientific concepts of both elements and atoms came to us from the ancient Greeks, who had a rather erroneous notion of the element and—for their time, at least—a highly advanced idea of the atom.

Unfortunately, atomic theory died away in later centuries, while the mistaken notion of four "elements" (earth, air, fire, and water) survived

DMITRI MENDELEEV. *(Bettmann/Corbis. Reproduced by permission.)*

virtually until the seventeenth century, an era that witnessed the birth of modern science. Yet the ancients did know of substances later classified as elements, even if they did not understand them as such. Among these were gold, tin, copper, silver, lead, and mercury. These, in fact, are such an old part of human history that their discoverers are unknown. The first individual credited with discovering an element was German chemist Hennig Brand (c. 1630-c. 1692), who discovered phosphorus in 1674.

MATURING CONCEPTS OF ATOMS, ELEMENTS, AND MOLECULES. The work of English physicist and chemist Robert Boyle (1627-1691) greatly advanced scientific understanding of the elements. Boyle maintained that no substance was an element if it could be broken down into other substances: thus air, for instance, was not an element. Boyle's studies led to the identification of numerous elements in the years that followed, and his work influenced French chemists Antoine Lavoisier (1743-1794) and Joseph-Louis Proust (1754-1826), both of whom helped define an element in the modern sense. These men in turn influenced English chemist John Dalton (1766-1844), who reintroduced atomic theory to the language of science.

In *A New System of Chemical Philosophy* (1808), Dalton put forward the idea that nature is composed of tiny particles, and in so doing he adopted the Greek word *atomos* to describe these basic units. Drawing on Proust's law of constant composition, Dalton recognized that the structure of atoms in a particular element or compound is uniform, but maintained that compounds are made up of compound "atoms." In fact, these compound atoms are really molecules, or groups of two or more atoms bonded to one another, a distinction clarified by Italian physicist Amedeo Avogadro (1776-1856).

Dalton's and Avogadro's contemporary, Swedish chemist Jons Berzelius (1779-1848), developed a system of comparing the mass of various atoms in relation to the lightest one, hydrogen. Berzelius also introduced the system of chemical symbols—H for hydrogen, O for oxygen, and so on—in use today. Thus, by the middle of the nineteenth century, scientists understood vastly more about elements and atoms than they had just a few decades before, and the need for a system of organizing elements became increasingly clear. By mid-century, a number of chemists had attempted to create just such an organizational system, and though Mendeleev's was not the first, it proved the most useful.

MENDELEEV CONSTRUCTS HIS TABLE

By the time Mendeleev constructed his periodic table in 1869, there were 63 known elements. At that point, he was working as a chemistry professor at the University of St. Petersburg, where he had become acutely aware of the need for a way of classifying the elements to make their relationships more understandable to his students. He therefore assembled a set of 63 cards, one for each element, on which he wrote a number of identifying characteristics for each.

Along with the element symbol, discussed below, he included the atomic mass for the atoms of each. In Mendeleev's time, atomic mass was understood simply to be the collective mass of a unit of atoms—a unit developed by Avogadro, known as the mole—divided by Avogadro's number, the number of atoms or molecules in a mole. With the later discovery of subatomic particles, which in turn made possible the discovery of isotopes, figures for atomic mass were clarified, as will also be discussed.

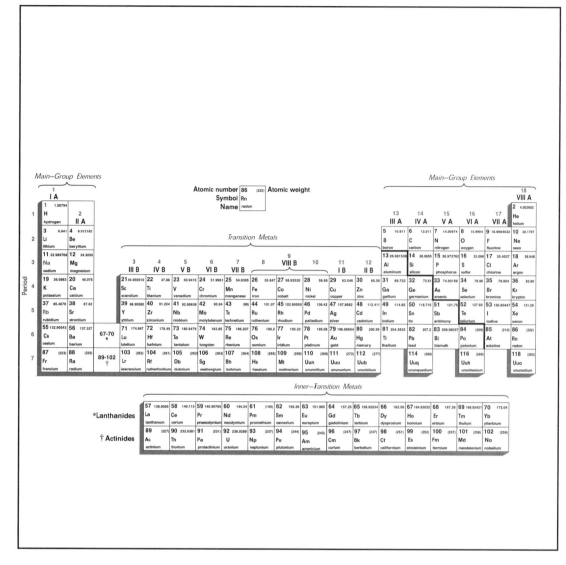

THE PERIODIC TABLE (IUPAC SYSTEM)

In addition, Mendeleev also included figures for specific gravity—the ratio between the density of an element and the density of water—as well as other known chemical characteristics of an element. Today, these items are typically no longer included on the periodic table, partly for considerations of space, but partly because chemists' much greater understanding of the properties of atoms makes it unnecessary to clutter the table with so much detail.

Again, however, in Mendeleev's time there was no way of knowing about these factors. As far as chemists knew in 1869, an atom was an indivisible little pellet of matter that could not be characterized by terms any more detailed than its mass and the ways it interacted with atoms of other elements. Mendeleev therefore ar-

ranged his cards in order of atomic mass, then grouped elements that showed similar chemical properties.

CONFIDENT PREDICTIONS. As Mendeleev observed, every eighth element on the chart exhibits similar characteristics, and thus, he established columns whereby element number x was placed above element number $x + 8$—for instance, helium (2) above neon (10). The patterns he observed were so regular that for any "hole" in his table, he predicted that an element to fill that space would be discovered.

Indeed, Mendeleev was so confident in the basic soundness of his organizational system that in some instances, he changed the figures for the atomic mass of certain elements because he was convinced they belonged elsewhere on the table.

Later discoveries of isotopes, which in some cases affected the average atomic mass considerably, confirmed his suppositions. Likewise the undiscovered elements he named "eka-aluminum," "eka-boron," and "eka-silicon" were later identified as gallium, scandium, and germanium, respectively.

REAL-LIFE APPLICATIONS

SUBATOMIC STRUCTURES CLARIFY THE PERIODIC TABLE

Over a period of 35 years, between the discovery of the electron in 1897 and the discovery of the neutron in 1932, chemists' and physicists' understanding of atomic structure changed completely. The man who identified the electron was English physicist J. J. Thomson (1856-1940). The electron is a negatively charged particle that contributes little to an atom's mass; however, it has a great deal to do with the energy an atom possesses. Thomson's discovery made it apparent that something else had to account for atomic mass, as well as the positive electric charge offsetting the negative charge of the electron.

Thomson's student Ernest Rutherford (1871-1937)—for whom, incidentally, rutherfordium (104 on the periodic table) is named—identified that "something else." In a series of experiments, he discovered that the atom has a nucleus, a center around which electrons move, and that the nucleus contains positively charged particles called protons. Protons have a mass 1,836 times as great as that of an electron, and thus, this seemed to account for the total atomic mass.

ISOTOPES AND ATOMIC MASS.

Later, working with English chemist Frederick Soddy (1877-1956), Rutherford discovered that when an atom emitted certain types of particles, its atomic mass changed. Rutherford and Soddy named these atoms of differing mass isotopes, though at that point—because the neutron had yet to be discovered—they did not know exactly what change had caused the change in mass. Certain types of isotopes, Soddy and Rutherford went on to conclude, had a tendency to decay by emitting particles or gamma rays, moving (sometimes over a great period of time) toward stabilization. In the process, these radioactive isotopes changed into other isotopes of the same element—and sometimes even to isotopes of other elements.

Soddy concluded that atomic mass, as measured by Berzelius, was actually an average of the mass figures for all isotopes within that element. This explained a problem with Mendeleev's periodic table, in which there seemed to be irregularities in the increase of atomic mass from element to element. The answer to these variations in mass, it turned out, related to the number of isotopes associated with a given element: the greater the number of isotopes, the more these affected the overall measure of the element's mass.

A CLEARER DEFINITION OF ATOMIC NUMBER.

Just a few years after Rutherford and Soddy discovered isotopes, Welsh physicist Henry Moseley (1887-1915) uncovered a mathematical relationship between the amount of energy a given element emitted and its atomic number. Up to this point, the periodic table had assigned atomic number in order of mass, beginning with the lightest element, hydrogen. Using atomic mass and other characteristics as his guides, Mendeleev had been able to predict the discovery of new elements, but such predictions had remained problematic. Thanks to Moseley's work, it became possible to predict the existence of undiscovered elements with much greater accuracy.

As Moseley discovered, the atomic number corresponds to the number of positive charges in the nucleus. Thus carbon, for instance, has an atomic number of 6 not because there are five lighter elements—though this is also true—but because it has six protons in its nucleus. The ordering by atomic number happens to correspond to the ordering by atomic mass, but atomic number provides a much more precise means of distinguishing elements. For one thing, atomic number is always a whole integer—1 for hydrogen, for instance, or 17 for chlorine, or 92 for uranium. Figures for mass, on the other hand, are almost always rendered with whole numbers and decimal fractions (for example, 1.008 for hydrogen).

If atoms have no electric charge, meaning that they have the same number of protons as electrons, then why do chemists not say that atomic number represents the number of protons or electrons? The reason is that electrons can easily be lost or gained by atoms to form ions, which have an electric charge. However, protons are very hard to remove.

NEUTRONS AND ATOMIC MASS.
By 1932, scientists had come a long way toward understanding the structure of the atom. Not only had the electron, nucleus, and proton been discovered, but the complex model of electron configuration (described later in this essay) had begun to evolve. Yet, one nagging question remained: the mass of the protons in the nucleus simply could not account for the entire mass of the atom. Neither did the electrons make a significant contribution to mass.

Suppose a proton was "worth" $1,836, while an electron had a value of only $1. In the "bank account" for deuterium, an isotope of hydrogen, there is $3,676, which poses a serious discrepancy in accounting. Because deuterium is a form of hydrogen, it has one proton as well as one electron, but that only accounts for $1,837. Where does deuterium get the other $1,839? These numbers are not chosen at random, as we shall see.

The answer to the problem of atomic mass came when English physicist James Chadwick (1891-1974) identified the neutron, a particle with no electric charge, residing in the nucleus alongside the protons. Whereas the proton has a mass 1,836 times as large as that of the electron, the neutron's mass is slightly larger—1,839 times that of an electron. This made it possible to clarify the values of atomic mass, which up to that time had been problematic, because a mole of atoms representing one element is likely to contain numerous isotopes.

Average Atomic Mass

Today, the periodic table lists, along with chemical symbol and atomic number, the average atomic mass of each element. As its name suggests, the average atomic mass provides the average value of mass—in atomic mass units (amu)—for a large sample of atoms. According to Berzelius's system for measuring atomic mass, 1 amu should be equal to the mass of a hydrogen atom, even though that mass had yet to be measured, since hydrogen almost never appears alone in nature. Today, in accordance with a 1960 agreement among members of the international scientific community, measurements of atomic

mass take carbon-12, an isotope found in all living things, as their reference point.

It is inconvenient, to say the least, to measure the mass of a single carbon-12 atom, or indeed of any other atom. Instead, chemists use a large number of atoms, a value known as Avogadro's number, which in general is the number of atoms in a mole (abbreviated mol). Avogadro's number is defined as $6.02214199 \cdot 10^{23}$, with an uncertainty of $4.7 \cdot 10^{16}$. In other words, the number of particles in a mole could vary by as much as 47,000,000,000,000,000 on either side of the value for Avogadro's number. This might seem like a lot, but in fact it is equal to only about 80 parts per billion.

When 1 is divided by Avogadro's number, the result is $1.66 \cdot 10^{-24}$—the value, in grams, of 1 amu. However, according to the 1960 agreement, 1 amu is officially 1/12 the mass of a carbon-12 atom, whose exact value (re-tested in 1998), is $1.6653873 \times 10^{-24}$ g. Carbon-12, sometimes represented as (12/6)C, contains six protons and six neutrons, so the value of 1 amu thus obtained is, in effect, an average of the mass for a proton and neutron.

Though atoms differ, subatomic particles do not. There is no such thing, for instance, as a "hydrogen proton"—otherwise, these subatomic particles, and not atoms, would constitute the basic units of an element. Given the unvarying mass of subatomic particles, combined with the fact that the neutron only weighs 0.16% more than a proton, the established value of 1 amu provides a convenient means of comparing mass. This is particularly useful in light of the large numbers of isotopes—and hence of varying figures for mass—that many elements have.

ATOMIC MASS UNITS AND THE PERIODIC TABLE. The periodic table as it is used today includes figures in atomic mass units for the average mass of each atom. As it turns out, Berzelius was not so far off in his use of hydrogen as a standard, since its mass is almost exactly 1 amu—but not quite. The value is actually 1.008 amu, reflecting the presence of slightly heavier deuterium isotopes in the average sample of hydrogen

Figures increase from hydrogen along the periodic table, though not by a regular pattern. Sometimes the increase from one element to the next is by just over 1 amu, and in other cases, the increase is by more than 3 amu. This only serves to prove that atomic number, rather than atomic mass, is a more straightforward means of ordering the elements.

Mass figures for many elements that tend to appear in the form of radioactive isotopes are usually shown in parentheses. This is particularly true for elements with very, very high atomic numbers (above 92), because samples of these elements do not stay around long enough to be measured. Some have a half-life—the period in which half the isotopes decay to a stable form—of just a few minutes, and for others, the half-life is a fraction of a second. Therefore, atomic mass figures represent the mass of the longest-lived isotope.

ELEMENTS

As of 2001, there were 112 known elements, of which about 90 occur naturally on Earth. Uranium, with an atomic number of 92, was the last naturally occurring element discovered: hence some sources list 92 natural elements. Other sources, however, subtract those elements with a lower atomic number than uranium that were first created in laboratories rather than discovered in nature. In any case, all elements with atomic numbers higher than 92 are synthetic, meaning that they were created in laboratories. Of these 20 elements—all of which have appeared only in the form of radioactive isotopes with short half-lives—the last three have yet to receive permanent names.

In addition, three other elements—designated by atomic numbers 114, 116, and 118, respectively—are still on the drawing board, as it were, and do not yet even have temporary names. The number of elements thus continues to grow, but these "new" elements have little to do with the daily lives of ordinary people. Indeed, this is true even for some of the naturally occurring elements: for example, few people who are not chemically trained would be able to identify yttrium, which has an atomic number of 39.

Though an element can exist theoretically as a gas, liquid, or a solid, in fact, the vast majority of elements are solids. Only 11 elements exist in the gaseous state at a normal temperature of about 77°F (25°C). These are the six noble gases; fluorine and chlorine from the halogen family; as well as hydrogen, nitrogen, and oxygen. Just two

are liquids at normal temperature: mercury, a metal, and the nonmetal halogen bromine. It should be noted that the metal gallium becomes liquid at just 85.6°F (29.76°C); below that temperature, however, it— like the elements other than those named in this paragraph—is a solid.

CHEMICAL NAMES AND SYMBOLS

For the sake of space and convenience, elements are listed on the periodic table by chemical symbol or element symbol—a one-or two-letter abbreviation for the name of the element according to the system first developed by Berzelius. These symbols, which are standardized and unvarying for any particular element, greatly aid the chemist in writing out chemical formulas, which could otherwise be quite cumbersome.

Many of the chemical symbols are simple one-letter designations: H for hydrogen, O for oxygen, and F for fluorine. Others are two-letter abbreviations, such as He for helium, Ne for neon, and Si for silicon. Note that the first letter is always capitalized, and the second is always lowercase. In many cases, the two-letter symbols indicate the first and second letters of the element's name, but this is not nearly always the case. Cadmium, for example, is abbreviated Cd, while platinum is Pt.

Many of the one-letter symbols indicate elements discovered early in history. For instance, carbon is represented by C, and later "C" elements took two-letter designations: Ce for cerium, Cr for chromium, and so on. Likewise, krypton had to take the symbol Kr because potassium had already been assigned K. The association of potassium with K brings up one of the aspects of chemical symbols most confusing to students just beginning to learn about the periodic table: why K and not P? The latter had in fact already been taken by phosphorus, but then why not Po, assigned many years later instead to polonium?

CHEMICAL SYMBOLS BASED IN OTHER LANGUAGES. In fact, potassium's symbol is one of the more unusual examples of a chemical symbol, taken from an ancient or non-European language. Soon after its discovery in the early nineteenth century, the element was named kalium, apparently after the Arabic *qali* or "alkali." Hence, though it is known as potassium today, the old symbol still stands.

The use of Arabic in naming potassium is unusual in the sense that "strange" chemical symbols usually refer to Latin and Greek names. Latin names include *aurum,* or "shining dawn" for gold, symbolized as Au; or *ferrum,* the Latin word for iron, designated Fe. Likewise, lead (Pb) and sodium (Na) are represented by letters from their Latin names, *plumbum* and *natrium,* respectively.

Some chemical elements are named for Greek or German words describing properties of the element. Consider, for instance, the halogens, collectively named for a Greek term meaning "salt producing." *Chloros,* in Greek, describes a sickly yellow color, and was assigned to chlorine; the name of bromine comes from a Greek word meaning "stink"; and that of iodine is a form of a Greek term meaning "violet-colored." Astatine, last-discovered of the halogens and the rarest of all natural elements, is so radioactive that it was given a name meaning "unstable." Another Greek-based example outside the halogen family is phosphorus, or "I bring light"—appropriate enough, in view of its phosphorescent properties.

NAMES OF LATER ELEMENTS. The names of several elements with high atomic numbers—specifically, the lanthanides, the transuranium elements of the actinide series, and some of the later transition metals—have a number of interesting characteristics. Several reflect the places where they were originally discovered or created: for example, germanium, americium, and californium. Other elements are named for famous or not-so-famous scientists. Most people could recognize einsteinium as being named after Albert Einstein (1879-1955), but the origin of the name gadolinium—Finnish chemist Johan Gadolin (1760-1852)—is harder for the average person to identify. Then of course there is element 101, named mendelevium in honor of the man who created the periodic table.

Two elements are named after women: curium after French physicist and chemist Marie Curie (1867-1934), and meitnerium after Austrian physicist Lise Meitner (1878-1968). Curie, the first scientist to receive two Nobel Prizes—in both physics and chemistry—herself discovered two elements, radium and polonium. In keeping with the trend of naming transuranium elements after places, she commemorated the land of her birth, Poland, in the name of polonium. One of Curie's students, French physicist Marguerite Perey (1909-1975), also discovered an element and named it after her own homeland: francium.

Meitnerium, the last element to receive a name, was created in 1982 at the Gesellschaft für Schwerionenforschung, or GSI, in Darmstadt, Germany, one of the world's three leading centers of research involving transuranium elements. The other two are the Joint Institute for Nuclear Research in Dubna, Russia, and the University of California at Berkeley, for which berkelium is named.

THE IUPAC AND THE NAMING OF ELEMENTS. One of the researchers involved with creating berkelium was American nuclear chemist Glenn T. Seaborg (1912-1999), who discovered plutonium and several other transuranium elements. In light of his many contributions, the scientists who created element 106 at Dubna in 1974 proposed that it be named seaborgium, and duly submitted the name to the International Union of Pure and Applied Chemistry (IUPAC).

Founded in 1919, the IUPAC is, as its name suggests, an international body, and it oversees a number of matters relating to the periodic table: the naming of elements, the assignment of chemical symbols to new elements, and the certification of a particular research team as the discoverers of that element. For many years, the IUPAC refused to recognize the name seaborgium, maintaining that an element could not be named after a living person. The dispute over the element's name was not resolved until the 1990s, but finally the IUPAC approved the name, and today seaborgium is included on the international body's official list.

Elements 110 through 112 had yet to be named in 2001, and hence were still designated by the three-letter symbols Uun, Uuu, and Uub respectively. These are not names, but alphabetic representations of numbers: un for 1, nil for 0, and bium for 2. Thus, the names are rendered as ununnilium, unununium, and ununbium; the undiscovered elements 114, 116, and 118 are respectively known as ununquadium, ununhexium, and ununoctium.

LAYOUT OF THE PERIODIC TABLE

TWO SYSTEMS FOR LABELING GROUPS. Having discussed the three items of information contained in the boxes of the periodic table—atomic number, chemical symbol/name, and average atomic mass—it is now possible to step back from the chart and look at its overall layout. To reiterate what was stated in the introduction to the periodic table above, the table is arranged in rows called periods, and columns known as groups. The deeper meaning of the periods and groups, however—that is, the way that chemists now understand them in light of what they know about electron configurations—will require some explanation.

All current versions of the periodic table show seven rows—in other words, seven periods—as well as 18 columns. However, the means by which columns are assigned group numbers varies somewhat. According to the system used in North America, only eight groups are numbered. These are the two "tall" columns on the left side of the "dip" in the chart, as well as the six "tall" columns to the right of it. The "dip," which spans 10 columns in periods 4 through 7, is the region in which the transition metals are listed. The North American system assigns no group numbers to these, or to the two rows set aside at the bottom, representing the lanthanide and actinide series of transition metals.

As for the columns that the North American system does number, this numbering may appear in one of four forms: either by Roman numerals; Roman numerals with the letter A (for example, IIIA); Hindu-Arabic numbers (for example, 3); or Hindu-Arabic numerals with the letter A. Throughout this book, the North American system of assigning Hindu-Arabic numerals without the letter A has been used. However, an attempt has been made in some places to include the group designation approved by the IUPAC, which is used by scientists in Europe and most parts of the world outside of North America. (Some scientists in North America are also adopting the IUPAC system.)

The IUPAC numbers all columns on the chart, so that instead of eight groups, there are 18. The table below provides a means of comparing the North American and IUPAC systems. Columns are designated in terms of the element family or families, followed in parentheses by the atomic numbers of the elements that appear at the top and bottom of that column. The first number following the colon is the number in the North American system (as described above, a

Hindu-Arabic numerical without an "A"), and the second is the number in the IUPAC system.

Element Family	North American	IUPAC
Hydrogen and alkali metals (1, 87)	1	1
Alkaline metals (4, 88)	2	2
Transition metals (21,89)		3
Transition metals (22,104)		4
Transition metals (23,105)		5
Transition metals (24,106)		6
Transition metals (25,107)		7
Transition metals (26,108)		8
Transition metals (27,109)		9
Transition metals (28,110)		10
Transition metals (29,111)		11
Transition metals (30,112)		12
Nonmetals and metals (5,81)	3	13
Nonmetals, metalloids, and metal (6,82)	4	14
Nonmetals, metalloids, and metal (7,83)	5	15
Nonmetals, metalloids, (8,84)	6	16
Halogens (9,85)	7	17
Noble gases (2,86)	8	18
Lanthanides (58,71)	No number group assigned in either system	
Actinides (90,103)	No number group assigned in either system	

VALENCE ELECTRONS, PERIODS, AND GROUPS

The merits of the IUPAC system are easy enough to see: just as there are 18 columns, the IUPAC lists 18 groups. Yet the North American system is more useful than it might seem: the group number in the North American system indicates the number of valence electrons, the electrons that are involved in chemical bonding. Valence electrons also occupy the highest energy level in the atom—which might be thought of as the orbit farthest from the nucleus, though in fact the reality is more complex.

A more detailed, though certainly far from comprehensive, discussion of electrons and energy levels, as well as the history behind these discoveries, appears in the Electrons essay. In what follows, the basics of electron configuration will be presented with the specific aim of making it clear exactly why elements appear in particular columns of the periodic table.

PRINCIPAL ENERGY LEVELS AND PERIODS. At one time, scientists thought that electrons moved around a nucleus in regular orbits, like planets around the Sun. In fact the paths of an electron are much more complicated, and can only be loosely defined in terms of orbitals, a set of probabilities regarding the positions that an electron is likely to occupy as it moves around the nucleus. The pattern of orbitals is determined by the principal energy level of the atom, which indicates a distance that an electron may move away from the nucleus.

Principal energy level is designated by a whole-number integer, beginning with 1 and moving upward: the higher the number, the further the electron is from the nucleus, and hence the greater the energy in the atom. Each principal energy level is divided into sublevels corresponding to the number n of the principal energy level: thus, principal energy level 1 has one sublevel, principal energy level 2 has two, and so on.

The relationship between principal energy level and period is relatively easy to demonstrate: the number n of a period on the periodic table is the same as the number of the highest principal energy level for the atoms on that row—that is, the principal energy level occupied by its valence electrons. Thus, elements on period 4 have a highest principal energy level of 4, whereas the valence electrons of elements on period 7 are at principal energy level 7. Note the conclusion that this allows us to draw: the further down the periodic table an element is positioned, the greater the energy in a single atom of that element. Not surprisingly, most of the elements used in nuclear power come from period 7, which includes the actinides.

VALENCE ELECTRON CONFIGU-RATIONS AND GROUPS. Now to a more involved subject, whereby group number is related to valence electron configuration. As mentioned earlier, the principal energy levels are divided into sublevels, which are equal in number to the principal energy level number: principal energy level 1 has one sublevel, level 2 has two sublevels, and so on. As one might expect, with an increase in principal energy levels and sublevels, there are increases in the complexity of the orbitals.

The four types of orbital patterns are designated as s, p, d, and f. Two electrons can move in an s orbital pattern or shell, six in a p, 10 in a d,

and 14 in an f orbital pattern or shell. This says nothing about the number of electrons that are actually in a particular atom; rather, the higher the principal energy level and the larger the number of sublevels, the greater the number of ways that the electrons can move. It does happen to be the case, however, that with higher atomic numbers—which means more electrons to offset the protons—the higher the energy level, the larger the number of orbitals for those electrons.

Let us now consider a few examples of valence shell configurations. Hydrogen, with the simplest of all atomic structures, has just one electron on principal energy level 1, so in effect its valence electron is also a core electron. The valence configuration for hydrogen is thus written as $1s^1$. Moving straight down the periodic table to francium (atomic number 87), which is in the same column as hydrogen, one finds that it has a valence electron configuration of $7s^1$. Thus, although francium is vastly more complex and energy-filled than hydrogen, the two elements have the same valence-shell configuration; only the number of the principal energy level is different.

Now look at two elements in Group 3 (Group 13 in the IUPAC system): boron and thallium, which respectively occupy the top and bottom of the column, with atomic numbers of 5 and 81. Boron has a valence-shell configuration of $2s^22p^1$. This means its valence shell is at principal energy level 2, where there are two electrons in an s orbital pattern, and 2 in a p orbital pattern. Thallium, though it is on period 6, nonetheless has the same valence-shell configuration: $6s^26p^1$.

Notice something about the total of the superscript figures for any element in Group 3 of the North American system: it is three. The same is true in the other columns numbered on North American charts, in which the total number of electrons equals the group number. Thus in Group 7, the valence shell configuration is ns^2np^5, where n is the principal energy level. There is only one exception to this: helium, in Group 8 (the noble gases), has a valence shell configuration of $1s^2$. Were it not for the fact that it clearly fits with the noble gases due to shared properties, helium would be placed next to hydrogen at the top of Group 2, where all the atoms have a valence-shell configuration of ns^2.

Obviously the group numbers in the IUPAC system do not correspond to the number of valence electrons, because the IUPAC chart includes numbers for the columns of transition metals, which are not numbered in the North American system. In any case, in both systems the columns contain elements that all have the same number of electrons in their valence shells. Thus the term "group" can finally be defined in accordance with modern chemists' understanding, which incorporates electron configurations of which Mendeleev was unaware. All the members of a group have the same number of valence electrons in the same orbital patterns, though at different energy levels. (Once again, helium is the lone exception.)

SOME CHALLENGES OF THE PERIODIC TABLE

IRREGULAR PATTERNS. The groups that are numbered in the North American system are referred to as "representative" elements, because they follow a clearly established pattern of adding valence shell electrons. By contrast, the 40 elements listed in the "dip" at the middle of the chart—the transition elements—do not follow such a pattern. This is why the North American system does not list them by group number, and also why neither system lists two "branches" of the transition-metal family, the lanthanides and actinides.

Even within the representative elements, there are some challenges as far as electron configuration. For the first 18 elements—1 (hydrogen) to 18 (argon)—there is a regular pattern of orbital filling. Beginning with helium (2) onward, all of principal level 1 is filled; then, beginning with beryllium (4), sublevel 2s begins to fill. Sublevel 2p—and hence principal level 2 as a whole—becomes filled at neon (10).

After argon, as one moves to the element occupying the nineteenth position on the periodic table—potassium—the rules change. Argon, in Group 8 of the North American system, has a valence shell of $3s^2 3p^6$, and by the pattern established with the first 18 elements, potassium should begin filling principal level 3d. Instead, it "skips" 3d and moves on to 4s. The element following argon, calcium, adds a second electron to the 4s sublevel.

After calcium, as the transition metals begin with scandium (21), the pattern again changes:

indeed, the transition elements are defined by the fact that they fill the d orbitals rather than the p orbitals, as was the pattern up to that point. After the first period of transition metals ends with zinc (30), the next representative element—gallium (31)—resumes the filling of the p orbital rather than the d. And so it goes, all along the four periods in which transition metals break up the steady order of electron configurations.

As for the lanthanide and actinide series of transitions metals, they follow an even more unusual pattern, which is why they are set apart even from the transition metals. These are the only groups of elements that involve the highly complex f sublevels. In the lanthanide series, the seven 4f orbital shells are filled, while the actinide series reflects the filling of the seven 5f orbital shells.

Why these irregularities? One reason is that as the principal energy level increases, the energy levels themselves become closer—i.e., there is less difference between the energy levels. The atom is thus like a bus that fills up: when there are just a few people on board, those few people (analogous to electrons) have plenty of room, but as more people get on, the bus becomes increasingly more crowded, and passengers jostle against one another. In the atom, due to differences in energy levels, the 4s orbital actually has a lower energy than the 3d, and therefore begins to fill first. This is also true for the 6s and 4f orbitals.

CHANGES IN ATOMIC SIZE. The subject of element families is a matter unto itself, and therefore a separate essay in this book has been devoted to it. The reader is encouraged to consult the Families of Elements essay, which discusses aspects of electron configuration as well as the properties of various element families.

One last thing should be mentioned about the periodic table: the curious fact that the sizes of atoms decreases as one moves from left to right across a row or period, even though the sizes increase as one moves from top to bottom along a group. The increase of atomic size in a group, as a function of increasing atomic number, is easy enough to explain. The higher the atomic number, the higher the principal energy level, and the greater the distance from the nucleus to the furthest probability range for the electron.

On the other hand, the decrease in size across a period is a bit more challenging to com-

ATOM: The smallest particle of an element that retains all the chemical and physical properties of the element.

ATOMIC MASS UNIT: An SI unit (abbreviated amu), equal to $1.66 \cdot 10^{-24}$ g, for measuring the mass of atoms.

ATOMIC NUMBER: The number of protons in the nucleus of an atom. Since this number is different for each element, elements are listed on the periodic table of elements in order of atomic number.

AVERAGE ATOMIC MASS: A figure used by chemists to specify the mass—in atomic mass units—of the average atom in a large sample.

AVOGADRO'S NUMBER: A figure, named after Italian physicist Amedeo Avogadro (1776-1856), equal to 6.022137×10^{23}. Avogadro's number indicates the number of atoms or molecules in a mole.

CHEMICAL SYMBOL: A one- or two-letter abbreviation for the name of an element.

COMPOUND: A substance made of two or more elements that have bonded chemically. These atoms are usually, but not always, joined in molecules.

ELECTRON: A negatively charged particle in an atom. The configurations of valence electrons define specific groups on the periodic table of elements, while the principal energy levels of those valence electrons define periods on the table.

ELEMENT: A substance made up of only one kind of atom, which cannot be chemically broken into other substances.

ELEMENT SYMBOL: Another term for chemical symbol.

GROUPS: Columns on the periodic table of elements. These are ordered according to the numbers of valence electrons in the outer shells of the atoms for the elements represented.

HALF-LIFE: The length of time it takes a substance to diminish to one-half its initial amount.

ION: An atom or atoms that has lost or gained one or more electrons, thus acquiring a net electric charge.

ISOTOPES: Atoms that have an equal number of protons, and hence are of the same element, but differ in their number of neutrons. This results in a difference of mass. Isotopes may be either stable or unstable. The latter type, known as radioisotopes, are radioactive.

MOLE: The SI fundamental unit for "amount of substance." A mole is, generally speaking, Avogadro's number of atoms, molecules, or other elementary particles; however, in the more precise SI definition, a mole is equal to the number of carbon atoms in 12.01 g of carbon.

MOLECULE: A group of atoms, usually but not always representing more than one element, joined by chemical bonds. Compounds are typically made of up molecules.

NEUTRON: A subatomic particle that has no electric charge. Neutrons, together with protons, account for the majority of average atomic mass. When atoms have the same number of protons—and hence are the same element—but differ in their number of neutrons, they are called isotopes.

NUCLEUS: The center of an atom, a region where protons and neutrons are located. The nucleus accounts for the vast majority of the average atomic mass.

ORBITAL: A pattern of probabilities regarding the regions that an electron can occupy within an atom in a particular energy state. The higher the principal energy level, the more complex the pattern of orbitals.

PERIODIC TABLE OF ELEMENTS: A chart that shows the elements arranged in order of atomic number, along with chemical symbol and the average atomic mass (in atomic mass units) for that particular element.

PERIODS: Rows of the periodic table of elements. These represent successive principal energy levels for the valence electrons in the atoms of the elements involved.

PRINCIPAL ENERGY LEVEL: A value indicating the distance that an electron may move away from the nucleus of an atom. This is designated by a whole-number integer, beginning with 1 and moving upward. The higher the principal energy level, the greater the energy in the atom, and the more complex the pattern of orbitals.

PROTON: A positively charged particle in an atom. The number of protons in the nucleus of an atom is the atomic number of an element.

RADIOACTIVITY: A term describing a phenomenon whereby certain isotopes known as radioisotopes are subject to a form of decay brought about by the emission of high-energy particles. "Decay" does not mean that the isotope "rots"; rather, it decays to form another isotope— either of the same element or another—until eventually it becomes stable. This stabilizing process may take a few seconds, or many years.

VALENCE ELECTRONS: Electrons that occupy the highest energy levels in an atom. These are the electrons involved in chemical bonding.

prehend; however, it just takes a little explaining. As one moves along a period from left to right, there is a corresponding increase in the number of protons within the nucleus. This means a stronger positive charge pulling the electrons inward. Therefore, the "cloud" of electrons is drawn ever closer toward the increasingly powerful charge at the center of the atom, and the size of the atom decreases because the electrons cannot move as far away from the nucleus.

WHERE TO LEARN MORE

Challoner, Jack. *The Visual Dictionary of Chemistry.* New York: DK Publishing, 1996.

"Elementistory" (Web site). <http://smallfry.dmu.ac.uk/chem/periodic/elementi.html> (May 22, 2001).

International Union of Pure and Applied Chemistry (Web site). <http://www.iupac.org> (May 22, 2001).

Knapp, Brian J. and David Woodroffe. *The Periodic Table.* Danbury, CT: Grolier Educational, 1998.

Oxlade, Chris. *Elements and Compounds.* Chicago: Heinemann Library, 2001.

"A Periodic Table of the Elements" Los Alamos National Laboratory (Web site). <http://pearl1.lanl.gov/periodic/> (May 22, 2001).

"The Pictorial Periodic Table" (Web site). <http://chem-lab.pc.maricopa.edu/periodic/periodic.html> (May 22, 2001).

Stwertka, Albert. *A Guide to the Elements.* New York: Oxford University Press, 1998.

"*Visual Elements*" (Web site). <http://www.chemsoc. org/viselements/> (May 22, 2001).

WebElements (Web site). <http://www.webelements.com (May 22, 2001).

FAMILIES OF ELEMENTS

CONCEPT

The term "family" is used to describe elements that share certain characteristics—not only in terms of observable behavior, but also with regard to atomic structure. All noble gases, for instance, tend to be highly nonreactive: only a few of them combine with other elements, and then only with fluorine, the most reactive of all substances. Fluorine is a member of another family, the halogens, which have so many shared characteristics that they are grouped together, despite the fact that two are gases, two are solids, and one—bromine—is one of only two elements that appears at room temperature as a solid. Despite these apparent differences, common electron configurations identify the halogens as a family. Families on the periodic table include, in addition to noble gases and halogens, the alkali metals, alkaline earth metals, transition metals, lanthanides, and actinides. The nonmetals form a loosely defined cross-family grouping, as do the metalloids.

HOW IT WORKS

The Basics of the Periodic Table

Created in 1869, and modified several times since then, the periodic table of the elements developed by Russian chemist Dmitri Ivanovitch Mendeleev (1834-1907) provides a highly useful means of organizing the elements. Certainly other organizational systems exist, but Mendeleev's table is the most widely used—and with good reason. For one thing, it makes it possible to see at a glance families of elements, many of which either belong to the same group (column) or the same period (row) on the table.

The periodic table is examined in depth within the essay devoted to that subject, and among the specifics discussed in that essay are the differing systems used for periodic-table charts in North America and the rest of the world. In particular, the North American system numbers only eight groups, leaving 10 columns unnumbered, whereas the other system—approved by the International Union of Pure and Applied Chemistry (IUPAC)—numbers all 18 columns. Both versions of the periodic table show seven periods.

The groups numbered in the North American system are the two "tall" columns on the left side of the "dip" in the chart, as well as the six "tall" columns to the right of it. Group 1 in this system consists of hydrogen and the alkali metals; Group 2, the alkaline earth metals; groups 3 through 6, an assortment of metals, nonmetals, and metalloids; Group 7, halogens; and Group 8, noble gases. The "dip," which spans 10 columns in periods 4 through 7, is the region in which the transition metals are listed. The North American system assigns no group numbers to these, or to the two rows set aside at the bottom, representing the lanthanide and actinide series of transition metals.

The IUPAC system, on the other hand, offers the obvious convenience of providing a number for each column. (Note that, like its North American counterpart, the IUPAC chart provides no column numbers for the lanthanides or actinides.) Furthermore, the IUPAC has behind it the authority of an international body, founded in 1919, which oversees a number of matters

	I A	II A	III A	IV A	VA	VIA	VIIA	VIIIA
Period 1	H							He
Period 2	Li	Be	B	C	N	O	F	Ne
Period 3	Na	Mg	Al	Si	P	S	Cl	Ar
Period 4	K	Ca	Ga	Ge	As	Se	Br	Kr
Period 5	Rb	Sr	In	Sn	Sb	Te	I	Xe
Period 6	Cs	Ba	Tl	Pb	Bi	Po	At	Rn

SIZE REPRESENTATION OF THE ATOMIC RADII OF THE MAIN-GROUP ELEMENTS.

relating to the periodic table: the naming of elements, the assignment of chemical symbols to new elements, and the certification of a particular individual or research team as the discoverers of that element. For these reasons, the IUPAC system is coming into favor among North American chemists as well.

Despite the international acceptance of the IUPAC system, as well as its merits in terms of convenience, the North American system is generally the one used in this book. The reason, in

part, is that most American schools still use this system; furthermore, there is a reasoning behind the assignment of numbers to only eight groups, as will be discussed. Where necessary or appropriate, however, group numbers in the IUPAC system will be provided as well.

PRINCIPAL ENERGY LEVELS

Group numbers in the North American system indicate the number of valence electrons, or the electrons that are involved in chemical bonding.

Valence electrons also occupy the highest energy level in the atom—which might be thought of as the orbit farthest from the nucleus, though in fact the term "orbit" is misleading when applied to the ways an electron moves.

Electrons do not move around the nucleus of an atom in regular orbits, like planets around the Sun; rather, their paths can only be loosely defined in terms of orbitals, a pattern of probabilities regarding the areas through which an electron is likely to move. The pattern of orbitals is determined by the principal energy level of the atom, which indicates the distance an electron may move away from the nucleus.

Principal energy level is designated by a whole-number integer, beginning with 1 and moving upward to 7: the higher the number, the further the electron is from the nucleus, and hence the greater the energy in the atom. The relationship between principal energy level and period is relatively easy to demonstrate. The number n of a period on the periodic table is the same as the number of the highest principal energy level for the atoms on that row—that is, the principal energy level occupied by its valence electrons. Thus, elements on period 1 have a highest principal energy level of 1, and so on.

VALENCE ELECTRON CONFIGURATIONS

When discussing families of elements, however, the periods or rows on the periodic table are not as important as the groups or columns. These are defined by the valence electron configurations, a subject more complicated than principal energy levels—though the latter requires a bit more explanation in order to explain electron configurations.

Each principal energy level is divided into sublevels corresponding to the number n of the principal energy level: thus, principal energy level 1 has one sublevel, principal energy level 2 has two, and so on. As one might expect, with an increase in principal energy levels and sublevels, there are increases in the complexity of the orbitals.

ORBITAL PATTERNS. The four basic types of orbital patterns are designated as s, p, d, and f. The s shape might be described as spherical, though when talking about electrons, nothing is quite so neat: orbital patterns, remember, only identify regions of probability for the elec-

tron. In other words, in an s orbital, the total electron cloud will probably end up being more or less like a sphere.

The p shape is like a figure eight around the nucleus, and the d like two figure eights meeting at the nucleus. Again, these and other orbital patterns do not indicate that the electron will necessary follow that path. What it means is that, if you could take millions of photographs of the electron during a period of a few seconds, the resulting blur of images in a p orbital would somewhat describe the shape of a figure eight.

The f orbital pattern is so complex that most basic chemistry textbooks do not even attempt to explain it, and beyond f are other, even more complicated, patterns designated in alphabetical order: g, h, and so on. In the discussion that follows, we will not be concerned with these, since even for the lanthanides and the actinides, an atom at the ground state does not fill orbital patterns beyond an f.

SUBLEVELS AND ORBITAL FILLING. Principal energy level 1 has only an s sublevel; 2 has an s and a p, the latter with three possible orientations in space; 3 has an s, p, and d (five possible spatial orientations); and 4 has an s, p, d, and f (seven possible spatial orientations.)

According to the Pauli exclusion principle, only two electrons can occupy a single orbital pattern—that is, the s sublevel or any one of the spatial orientations in p, d, and f—and those two electrons must be spinning in opposite directions. Thus, two electrons can move in an s orbital pattern or shell, six in a p, 10 in a d, and 14 in an f orbital pattern or shell. Valence shell configurations are therefore presented with superscript figures indicating the number of electrons in that orbital pattern—for instance, s^1 for one electron in the s orbital, or d^{10}, indicating a d orbital that has been completely filled.

REAL-LIFE APPLICATIONS

REPRESENTATIVE ELEMENTS

Hydrogen (atomic number 1), with the simplest of all atomic structures, has just one electron on principal energy level 1, so, in effect, its valence electron is also a core electron. The valence configuration for hydrogen is thus written as $1s^1$. It should be noted, as described in the Electrons essay, that if a hydrogen atom (or any other atom) is in an excited state, it may reach energy levels beyond its normal, or ground, state.

Moving straight down the periodic table to francium (atomic number 87), which is in the same column as hydrogen, one finds that it has a valence electron configuration of $7s^1$. Thus, although francium is vastly more complex and energy-filled than hydrogen, the two elements have the same valence shell configuration; only the number of the principal energy level is different. All the elements listed below hydrogen in Group 1 are therefore classified together as alkali metals. Obviously, hydrogen—a gas—is not part of the alkali metal family, nor does it clearly belong to any other family: it is the "lone wolf" of the periodic table.

Now look at two elements in Group 2, with beryllium (atomic number 4) and radium (88) at the top and bottom respectively. Beryllium has a valence shell configuration of $2s^2$. This means its valence shell is at principal energy level 2, where there are two electrons on an s orbital pattern. Radium, though it is on period 7, nonetheless has the same valence shell configuration: $7s^2$. This defines the alkaline earth metals family in terms of valence shell configuration.

For now, let us ignore groups 3 through 6—not to mention the columns between groups 2 and 3, unnumbered in the North American system—and skip over to Group 7. All the elements in this column, known as halogens, have valence shell configurations of ns^2np^5. Beyond Group 7 is Group 8, the noble gases, all but one of whom have valence shell configurations of ns^2np^6. The exception is helium, which has an s^2 valence shell. This seems to put it with the alkaline earth metals, but of course helium is not a metal. In terms of its actual behavior, it clearly belongs to the noble gases family.

The configurations of these valence shells have implications with regard to the ways in which elements bond, a subject developed at some length in the Chemical Bonding essay. Here we will consider it only in passing, to clarify the fact that electron configuration produces observable results. This is most obvious with the noble gases, which tend to resist bonding with most other elements because they already have eight electrons in their valence shell—the same number of valence electrons that most other atoms achieve only after they have bonded.

FROM THE REPRESENTATIVE ELEMENTS TO THE TRANSITION ELEMENTS

Groups 3 through 6, along with hydrogen and the four families so far identified, constitute the 44 representative or main-group elements. In 43 of these 44, the number of valence shell electrons is the same as the group number in the North American system. (Helium, which is in Group 8 but has two valence electrons, is the lone exception.) By contrast, the 40 elements listed in the "dip" at the middle of the chart—the transition metals—follow a less easily defined pattern. This is part of the reason why the North American system does not list them by group number, and also why neither system lists the two other families within the transition elements—the lanthanides and actinides.

Before addressing the transition metals, however, let us consider patterns of orbital filling, which also differentiate the representative elements from the transition elements. Each successive representative element fills all the orbitals of the elements that precede it (with some exceptions that will be explained), then goes on to add one more possible electron configuration. The total number of electrons—not just valence shell electrons—is the same as the atomic number. Thus fluorine, with an atomic number of 9, has a complete configuration of $1s^22s^22p^5$. Neon, directly following it with an atomic number of 10, has a total configuration of $1s^22s^22p^6$. (Again, this is not the same as the valence shell configuration, which is contained in the last two sublevels represented: for example, $2s^22p^6$ for neon.)

The chart that follows shows the pattern by which orbitals are filled. Note that in several places, the pattern of filling becomes "out of order," something that will be explained below.

Orbital Filling by Principal Energy Level
- $1s$ (2)
- $2s$ (2)
- $2p$ (6)
- $3s$ (2)
- $3p$ (6)
- $4s$ (2)
- $3d$ (10)
- $4p$ (6)
- $5s$ (2)
- $4d$ (10)
- $5p$ (6)

- 6s (2)
- 4f (14)
- 5d (10)
- 6p (6)
- 7s (2)
- 5f (14)
- 6d (10)

PATTERNS OF ORBITAL FILL-ING. Generally, the 44 representative elements follow a regular pattern of orbital filling, and this is particularly so for the first 18 elements. Imagine a small amphitheater, shaped like a cone, with smaller rows of seats at the front. These rows are also designated by section, with the section number being the same as the number of rows in that section.

The two seats in the front row comprise a section labeled 1 or 1s, and this is completely filled after helium (atomic number 2) enters the auditorium. Now the elements start filling section 2, which contains two rows. The first row of section 2, labeled 2s, also has two seats, and after beryllium (4), it too is filled. Row 2p has 6 seats, and it is finally filled with the entrance of neon (10). Now, all of section 2 has been filled; therefore, the eleventh element, sodium, starts filling section 3 at the first of its three rows. This row is 3s—which, like all s rows, has only two seats. Thus, when element 13, aluminum, enters the theatre, it takes a seat in row 3p, and eventually argon (18), completes that six-seat row.

By the pattern so far established, element 19 (potassium) should begin filling row 3d by taking the first of its 10 seats. Instead, it moves on to section 4, which has four rows, and it takes the first seat in the first of those rows, 4s. Calcium (20) follows it, filling the 4s row. But when the next element, scandium (21), comes into the theatre, it goes to row 3d, where potassium "should have" gone, if it had continued filling sections in order. Scandium is followed by nine companions (the first row of transition elements) before another representative element, gallium (31), comes into the theatre. (For reasons that will not be discussed here, chromium and copper, elements 24 and 29, respectively, have valence electrons in 4s—which puts them slightly off the transition metal pattern.)

According to the "proper" order of filling seats, now that 3d (and hence all of section 3) is filled, gallium should take a seat in 4s. But those seats have already been taken by the two preced-

ing representative elements, so gallium takes the first of six seats in 4p. After that row fills up at krypton (36), it is again "proper" for the next representative element, rubidium (37), to take a seat in 4d. Instead, just as potassium skipped 3d, rubidium skips 4d and opens up section 5 by taking the first of two seats in 5s.

Just as before, the next transition element—yttrium (39)—begins filling up section 4d, and is followed by nine more transition elements until cadmium (48) fills up that section. Then, the representative elements resume with indium (49), which, like gallium, skips ahead to section 5p. And so it goes through the remainder of the periodic table, which ends with two representative elements followed by the last 10 transition metals.

TRANSITION METALS

Given the fact that it is actually the representative elements that skip the d sublevels, and the transition metals that go back and fill them, one might wonder if the names "representative" and "transition" (implying an interruption) should be reversed. However, remember the correlation between the number of valence shell electrons and group number for the representative elements. Furthermore, the transition metals are the only elements that fill the d orbitals.

This brings us to the reason why the lanthanides and actinides are set apart even from the transition metals. In most versions of the periodic table, lanthanum (57) is followed by hafnium (72) in the transition metals section of the chart. Similarly, actinium (89) is followed by rutherfordium (104). The "missing" metals—lanthanides and actinides, respectively—are listed at the bottom of the chart. There are reasons for this, as well as for the names of these groups.

After the 6s orbital fills with the representative element barium (56), lanthanum does what a transition metal does—it begins filling the 5d orbital. But after lanthanum, something strange happens: cerium (58) quits filling 5d, and moves to fill the 4f orbital. The filling of that orbital continues throughout the entire lanthanide series, all the way to lutetium (71). Thus, lanthanides can be defined as those metals that fill the 4f orbital; however, because lanthanum exhibits similar properties, it is usually included with the lanthanides. Sometimes the term "lan-

KEY TERMS

ACTINIDES: Those transition metals that fill the 5f orbital. Because actinium—which does not fill the 5f orbital—exhibits characteristics similar to those of the actinides, it is usually considered part of the actinides family.

ALKALI METALS: All members, except hydrogen, of Group 1 on the periodic table of elements, with valence electron configurations of ns^1.

ALKALINE EARTH METALS: Group 2 on the periodic table of elements, with valence electron configurations of ns^2.

ELECTRON CLOUD: A term used to describe the pattern formed by orbitals.

FAMILIES OF ELEMENTS: Related elements, including the noble gases, halogens, alkali metals, alkaline earth metals, transition metals, lanthanides, and actinides. In addition, metals, nonmetals, and metalloids form loosely defined families. Other family designations—such as carbon family—are sometimes used.

GROUND STATE: A term describing the state of an atom at its ordinary energy level.

GROUPS: Columns on the periodic table of elements. These are ordered according to the numbers of valence electrons in the outer shells of the atoms for the elements represented.

HALOGENS: Group 7 of the periodic table of elements, with valence electron configurations of ns^2np^5.

ION: An atom or atoms that has lost or gained one or more electrons, and thus has a net electric charge.

LANTHANIDES: The transition metals that fill the 4f orbital. Because lanthanum—which does not fill the 4f orbital—exhibits characteristics similar to those of the lanthanides, it is usually considered part of the lanthanide family.

MAIN-GROUP ELEMENTS: The 44 elements in Groups 1 through 8 on the periodic table of elements, for which the number of valence electrons equals the group number. (The only exception is helium.) The main-group elements, also called representative elements, include the families of alkali metals, alkali earth metals, halogens, and noble gases, as well as other metals, nonmetals, and metalloids.

METALLOIDS: Elements which exhibit characteristics of both metals and nonmetals. Metalloids are all solids, but are not lustrous or shiny, and they conduct heat and electricity moderately well. The six metalloids occupy a diagonal region between the metals and nonmetals on the right side of the periodic table. Sometimes astatine is included with the metalloids, but in this book it is treated within the context of the halogens family.

METALS: A collection of 87 elements that includes numerous families—the alkali metals, alkaline earth metals, transition metals, lanthanides, and actinides, as well as seven elements in groups 3 through 5. Metals, which occupy the left, center, and part of the right-hand side of the periodic table, are lustrous or shiny in appearance, and malleable, meaning that they can be molded into different shapes without breaking. They are excellent conductors of heat and electricity, and tend to form positive ions by losing electrons.

NOBLE GASES: Group 8 of the periodic table of elements, all of whom (with the exception of helium) have valence electron configurations of ns^2np^6.

NONMETALS: Elements that have a dull appearance; are not malleable; are poor conductors of heat and electricity; and tend to gain electrons to form negative ions. They are thus the opposite of metals in most regards, as befits their name. Aside from hydrogen, the other 18 nonmetals occupy the upper right-hand side of the periodic table, and include the noble gases, halogens, and seven elements in groups 3 through 6.

ORBITAL: A pattern of probabilities regarding the position of an electron for an atom in a particular energy state. The higher the principal energy level, the more complex the pattern of orbitals. The four types of orbital patterns are designated as *s, p, d,* and *f*—each of which is more complex than the one before.

PERIODIC TABLE OF ELEMENTS: A chart that shows the elements arranged in order of atomic number, along with chemical symbol and the average atomic mass

(in atomic mass units) for that particular element.

PERIODS: Rows of the periodic table of elements. These represent successive energy levels in the atoms of the elements involved.

PRINCIPAL ENERGY LEVEL: A value indicating the distance that an electron may move away from the nucleus of an atom. This is designated by a whole-number integer, beginning with 1 and moving upward. The higher the principal energy level, the greater the energy in the atom, and the more complex the pattern of orbitals.

REPRESENTATIVE ELEMENTS: See main-group elements.

TRANSITION METALS: A group of 40 elements, which are not assigned a group number in the North American version of the periodic table. These are the only elements that fill the *d* orbitals.

VALENCE ELECTRONS: Electrons that occupy the highest energy levels in an atom. These are the electrons involved in chemical bonding.

thanide series" is used to distinguish the other 14 lanthanides from lanthanum itself.

A similar pattern occurs for the actinides. The 7*s* orbital fills with radium (88), after which actinium (89) begins filling the 6*d* orbital. Next comes thorium, first of the actinides, which begins the filling of the 5*f* orbital. This is completed with element 103, lawrencium. Actinides can thus be defined as those metals that fill the 5*f* orbital; but again, because actinium exhibits similar properties, it is usually included with the actinides.

METALS, NONMETALS, AND METALLOIDS

The reader will note that for the seven families so far identified, we have generally not discussed them in terms of properties that can more easily be discerned—such as color, phase of matter, bonding characteristics, and so on. Instead, they have been examined primarily from the standpoint of orbital filling, which provides a solid chemical foundation for identifying families. Macroscopic characteristics, as well as the ways that the various elements find application in daily life, are discussed within essays devoted to the various groups.

Note, also, that the families so far identified account for only 92 elements out of a total of 112 listed on the periodic table: hydrogen; six alkali metals; six alkaline earth metals; five halogens; six noble gases; 40 transition metals; 14 lanthanides; and 14 actinides. What about the other 20? Some discussions of element families assign these elements, all of which are in groups 3 through 6, to families of their own, which will be mentioned briefly. However, because these "families" are not recognized by all chemists, in this book the 20 elements of groups 3 through 6 are described generally as metals, nonmetals, and metalloids.

METALS AND NONMETALS. Metals are lustrous or shiny in appearance, and malleable, meaning that they can be molded into different shapes without breaking. They are excellent conductors of heat and electricity, and tend to form positive ions by losing electrons. On the periodic table, metals fill the left, center, and part of the right-hand side of the chart. Thus it should not come as a surprise that most elements (87, in fact) are metals. This list includes alkali metals, alkaline earth metals, transition metals, lanthanides, and actinides, as well as seven elements in groups 3 through 6—aluminum, gallium, indium, thallium, tin, lead, and bismuth.

Nonmetals have a dull appearance; are not malleable; are poor conductors of heat and electricity; and tend to gain electrons to form negative ions. They are thus the opposite of metals in most regards, as befits their name. Nonmetals, which occupy the upper right-hand side of the periodic table, include the noble gases, halogens, and seven elements in groups 3 through 5. These nonmetal "orphans" are boron, carbon, nitrogen, oxygen, phosphorus, sulfur, and selenium. To these seven orphans could be added an eighth, from Group 1: hydrogen. As with the metals, a separate essay—with a special focus on the "orphans"—is devoted to nonmetals.

METALLOIDS AND OTHER "FAMILY" DESIGNATIONS. Occupying a diagonal region between the metals and nonmetals are metalloids, elements which exhibit characteristics of both metals and nonmetals. They are all solids, but are not lustrous, and conduct heat and electricity moderately well. The six metalloids are silicon, germanium, arsenic, antimony, tellurium, and polonium. Astatine is sometimes identified as a seventh metalloid; however, in this book, it is treated as a member of the halogen family.

Some sources list "families" rather than collections of "orphan" metals, metalloids, and nonmetals, in groups 3 through 6. These designations are not used in this book; however, they should be mentioned briefly. Group 3 is sometimes called the boron family; Group 4, the carbon family; Group 5, the nitrogen family; and Group 6, the oxygen family. Sometimes Group 5 is designated as the pnictogens, and Group 6 as the chalcogens.

WHERE TO LEARN MORE

Bankston, Sandy. "Explore the Periodic Table and Families of Elements" The Rice School Science Department (Web site). <http://www.ruf.rice.edu/~sandyb/Lessons/chem.html> (May 23, 2001).

Challoner, Jack. The Visual Dictionary of Chemistry. New York: DK Publishing, 1996.

"Elementistory" (Web site). <http://smallfry.dmu.ac.uk/chem/periodic/elementi.html> (May 22, 2001).

"Families of Elements" (Web site). <http://homepages.stuy.edu/~bucherd/ch23/families.html> (May 23, 2001).

Knapp, Brian J. and David Woodroffe. The Periodic Table. Danbury, CT: Grolier Educational, 1998.

Maton, Anthea. Exploring Physical Science. Upper Saddle River, N.J.: Prentice Hall, 1997.

Oxlade, Chris. Elements and Compounds. Chicago: Heinemann Library, 2001.

"The Pictorial Periodic Table" (Web site). <http://chemlab.pc.maricopa.edu/periodic/periodic.html> (May 22, 2001).

Stwertka, Albert. A Guide to the Elements. New York: Oxford University Press, 1998.

"Visual Elements" (Web site). <http://www.chemsoc.org/viselements/> (May 22, 2001).

METALS

METALS

CONCEPT

A number of characteristics distinguish metals, including their shiny appearance, as well as their ability to be bent into various shapes without breaking. In addition, metals tend to be highly efficient conductors of heat and electricity. The vast majority of elements on the periodic table are metals, and most of these fall into one of five families: alkali metals; alkaline earth metals; the very large transition metals family; and the inner transition metal families, known as the lanthanides and actinides. In addition, there are seven "orphans," or metals that do not belong to a larger family of elements: aluminum, gallium, indium, thallium, tin, lead, and bismuth. Some of these may be unfamiliar to most readers; on the other hand, a person can hardly spend a day without coming into contact with aluminum. Tin is a well-known element as well, though much of what people call "tin" is not really tin. Lead, too, is widely known, and once was widely used as well; today, however, it is known primarily for the dangers it poses to human health.

HOW IT WORKS

GENERAL PROPERTIES OF METALS

Metals are lustrous or shiny in appearance, and malleable or ductile, meaning that they can be molded into different shapes without breaking. Despite their ductility, metals are extremely durable and have high melting and boiling points. They are excellent conductors of heat and electricity, and tend to form positive ions by losing electrons.

The bonds that metals form with each other, or with nonmetals, are known as ionic bonds, which are the strongest type of chemical bond. Even within a metal, however, there are extremely strong bonds. Therefore, though it is easy to shape metals (that is, to slide the atoms in a metal past one another), it is very difficult to separate metal atoms. Their internal bonding is thus very strong, but nondirectional.

Internal bonding in metals is described by the electron sea model, which depicts metal atoms as floating in a "sea" of valence electrons, the electrons involved in bonding. These valence electrons are highly mobile within the crystalline structure of the metal, and this mobility helps to explain metals' high conductivity. The ease with which metal crystals allow themselves to be rearranged explains not only metals' ductility, but also their ability to form alloys, a mixture containing one or more metals.

ABUNDANCE OF METALS

Metals constitute a significant portion of the elements found in Earth's crust, waters, and atmosphere.

The following list shows the most abundant metals on Earth, with their ranking among all elements in terms of abundance. Many other metals are present in smaller, though still significant, quantities.

- 3. Aluminum (7.5%)
- 4. Iron (4.71%)
- 5. Calcium (3.39%)
- 6. Sodium (2.63%)
- 7. Potassium (2.4%)
- 8. Magnesium (1.93%)

ALUMINUM IS THE MOST ABUNDANT METAL ON EARTH. THIS PHOTO SHOWS THE FIRST BLOBS OF ALUMINUM EVER ISOLATED, A FEAT PERFORMED BY CHARLES HALL IN 1886. *(James L. Amos/Corbis. Reproduced by permission.)*

- 10. Titanium (0.58%)
- 13. Manganese (0.09%)

In addition, metals are also important components of the human body. The following list shows the ranking of the most abundant metals among the elements in the body, along with the percentage that each constitutes. As with the quantities of metals on Earth, other metals are present in much smaller, but still critical, amounts.

- 5. Calcium (1.4%)
- 7. Magnesium (0.50%)
- 8. Potassium (0.34%)
- 10. Sodium (0.14%)
- 12. Iron (0.004%)
- 13. Zinc (0.003%)

METALS ON THE PERIODIC TABLE

On the periodic table of elements, metals fill the left, center, and part of the right-hand side of the chart. The remainder of the right-hand section is taken by nonmetals, which have properties quite different from (and often opposite to) those of metals, as well as metalloids, which display characteristics both of metals and nonmetals. Nonmetallic elements are covered in essays on Nonmetals; Metalloids; Halogens; Noble Gases; and Carbon. In addition, hydrogen, the one nonmetal on the left side of the periodic table—first among the elements, in the upper left corner—is also discussed in its own essay.

The total number of metals on the periodic table is 87—in other words, about 80% of all known elements. These include the six alkali metals, which occupy Group 1 of the periodic table, as well as the six alkaline earth metals, or Group 2. To the right of these are the 40 transition metals, whose groups are unnumbered in the North American version of the periodic table. Among the transition metals are two elements, lanthanum and actinium, often lumped in with the families of inner transition metals that exhibit similar properties. These are, respectively, the 14 lanthanides and 14 actinides. In addition, there are the seven "orphan" metals of periods 3 through 5, mentioned in the introduction to this essay, which will be discussed below.

The various families of elements are defined, not so much by external characteristics as by the configurations of valence electrons in their atoms' shells. This subject will not be discussed in any depth here; instead, the reader is encouraged to consult essays on the specific metal families, which discuss the electron configurations of each.

BEFORE THEY WERE DESTROYED BY TERRORIST ATTACKS ON SEPTEMBER 11, 2001, THE TWIN TOWERS OF THE WORLD TRADE CENTER IN NEW YORK CITY GLEAMED WITH A CORROSION-RESISTANT COVERING OF ALUMINUM. *(Bill Ross/Corbis. Reproduced by permission.)*

REAL-LIFE APPLICATIONS

ALKALI METALS

The members of the alkali metal family are distinguished by the fact that they have a valence electron configuration of s^1 This means that they tend to bond easily with other substances. Alkali metals tend to form positive ions, or cations, with a charge of 1+.

In contact with water, alkali metals form a negatively charged hydroxide ion (OH^-). When alkali metals react with water, one hydrogen atom splits off from the water molecule to form hydrogen gas, while the other hydrogen atom remains bonded to the oxygen, forming hydroxide. Where the heavier members of the alkali metal family are concerned, reactions can often be so vigorous that the result is combustion or even explosion. Alkali metals also react with oxygen to produce either an oxide, peroxide, or superoxide, depending on the particular member of the alkali metal family involved.

Shiny and soft enough to be cut with a knife, the alkali metals are usually white, though cesium is a yellowish white. When placed in a flame, most of these substances produce charac-teristic colors: lithium, for instance, glows bright red, and sodium an intense yellow. Heated potassium produces a violet color, rubidium a dark red, and cesium a light blue. This makes it possible to identify the metals by color when heated— a useful trait, since they so often tend to be bonded with other elements.

The alkali metals, listed below by atomic number and chemical symbol, are:

- 3. Lithium (Li)
- 11. Sodium (Na)
- 19. Potassium (K)
- 37. Rubidium (Rb)
- 55. Cesium (Cs)
- 87. Francium (Fr)

ALKALINE EARTH METALS

Occupying Group 2 of the periodic table are the alkaline earth metals, which have a valence electron configuration of s^2. Like the alkali metals, the alkaline earth metals are known for their high reactivity—a tendency for bonds between atoms or molecules to be made or broken in such a way that materials are transformed. Thus they are sel-

MOST "TIN" ROOFS NOW CONSIST OF AN IRON OR STEEL ROOF COVERED WITH ZINC, NOT TIN. *(John Hulme; Eye Ubiquitous/Corbis. Reproduced by permission.)*

dom found in pure form, but rather in compounds with other elements—for example, salts, a term that generally describes a compound consisting of a metal and a nonmetal.

The alkaline earth metals are also like the alkali metals in that they have the properties of a base as opposed to an acid. An acid is a substance that, when dissolved in water, produces positive ions of hydrogen, designated symbolically as H^+ ions. It is thus known as a proton donor. A base, on the other hand, produces negative hydroxide ions when dissolved in water. These are designated by the symbol OH^-. Bases are therefore characterized as proton acceptors.

The alkaline earth metals are shiny, and most are white or silvery in color. Like their "cousins" in the alkali metal family, they glow with characteristic colors when heated. Calcium glows orange, strontium a very bright red, and barium an apple green. Physically they are soft, though not as soft as the alkali metals; nor are their levels of reactivity as great as those of their neighbors in Group 1.

The alkaline earth metals, listed below by atomic number and chemical symbol, are:

- 4. Beryllium (Be)
- 12. Magnesium (Mg)
- 20. Cadmium (Ca)
- 38. Strontium (Sr)
- 56. Barium (Ba)
- 88. Radium (Ra)

TRANSITION METALS

The transition metals are the only elements that fill the dorbitals. They are also the only elements that have valence electrons on two different principal energy levels. This sets them apart from the representative or main-group elements, of which the alkali metals and alkaline earth metals are a part. The transition metals also follow irregular patterns of orbital filling, which further distinguish them from the representative elements.

Other than that, there is not much that differentiates the transition metals, a broad grouping that includes precious metals such as gold, silver, and platinum, as well as highly functional metals such as iron, manganese, and zinc. Many of the transition metals, particularly those on periods 4, 5, and 6, form useful alloys with one another, and with other elements. Because of their differences in valence electron configuration, however, they do not always combine in the same ways, even within an element. Iron, for instance, sometimes releases two electrons in chemical bonding, and at other times three.

GROUPING THE TRANSITION METALS. There is no easy way to group the transition metals, though some of these elements are traditionally categorized together. These do not constitute "families" as such, but they do provide useful ways to break down the otherwise rather daunting lineup of the transition metals. In two cases, there is at least a relationship between group number on the periodic table and the categories loosely assigned to a collection of transition metals. Thus the "coinage metals"—copper, silver, and gold—all occupy Group 9 on the IUPAC version of the periodic table. (These have traditionally been associated with one another because their resistance to oxidation, combined with their malleability and beauty, has made them useful materials for fashioning coins.

Likewise the members of the "zinc group"—zinc, cadmium, and mercury—on Group 10 on the IUPAC periodic table have often been associated as a miniature unit due to common properties. Members of the "platinum group"—platinum, iridium, osmium, palladium, rhodium, and ruthenium—occupy a rectangle on the table, corresponding to periods 5 and 6, and groups 6 through 8. What actually makes them a "group," however, is the fact that they tend to appear together in nature. Iron, nickel, and cobalt, found alongside one another on Period 4, may be grouped together because they are all magnetic to some degree.

The transition metals, listed below by atomic number and chemical symbol, are:

- 21. Scandium (Sc)
- 22. Titanium (Ti)
- 23. Vanadium (V)
- 24. Chromium (Cr)
- 25. Manganese (Mn)
- 26. Iron (Fe)
- 27. Cobalt (Co)
- 28. Nickel (Ni)
- 29. Copper (Cu)
- 30. Zinc (Zn)
- 39. Yttrium (Y)
- 40. Zirconium (Zr)
- 41. Niobium (Nb)
- 42. Molybdenum (Mo)
- 43. Technetium (Tc)
- 44. Ruthenium (Ru)

- 45. Rhodium (Rh)
- 46. Palladium (Pd)
- 47. Silver (Ag)
- 48. Cadmium (Cd)
- 57. Lanthanum (La)
- 72. Hafnium (Hf)
- 73. Tantalum (Ta)
- 74. Tungsten (W)
- 75. Rhenium (Re)
- 76. Osmium (Os)
- 77. Iridium (Ir)
- 78. Platinum (Pt)
- 79. Gold (Au)
- 80. Mercury (Hg)
- 89. Actinium (Ac)
- 104. Rutherfordium (Rf)
- 105. Dubnium (Db)
- 106. Seaborgium (Sg)
- 107. Bohrium (Bh)
- 108. Hassium (Hs)
- 109. Meitnerium (Mt)
- 110. Ununnilium (Uun)
- 111. Unununium (Uuu)
- 112. Ununbium (Uub)

The last nine, along with 11 of the actinides, are part of the list of metals known collectively as the transuranium elements—that is, elements with atomic numbers higher than 92. These have all been produced artificially, in most cases within a laboratory. Note that the last three had not been named as of 2001: the "names" by which they are known simply mean, respectively, 110, 111, and 112.

LANTHANIDES

The lanthanides are the first of two inner transition metal groups. Note that in the list of transition metals above, lanthanum (57) is followed by hafnium (72). The "missing" group of 14 elements, normally shown at the bottom of the periodic table, is known as the lanthanides, a family which can be defined by the fact that all fill the 4f orbital. However, because lanthanum (which does not fill an f orbital) exhibits similar properties, it is usually included with the lanthanides.

Bright and silvery in appearance, many of the lanthanides—though they are metals—are so soft they can be cut with a knife. They also tend to be highly reactive. Though they were once known as the "rare earth metals," lanthanides only seemed rare because they were difficult to extract from compounds containing other substances—including other lanthanides. Because their properties are so similar, and because they tend to be found together in the same substances, the original isolation and identification of the lanthanides was an arduous task that took well over a century.

The lanthanides (in addition to lanthanum, listed earlier with the transition metals), are:

- 58. Cerium (Ce)
- 59. Praseodymium (Pr)
- 60. Neodymium (Nd)
- 61. Promethium (Pm)
- 62. Samarium (Sm)
- 63. Europium (Eu)
- 64. Gadolinium (Gd)
- 65. Terbium (Tb)
- 66. Dysprosium (Dy)
- 67. Holmium (Ho)
- 68. Erbium (Er)
- 69. Thulium (Tm)
- 70. Ytterbium (Yb)
- 71. Lutetium (Lu)

ACTINIDES

Just as lanthanum is followed on the periodic table by an element with an atomic number 15 points higher, so actinium (89) is followed by rutherfordium (104). The elements in the "gap" are the other inner transition family, the actinides, which all fill the 5f orbital. These are placed below the lanthanides at the bottom of the chart, and just as lanthanum is included with the lanthanides, even though it does not fill an f orbital, so actinium is usually lumped in with the actinides due to similarities in properties.

Most of the actinides tend to be unstable or radioactive, the later ones highly so. The first three members of the group (not counting actinium) occur in nature, while the other 11 are transuranium elements created artificially. In most cases, these were produced with a cyclotron or other machine for accelerating atoms, generally at the research center of the University of California at Berkeley during a period from 1940 to 1961. Einsteinium and fermium, however, were by-products of nuclear testing at Bikini Atoll in the south Pacific in 1952. The principal use of the actinides is in nuclear applications—weaponry

or power plants—though some of them also have specialized uses as well.

The actinides (in addition to actinium, listed earlier with the transition metals), are:

- 90. Thorium (Th)
- 91. Protactinium (Pa)
- 92. Uranium (U)
- 93. Neptunium (Np)
- 94. Plutonium (Pu)
- 95. Americium (Am)
- 96. Curium (Cm)
- 97. Berkelium (Bk)
- 98. Californium (Cf)
- 99. Einsteinium (Es)
- 100. Fermium (Fm)
- 101. Mendelevium (Md)
- 102. Nobelium (No)
- 103. Lawrencium (Lr)

"ORPHAN" METALS

Seven other metals remain to be discussed. These are the "orphan" metals, which appear in groups 3, 4, and 5 of the North American periodic table (13, 14, and 15 of the IUPAC table), and they are called "orphans" simply because none belongs to a clearly defined family. Sometimes aluminum and the three elements below it in Group 3—gallium, indium, and thallium—are lumped together as the "aluminum family." This is not a widely recognized distinction, but will be used here only for the purpose of giving some form of organization to the "orphan" elements.

Though they do not belong to families, the "orphan" metals are nonetheless highly important. Aluminum, after all, is the most abundant of all metals on Earth, and has wide applications in daily life. Tin, too, is widely known, as is lead. Somewhat less recognizable is bismuth, but it appears in a product with which most Americans are familiar: Pepto-Bismol.

The seven "orphans" are listed below, along with atomic number and chemical symbol:

- 13. Aluminum (Al)
- 31. Gallium (Ga)
- 49. Indium (In)
- 50. Tin (Sn)
- 81. Thallium (Tl)
- 82. Lead (Pb)
- 83. Bismuth (Bi)

Named after alum, a salt known from ancient times and used by the Egyptians, Romans, and Greeks, aluminum is so reactive that it proved difficult to isolate. Only in 1825 did Danish physicist Hans Christian Ørsted (1777-1851) isolate an impure version of aluminum metal, using a complicated four-step process. Two years later, German chemist Friedrich Wöhler (1800-1882) obtained pure aluminum from a reaction of metallic potassium with aluminum chloride, or $AlCl_3$.

For years thereafter, aluminum continued to be so difficult to produce that it acquired the status of a precious metal. European kings displayed treasures of aluminum as though they were gold or silver, and by 1855, it was selling for about $100,000 a pound. Then in 1886, French metallurgist Paul-Louis-Toussaint Héroult (1863-1914) and American chemist Charles Martin Hall (1863-1914) independently developed a process that made aluminum relatively easy to extract by means of electrolysis. Thanks to what became known as the Hall-Héroult process, the price of aluminum dropped to around 30 cents a pound.

America's aluminum comes primarily from mines in Alabama, Arkansas, and Georgia, where it often appears in a clay called kaolin, used in making porcelain. Aluminum can also be found in deposits of feldspar, mica, granite, and other clays. The principal sources of aluminum outside the United States include France, Surinam, Jamaica, and parts of Africa.

USES FOR ALUMINUM. Though aluminum is highly reactive, it does not corrode in moist air. Instead of forming rust the way iron does, it forms a thin, hard, invisible coating of aluminum oxide. This, along with its malleability, makes it highly useful for producing cans and other food containers. The World Trade Center Towers in New York City gleam with a corrosion-resistant covering of aluminum. Likewise the element is used as a coating for mirrors: not only is it cheaper than silver, but unlike silver, it does not tarnish and turn black. Also, a thin coating of metallic aluminum gives Mylar balloons their silvery sheen.

Aluminum conducts electricity about 60% as well as copper, meaning that it is an extraordinarily good conductor, useful for transmitting electrical power. Because it is much more light-weight than copper and highly ductile, it is often used in high-voltage electric lines. Its conductivity also makes aluminum a favorite material for kitchen pots and pans. Unlike copper, it is not known to be toxic; however, because aluminum dissolves in strong acids, sometimes tomatoes and other acidic foods acquire a "metallic" taste when cooked in an aluminum pot.

Though it is soft in its pure form, when combined with copper, manganese, silicon, magnesium, and/or zinc, aluminum can produce strong, highly useful alloys. These find application in automobiles, airplanes, bridges, highway signs, buildings, and storage tanks. Aluminum's high ductility also makes it the metal of choice for foil wrappings—that is, aluminum foil. Compounds containing aluminum are used in a wide range of applications, including antacids; antiperspirants (potassium aluminum sulfate tends to close up sweat-gland ducts); and water purifiers.

OTHER ELEMENTS IN THE ALUMINUM GROUP

GALLIUM. When Russian chemist Dmitri Ivanovitch Mendeleev (1834-1907) created the periodic table in 1869, he predicted the existence of several missing elements on the basis of "holes" between elements on the table. Among these was what he called "eka-aluminum," discovered in the 1870s by French chemist Paul Emile Lecoq de Boisbaudran (1838-1912). Boisbaudran named the new element gallium, after the ancient name of his homeland, Gaul.

Gallium will melt if held in the hand, and remains liquid over a large range of temperatures—from 85.6°F (29.76°C) to 3,999.2°F (2,204°C). This makes it highly useful for thermometers with a large temperature range. Used in a number of applications within the electronics industry, gallium is present in the compounds gallium arsenide and gallium phosphide, applied in lasers, solar cells, transistors, and other solid-state devices.

INDIUM AND THALLIUM. As Boisbaudran would later do, German mineralogists Ferdinand Reich (1799-1882) and Hieronymus Theodor Richter (1824-1898) used a method called spectroscopy to discover one of the aluminum-group metals. Spectroscopy involves analyzing the frequencies of light (that is, the colors) emitted by an item of matter.

However, Reich faced a problem when studying the zinc ores that he believed contained a new element: he was colorblind. Therefore he called on the help of Richter, his assistant. In 1863, they discovered an element they called indium because of the indigo blue color it emitted. Reich later tried to claim that he made the discovery alone, but Richter's role is indisputable.

Indium is used in making alloys, one of which is a combination with gallium that is liquid at room temperature. Its alloys are used in bearings, dental amalgams, and nuclear control roles. Various compounds are applied in solid-state electronics devices, and in electronics research.

Two years before the discovery of indium, English physicist William Crookes (1832-1919) discovered another element using spectroscopic analysis. This was thallium, named after the Greek word thallos ("green twig"), a reference to a lime-green spectral line that told Crookes he had found a new element. Thallium is applied in electronic equipment, and the compound thallium sulfate is used for killing ants and rodents.

THREE OTHER "ORPHANS"

TIN. Known from ancient times, tin was combined with copper to make bronze—an alloy so significant it defines an entire stage of human technological development. Tin's chemical symbol, Sn, refers to the name the Romans gave it: stannum. Just as tin added to copper gave bronze its strength, alloys of tin with copper and antimony created pewter, a highly useful, malleable material for pots and dinnerware popular from the thirteenth century through the early nineteenth century.

Tin is found primarily in Bolivia, Malaysia, Indonesia, Thailand, Zaire, and Nigeria—a wide geographical range. Like aluminum today, it has been widely used as a coating for other metals to retard corrosion. Hence the term "tin can," referring to a steel can coated with tin. Also, tin has been used to coat iron or steel roofs, but because zinc is cheaper, it is now most often the coating of choice; nonetheless, these are still typically called "tin roofs."

LEAD. Due to knowledge of its toxic properties, lead is not frequently used today, but this was not always the case. The Romans, for instance, used plumbum as they called it (hence the chemical symbol Pb) as a material for making water pipes. This is the root of our own word plumber. (The same root explains the verb plumb or the noun plumb bob. A plumb bob is a lead weight, hung from a string, used to ensure that the walls of a structure are "plumb"—that is, perpendicular to the foundation.) Many historians believe that plumbum in the Romans' water supply was one of the reasons behind the decline and fall of the Roman Empire.

The human body can only excrete very small quantities of lead a day, and this is particularly true of children. Even in small concentrations, lead can cause elevation of blood pressure; reduction in the synthesis of hemoglobin, which carries oxygen from the lungs to the blood and organs; and decreased ability to utilize vitamin D and calcium for strengthening bones. Higher concentrations of lead can effect the central nervous system, resulting in decreased mental functioning and hearing damage. Prolonged exposure can result in a coma or even death.

Before these facts became widely known in the late twentieth century, lead was applied as an ingredient in paint. In addition, it was used in water pipes, and as an anti-knock agent in gasolines. Increased awareness of the health hazards involved have led to a discontinuation of these practices. (Note that pencil "lead" is actually graphite, a form of carbon.)

Lead, however, is used for absorbing radiation—for instance, as a shield against X rays. Another important use for lead today is in storage batteries. Yet another "application" of lead relates to its three stable isotopes (lead-206, -207, and -208), the end result of radioactive decay on the part of various isotopes of uranium and other radioactive elements.

BISMUTH. During the Middle Ages, when a number of unscientific ideas prevailed, scholars believed that lead eventually became tin, tin eventually turned into bismuth, and bismuth developed into silver. Hence they referred to bismuth as "roof of silver." Only in the sixteenth century did German metallurgist Georgius Agricola (George Bauer; 1494-1555) put forward the idea that bismuth was a separate element.

Bismuth appears in nature not only in its elemental form, but also in a number of compounds and ores. It is also obtained as a by-product from the smelting of gold, silver, copper, and lead. Because bismuth expands dramatically when it cools, early printers used a bismuth alloy

KEY TERMS

ACTINIDES: Those transition metals that fill the 5f orbital. Because actinium—which does not fill the *f* orbital—exhibits characteristics similar to those of the actinides, it is usually considered part of the actinides family.

ALKALI METALS: All members, ex-cept hydrogen, of Group 1 on the periodic table of elements, with valence electron configurations of ns^1.

ALKALINE EARTH METALS: Group 2 on the periodic table of elements, with valence electron configurations of ns^2.

ALLOY: A mixture containing more than one metal.

CONDUCTIVITY: The ability to conduct heat or electricity. Metals are noted for their high levels of conductivity.

CRYSTALLINE: A term describing an internal structure in which the constituent parts have a simple and definite geometric arrangement, repeated in all directions. Metals have a crystalline structure.

DUCTILE: Capable of being bent or molded into various shapes without breaking.

ELECTRON SEA MODEL: A widely accepted model of internal bonding within metals, which depicts metal atoms as floating in a "sea" of valence electrons.

GROUPS: Columns on the periodic table of elements. These are ordered according to the numbers of valence electrons in the outer shells of the atoms for the elements represented.

INNER TRANSITION METALS: The lanthanides and actinides, both of which fill the f orbitals. For this reason, they are usually treated as distinct from the transition metals.

ION: An atom or group of atoms that has lost or gained one or more electrons, and thus has a net electric charge.

ISOTOPES: Atoms that have an equal number of protons, and hence are of the same element, but differ in their number of neutrons. This results in a difference of mass. Isotopes may be either stable or unstable. The latter type, known as radioisotopes, are radioactive.

IUPAC SYSTEM: A version of the periodic table of elements, authorized by the International Union of Pure and Applied Chemistry (IUPAC), which numbers all groups on the table from 1 to 18. Thus in the IUPAC system, in use primarily outside of North America, both the representative or main-group elements and the transition metals are numbered.

LANTHANIDES: The transition metals that fill the 4f orbital. Because lanthanum—which does not fill the f orbital—exhibits characteristics similar to those of the lanthanides, it is usually considered part of the lanthanide family.

METALS: A collection of 87 elements that includes numerous families—the alkali metals, alkaline earth metals, transition metals, lanthanides, and actinides, as well as seven elements in groups 3 through 5 of the North American system periodic table. Metals are lustrous or shiny in appearance, and ductile. They are excellent conductors of heat and electricity, and tend to form positive ions by losing electrons.

NORTH AMERICAN SYSTEM: A version of the periodic table of elements that only numbers groups of elements in which the number of valence electrons equals the group number. Hence the transition metals are usually not numbered in this system.

ORBITAL: A pattern of probabilities regarding the position of an electron for an atom in a particular energy state. The higher the principal energy level, the more complex the pattern of orbitals.

PERIODIC TABLE OF ELEMENTS: A chart that shows the elements arranged in order of atomic number, along with chemical symbol and the average atomic mass for that particular element. Elements are arranged in groups or columns, as well as in rows or periods, according to specific aspects of their valence electron configurations.

PERIODS: Rows on the periodic table of elements. These represent successive principal energy levels for the valence electrons in the atoms of the elements involved.

PRINCIPAL ENERGY LEVEL: A value indicating the distance that an electron may move away from the nucleus of an atom. This is designated by a whole-number integer, beginning with 1 and moving upward. The higher the principal energy level, the greater the energy in the atom, and the more complex the pattern of orbitals.

RADIOACTIVITY: A term describing a phenomenon whereby certain isotopes known as radioisotopes are subject to a form of decay brought about by the emission of high-energy particles. "Decay" does not mean that the isotope "rots"; rather, it decays to form another isotope until eventually (though this may take a long time) it becomes stable.

REACTIVITY: The tendency for bonds between atoms or molecules to be made or broken in such a way that materials are transformed.

REPRESENTATIVE OR MAIN-GROUP ELEMENTS: The 44 elements in Groups 1 through 8 on the North American version of periodic table of elements, for which the number of valence electrons equals the group number. (The only exception is helium.) The alkali metals and alkaline earth metals are representative elements; all other metals are not.

SALT: Generally, a salt is any compound that brings together a metal and a nonmetal. Salts are produced, along with water, in the reaction of an acid and a base.

TRANSITION METALS: A group of 40 elements (counting lanthanum and actinium), which are not assigned a group number in the version of the periodic table of elements known as the North American system. These are the only elements that fill the d orbital. In addition, the transition metals—unlike the representative or main-group elements—have their valence electrons on two different principal energy levels. Though the lanthanides and actinides are known as inner transition metals, they are usually treated separately.

VALENCE ELECTRONS: Electrons that occupy the highest principal energy level in an atom. These are the electrons involved in chemical bonding.

to make metal type. When poured into forms cut with the shapes of letters and other characters, the alloy produced symbols with clear, sharp edges.

Wood's metal is a bismuth alloy involving cadmium, tin, and lead, which melts at a temperature of 158°F (70°C). This makes it useful for automatic sprinkler systems inside a building: when a fire breaks out, it melts the Wood's metal seal, and turns on the sprinklers. Bismuth compounds are applied in cosmetics, paints, and dyes, and prior to the twentieth century, compounds of bismuth were widely used for treating sexually transmitted diseases. Today, antibiotics have largely taken their place, but bismuth still has one well-known medicinal application: in bismuth subsalicylate, better known by the brand name Pepto-Bismol.

WHERE TO LEARN MORE

Aluminum Association (Web site). <http://www.aluminum.org> (May 28, 2001).

Kerrod, Robin. *Matter and Materials.* Illustrated by Terry Hadler. Tarrytown, NY: Benchmark Books, 1996.

"Lead Programs." Environmental Protection Agency (Web site). <http://www.epa.gov/opptintr/lead/> (May 28, 2001).

Mebane, Robert C. and Thomas R. Rybolt. *Metals.* Illustrated by Anni Matsick. New York: Twenty-First Century Books, 1995.

"Metals Industry Guide." About.com (Web site). <http://metals.about.com/industry/metals/mbody.htm> (May 28, 2001).

"Minerals and Metals." Natural Resources of Canada (Web site). <http://www.nrcan.gc.ca/mms/school/e_mine.htm> (May 28, 2001).

Oxlade, Chris. *Metal.* Chicago, IL: Heinemann Library, 2001.

Snedden, Robert. *Materials.* Des Plaines, IL: Heinemann Library, 1999.

WebElements (Web site). <http://www.webelements.com> (May 22, 2001).

Whyman, Kathryn. *Metals and Alloys.* Illustrated by Louise Nevett and Simon Bishop. New York: Gloucester Press, 1988.

Zumdahl, Steven S. *Introductory Chemistry: A Foundation,* 4th ed. Boston: Houghton Mifflin, 2000.

ALKALI METALS

CONCEPT

Group 1 of the periodic table of elements consists of hydrogen, and below it the six alkali metals: lithium, sodium, potassium, rubidium, cesium, and francium. The last three are extremely rare, and have little to do with everyday life; on the other hand, it is hard to spend a day without encountering at least one of the first three—particularly sodium, found in table salt. Along with potassium, sodium is an important component of the human diet, and in compounds with other substances, it has an almost endless array of uses. Lithium does not have as many applications, but to many people who have received it as a medication for bipolar disorder, it is quite literally a life-saver.

HOW IT WORKS

ELECTRON CONFIGURATION OF THE ALKALI METALS

In the essay on Families of Elements, there is a lengthy discussion concerning the relationship between electron configuration and the definition of a particular collection of elements as a "family." Here that subject will only be touched upon lightly, inasmuch as it relates to the alkali metals.

All members of Group 1 on the periodic table of elements have a valence electron configuration of s^1. This means that a single electron is involved in chemical bonding, and that this single electron moves through an orbital, or range of probabilities, roughly corresponding to a sphere.

Most elements bond according to what is known as the octet rule, meaning that when two or more atoms are bonded, each has (or shares) eight valence electrons. It is for this reason that the noble gases, at the opposite side of the periodic table from the alkali metals, almost never bond with other elements: they already have eight valence electrons.

The alkali metals, on the other hand, are quite likely to find "willing partners," since they each have just one valence electron. This brings up one of the reasons why hydrogen, though it is also part of Group 1, is not included as an alkali metal. First and most obviously, it is not a metal; additionally, it bonds according to what is called the duet rule, such that it shares two electrons with another element.

CHEMICAL AND PHYSICAL CHARACTERISTICS OF THE ALKALI METALS

The term "alkali" (essentially the opposite of an acid) refers to a substance that forms the negatively charged hydroxide ion (OH-) in contact with water. On their own, however, alkali metals almost always form positive ions, or cations, with a charge of +1.

When alkali metals react with water, one hydrogen atom splits off from the water molecule to form hydrogen gas, while the other hydrogen atom joins the oxygen to form hydroxide. Where the heavier members of the alkali metal family are concerned, reactions can often be so vigorous that the result is combustion or even explosion. Alkali metals also react with oxygen to produce

ONE OF THE USES OF LITHIUM IS IN BATTERIES. THE NISSAN HYPERMINI IS AN ELECTRIC VEHICLE POWERED BY A LITHIUM BATTERY. *(Reuters NewMedia Inc./Corbis. Reproduced by permission.)*

either an oxide, peroxide, or superoxide, depending on the particular member of the alkali metal family involved.

Shiny and soft enough to be cut with a knife, the alkali metals are usually white (though cesium is more of a yellowish white). When placed in a flame, most of these substances produce characteristic colors: lithium, for instance, glows bright red, and sodium an intense yellow. Heated potassium produces a violet color, rubidium a dark red, and cesium a light blue. This makes it possible to identify the metals, when heated, by color—a useful trait, since they are so often inclined to be bonded with other elements.

MELTING AND BOILING POINTS. As one moves down the rows or

periods of the periodic table, the mass of atoms increases, as does the energy each atom possesses. Yet the amount of energy required to turn a solid alkali metal into a liquid, or to vaporize a liquid alkali metal, actually decreases with higher atomic number. In other words, the higher the atomic number, the lower the boiling and melting points.

The list below lists the six alkali metals in order of atomic number, along with chemical symbol, atomic mass, melting point, and boiling point. Note that for francium, which is radioactive, the figures given are for its most stable isotope, francium-223 (^{223}Fr)—which has a half-life of only about 21 minutes.

ONE OF THE MOST STRIKING USES OF LITHIUM OCCURRED IN 1932, WITH THE DEVELOPMENT OF THE FIRST PARTI-
CLE ACCELERATOR. SHOWN HERE ARE ITS INVENTORS, ERNEST WALTON (LEFT), ERNEST RUTHERFORD (CENTER),
AND JOHN COCKCROFT. *(Hulton-Deutsch Collection/Corbis. Reproduced by permission.)*

Atomic Number, Mass, and Melting and Boiling Points of the Alkali Metals:

- 3. **Lithium** (Li), 6.941; Melting Point: 356.9°F (180.5°C); Boiling Point: 2,457°F (1,347°C)

- 11. **Sodium** (Na), 22.99; Melting Point: 208°F (97.8°C); Boiling Point: 1,621.4°F (883°C)

- 19. **Potassium** (K), 39.10; Melting Point: 145.9°F (63.28°C); Boiling Point: 1,398.2°F (759°C)

- 37. **Rubidium** (Rb), 85.47; Melting Point: 102.8°F (39.31°C); Boiling Point: 1,270.4°F (688°C)

- 55. **Cesium** (Cs), 132.9; Melting Point: 83.12°F (28.4°C); Boiling Point: 1,239.8°F (671°C)

- 87. **Francium** (Fr), (223); Melting Point: 80.6°F (27°C); Boiling Point: 1,250.6°F (677°C)

ABUNDANCE OF ALKALI METALS

Sodium and potassium are, respectively, the sixth and seventh most abundant elements on Earth, comprising 2.6% and 2.4% of the planet's known elemental mass. This may not seem like much, but considering the fact that just two elements—oxygen and silicon—make up about 75%, and that just 16 elements make up most of the remainder, it is an impressive share.

Lithium, on the other hand, is much less abundant, and therefore, figures for its part of Earth's known elemental mass are measured in parts per million (ppm). The total lithium in Earth's crust is about 17 ppm. Surprisingly, rubidium is more abundant, at 60 ppm; less surprisingly, cesium, with just 3 ppm, is very rare. Almost no francium is found naturally, except in very small quantities within uranium ores.

REAL-LIFE APPLICATIONS

LITHIUM

Swedish chemist Johan August Arfvedson (1792-1841) discovered lithium in 1817, and named it after the Greek word for "stone." Four years later, another scientist named W. T. Brande succeeded in isolating the highly reactive metal. Most of the lithium available on Earth's crust is bound up with aluminum and silica in minerals.

Since the time of its discovery, lithium has been used in lubricants, glass, and in alloys of lead, aluminum, and magnesium. In glass, it acts as a strengthening agent; likewise, metal alloys that contain lithium tend to be stronger, yet less dense. In 1994, physicist Jeff Dahn of Simon Fraser University in British Columbia, Canada, developed a lithium battery. Not only was the battery cheaper to produce than the traditional variety, Dahn and his colleagues announced, but the disposal of used lithium batteries presented less danger to the environment.

One of the most striking uses of lithium occurred in 1932, when English physicist John D. Cockcroft (1897-1967) and Irish physicist Ernest Walton (1903-1995) built the first particle accelerator. By bombarding lithium atoms, they produced highly energized alpha particles. This was the first nuclear reaction brought about by the use of artificially accelerated particles—in other words, without the need for radioactive materials such as uranium-235. Cockcroft's and Walton's experiment with lithium thus proved pivotal to the later creation of the atomic bomb.

LITHIUM IN PSYCHIATRIC TREATMENT.

The most important application of lithium, however, is in treatment for the psychiatric condition once known as manic depression, today identified as bipolar disorder. Persons suffering from bipolar disorder tend toward mood swings: during some periods the patient is giddy ("manic," or in a condition of "mania"), and during others the person is suicidal. Indeed, prior to the development of lithium as a treatment for bipolar disorder, as many as one in five patients with this condition committed suicide.

Doctors do not know exactly how lithium does what it does, but it obviously works: between 70% and 80% of patients with the bipolar condition respond well to treatment, and are able to go on with their lives in such a way that their condition is no longer outwardly evident. Lithium is also administered to patients who suffer unipolar depression and some forms of schizophrenia.

EARLY MEDICINAL USES OF LITHIUM.

It is said that the great Greco-Roman physician Galen (129-c. 199) counseled patients suffering from "mania" to bathe in, and even drink the water from, alkaline springs. If so, he was nearly 2,000 years ahead of his time. Even in the 1840s, not long after lithium was discovered, the mineral—mixed with carbonate or citrate—was touted as a cure for insomnia, gout, epilepsy, diabetes, and even cancer.

None of these alleged cures proved a success; nor did a lithium chloride treatment administered in the 1940s as a salt substitute for patients on low-sodium diets. As it turned out, when not enough sodium is present, the body experiences a buildup of sodium's sister element, lithium. The result was poisoning, which in some cases proved fatal.

CADE'S BREAKTHROUGH.

Then in 1949, Australian psychiatrist John Cade discovered the value of lithium for psychiatric treatment. He approached the problem from an entirely different angle, experimenting with uric acid, which he believed to be a cause of manic behavior. In administering the acid to guinea pigs, he added lithium salts merely to keep the uric acid soluble—and was very surprised by what he discovered. The uric acid did not make the guinea pigs manic, as he had expected; instead, they became exceedingly calm.

Cade changed the focus of his research, and tested lithium treatment on ten manic patients. Again, the results were astounding: one patient who had suffered from an acute bipolar disorder (as it is now known) for five years was released from the hospital after three months of lithium treatment, and went on to lead a healthy, normal life.

Encouraged by the changes he had seen in patients who received lithium, Cade published a report on his findings in the *Medical Journal of Australia*, but his work had little impact at the time. Nor did the idea of lithium treatment meet with an enthusiastic reception on the other side of the Pacific: in the aftermath of the failed experiments with lithium as a sodium substitute in the 1940s, stories of lithium poisoning were widespread in the United States.

LITHIUM TODAY.

Were it not for the efforts of Danish physician Mogens Schou, lithium might never have taken hold in the medical community. During the 1950s and 1960s, Schou campaigned tirelessly for recognition of lithium as a treatment for manic-depressive illness. Finally during the 1960s, the U.S. Food and Drug Administration began conducting trials of lithium, and approved its use in 1974. Today some 200,000 Americans receive lithium treatments.

A non-addictive and non-sedating medication, lithium—as evidenced by the failed experiment in the 1940s—may still be dangerous in large quantities. It is absorbed quickly into the bloodstream and carried to all tissues in the brain and body before passing through the kidneys. Both lithium and sodium are excreted through the kidneys, and since sodium affects lithium excretion, it is necessary to maintain a proper quantity of sodium in the body. For this reason, patients on lithium are cautioned to avoid a low-salt diet.

SODIUM

Sodium compounds had been known for some time prior to 1807, when English chemist Sir Humphry Davy (1778-1829) succeeded in isolating sodium itself. The element is represented by a chemical symbol (Na), reflecting its Latin name, *natrium*. In its pure form, sodium has a bright, shiny surface, but in order to preserve this appearance, it must be stored in oil: sodium reacts quickly with oxygen, forming a white crust of sodium oxide.

Pure sodium never occurs in nature; instead, it combines readily with other substances to form compounds, many of which are among the most widely used chemicals in industry. It is also highly soluble: thus whereas sodium and potassium occur in crystal rocks at about the same ratio, sodium is about 30 times more abundant in seawater than its sister element.

OBTAINING SODIUM CHLORIDE. Though the extraction of sodium involves the use of a special process, the metal is plentiful in the form of sodium chloride—better known as table salt. In fact, the term salt in chemistry refers generally to any combination of a metal with a nonmetal. More specifically, salts are (along with water) the product of reactions between acids and bases.

Sodium chloride is so easy to obtain, and therefore so cheap, that most industries making other sodium compounds use it, simply separating out the chloride (as described below) before adding other elements. The United States is the world's largest producer of sodium chloride, obtained primarily from brine, a term used to describe any solution of sodium chloride in water. Brine comes from seawater, subterranean wells, and desert lakes, such as the Great Salt Lake in Utah. Another source of sodium chloride is

rock salt, created underground by the evaporation of long-buried saltwater seas.

Other top sodium-chloride-producing nations include China, Germany, Great Britain, France, India, and various countries in the former Soviet Union. Salt may be cheap and plentiful for the world in general, but there are places where it is a precious commodity. One such place is the Sahara Desert, where salt caravans ply a brisk trade today, much as they have since ancient times.

ISOLATING SODIUM. Modern methods for the production of sodium represent an improvement in the technique Davy used in 1807, although the basic principle is the same. Though several decades passed before electricity came into widespread public use, scientists had been studying its properties for years, and Davy applied it in a process called electrolysis.

Electrolysis is the use of an electric current to produce a chemical reaction—in this case, to separate sodium from the other element or elements with which it is combined. Davy first fused or melted a sample of sodium chloride, then electrolyzed it. Using an electrode, a device that conducts electricity and is used to emit or collect electric charge, he separated the sodium chloride in such a way that liquid sodium metal collected on the cathode, or negatively charged end. Meanwhile, the gaseous chlorine was released through the anode, or the positively charged end.

The apparatus used for sodium separation today is known as the Downs cell, after its inventor, J. C. Downs. In a Downs cell, sodium chloride and calcium chloride are combined in a molten mixture in which the presence of calcium chloride lowers the melting point of the sodium chloride by more than 30%. When an electric current is passed through the mixture, sodium ions move to the cathode, where they pick up electrons to become sodium atoms. At the same time, ions of chlorine migrate to the anode, losing electrons to become chlorine atoms.

Sodium is a low-density material that floats on water, and in the Downs cell, the molten sodium rises to the top, where it is drawn off. The chlorine gas is allowed to escape through a vent at the top of the anode end of the cell, and the resulting sodium metal—that is, the elemental form of sodium—is about 99.8% pure.

USES FOR SODIUM CHLORIDE. As indicated earlier, sodium chloride is by far the most widely known and commonly used sodium compound—and this in itself is a distinction, given the fact that so many sodium compounds are a part of daily life. Today people think of salt primarily as a seasoning to enhance the taste of food, but prior to the development of refrigeration, it was vital as a preservative because it kept microbes away from otherwise perishable food items.

Salt does not merely improve the taste of food; it is an essential nutrient. Sodium compounds regulate transmission of signals through the nervous system, alter the permeability of membranes, and perform a number of other life-preserving functions. On the other hand, too much salt can aggravate high blood pressure. Thus, since the 1970s and 1980s, food manufacturers have increasingly offered products low in sodium, a major selling point for health-conscious consumers.

OTHER SODIUM COMPOUNDS. In addition to its widespread use in consumer goods, sodium chloride is the principal source of sodium used in making other sodium compounds. These include sodium hydroxide, for manufacturing cellulose products such as film, rayon, soaps, and paper, and for refining petroleum. In its application as a cleaning solution, sodium hydroxide is known as caustic soda or lye.

Another widely used sodium compound is sodium carbonate or, soda ash, applied in glassmaking, paper production, textile manufacturing, and other areas, such as the production of soaps and detergents. Sodium also can be combined with carbon to produce sodium bicarbonate, or baking soda. Sodium sulfate, sometimes known as salt cake, is used for making cardboard and kraft paper. Yet another widely used sodium compound is sodium silicate, or "water glass," used in the production of soaps, detergents, and adhesives; in water treatment; and in bleaching and sizing of textiles.

Still other sodium compounds used by industry and/or consumers include sodium borate, or borax; sodium tartrate, or sal tartar; the explosive sodium nitrate, or Chilean saltpeter; and the food additive monosodium glutamate (MSG). Perhaps ironically, there are few uses for pure metallic sodium. Once applied as an "anti-knock" additive in leaded gasoline, before those products were phased out for environmental reasons, metallic sodium is now used as a heat-exchange medium in nuclear reactors. But its widest application is in the production of the many other sodium compounds used around the world.

POTASSIUM

In some ways, potassium is a strange substance, as evidenced by its behavior in response to water. As everyone knows, water tends to put out a fire, and most explosives, when exposed to sufficient quantities of water, become ineffective. Potassium, on the other hand, explodes in contact with water and reacts violently with ice at temperatures as low as -148°F (-100°C). In a complete reversal of the procedures normally followed for most substances, potassium is stored in kerosene, because it might burst into flames if exposed to moist air!

Many aspects of potassium mirror those already covered with regard to sodium. The two have a number of the same applications, and in certain situations, potassium is used as a sodium substitute. Like sodium, potassium is never found alone in nature; instead, it comes primarily from sylvinite and carnalite, two ores containing potassium chloride. Also, like sodium, potassium was first isolated in 1807 by Davy, using the process of electrolysis described above. A few years later, a German chemist dubbed the newly isolated element "kalium," apparently a derivation of the Arabic *qali*, for "alkali"; hence the use of K as the chemical symbol for potassium.

USES FOR POTASSIUM. Potassium has another similarity with sodium; although it was not isolated until the early nineteenth century, its compounds have been in use for many centuries. The Romans, for instance, used potassium carbonate, or potash, obtained from the ashes of burned wood, to make soap. During the Middle Ages, the Chinese applied a form of saltpeter, potassium nitrate, in making gunpowder. And in colonial America, potash went into the production of soap, glass, and other products.

The production of just one ton of potash required the burning of several acres' worth of trees—a wasteful practice in more ways than one. Though there was no environmentalist movement in those days, financial concerns never go out of style. In order to save the money lost by using up vast acres of timber, American industry in the nineteenth century sought another means of making potash. The many similarities between

IN THE SAHARA DESERT, SALT CARAVANS PLY A BRISK TRADE TODAY, MUCH AS THEY HAVE SINCE ANCIENT TIMES. HERE, TRADERS IN MALI LOAD A CAMEL FOR A SALT CARAVAN. *(Nik Wheeler/Corbis. Reproduced by permission.)*

sodium and potassium provided a key, and the substitution of sodium carbonate for potassium carbonate saved millions of trees.

In 1847, German chemist Justus von Liebig (1803-1873) discovered potassium in living tis-sues. As a result, scientists became aware of the role this alkali metal plays in sustaining life: indeed, potassium is present in virtually all living cells. In the human body, potassium—which accounts for only 0.4% of the body's mass—is

KEY TERMS

ALKALI METALS: The elements in Group 1 of the periodic table of elements, with the exception of hydrogen. The alkali metals, which include lithium, sodium, potassium, rubidium, cesium, and francium, all have one valence electron in the s^1 orbital, and are highly reactive.

ANODE: An electrode at the positively charged end of a supply of electric current.

BRINE: A term used to describe any solution of sodium chloride in water.

CATHODE: An electrode at the negatively charged end of a supply of electric current.

CATION: The positive ion that results when an atom loses one or more electrons. All of the alkali metals tend to form +1 cations (pronounced KAT-ie-un).

ELECTRODE: A structure, often a metal plate or grid, that conducts electricity, and which is used to emit or collect electric charge.

ELECTROLYSIS: The use of an electric current to cause a chemical reaction.

HALF-LIFE: The length of time it takes a substance to diminish to one-half its initial amount.

ION: An atom or group of atoms that has lost or gained one or more electrons, and thus has a net electric charge.

ISOTOPES: Atoms that have an equal number of protons, and hence are of the same element, but differ in their number of neutrons. This results in a difference of mass. Isotopes may be either stable or unstable—that is, radioactive. Such is the case with francium, a radioactive member of the alkali metals family.

ORBITAL: A pattern of probabilities regarding the position of an electron for an atom in a particular energy state. The six alkali metals all have valence electrons in an s^1 orbital, which describes a more or less spherical shape.

PERIODIC TABLE OF ELEMENTS: A chart that shows the elements arranged in order of atomic number. Vertical columns within the periodic table indicate groups or "families" of elements with similar chemical characteristics; the alkali metals occupy Group 1.

RADIOACTIVITY: A term describing a phenomenon whereby certain materials are subject to a form of decay brought about by the emission of high-energy particles. "Decay" in this sense does not mean "rot"; instead, radioactive isotopes continue changing into other isotopes until they become stable.

SALT: A compound formed, along with water, by the reaction of an acid and a base. Generally, salts are any compounds that bring together a metal and a nonmetal.

VALENCE ELECTRONS: Electrons that occupy the highest energy levels in an atom, and which are involved in chemical bonding. The alkali metals all have one valence electron, in the s^1 orbital.

essential to the functioning of muscles. In larger quantities, however, it can be dangerous, causing a state of permanent relaxation known as potassium inhibition.

Since plants depend on potassium for growth, it was only logical that potassium, in the form of potassium chloride, was eventually applied as a fertilizer. This, at least, distinguishes it from its sister element: sodium, or sodium chloride, which can kill plants if administered to the soil in large enough quantities.

Another application of potassium is in the area pioneered by the Chinese about 800 years ago: the manufacture of fireworks and gunpowder from potassium nitrate. Like ammonium nitrate, made infamous by its use in the 1993 World Trade Center bombing and the Oklahoma City bombing in 1995, potassium nitrate doubles as a fertilizer.

RUBIDIUM, CESIUM, AND FRANCIUM

The three heaviest alkali metals are hardly household names, though one of them, cesium, does have several applications in industry. Rubidium and cesium, discovered in 1860 by German chemist R. W. Bunsen (1811-1899) and German physicist Gustav Robert Kirchhoff (1824-1887), were the first elements ever found using a spectroscope. Matter emits electromagnetic radiation along various spectral lines, which can be recorded using a spectroscope and then analyzed to discern the particular "fingerprint" of the substance in question.

When Bunsen and Kirchhoff saw the bluish spectral lines emitted by one of the two elements, they named it cesium, after a Latin word meaning "sky blue." Cesium, which is very rare, appears primarily in compounds such as pollucite. It is used today in photoelectric cells, military infrared lamps, radio tubes, and video equipment. During the 1940s, American physicist Norman F. Ramsey, Jr. (1915-) built a highly accurate atomic clock based on the natural frequencies of cesium atoms.

Rubidium, by contrast, has far fewer applications, and those are primarily in areas of scientific research. On Earth it is found in pollucite, lepidolite, and carnallite. It is considerably more abundant than cesium, and vastly more so than francium. Indeed, it is estimated that if all the francium in Earth's crust were combined, it would have a mass of about 25 grams.

Francium was discovered in 1939 by French physicist Marguerite Perey (1909-1975), student of the famous French-Polish physicist and chemist Marie Curie (1867-1934). For about four decades, scientists had been searching for the mysterious Element 87, and while studying the decay products of an actinium isotope, actinium-227, Perey discovered that one out of 100 such atoms decayed to form the undiscovered element. She named it francium, after her homeland. Though the discovery of francium solved a mystery, the element has no known uses outside of its applications in research.

WHERE TO LEARN MORE

"Alkali Metals" (Web site). <http://www.midlink.com/~jfromm/elements/alkali.htm> (May 24, 2001).

"Alkali Metals" ChemicalElements.com (Web site). <http://www.chemicalelements.com/groups/alkali.html> (May 24, 2001).

"Hydrogen and the Alkali Metals." University of Colorado Department of Physics (Web site). <http://www.colorado.edu/physics/2000/periodic_table/alkalimetals.html> (May 24, 2001).

Kerrod, Robin. Matter and Materials. Illustrated by Terry Hadler. Tarrytown, N.Y.: Benchmark Books, 1996.

Mebane, Robert C. and Thomas R. Rybolt. Metals. Illustrated by Anni Matsick. New York: Twenty-First Century Books, 1995.

Oxlade, Chris. Metal. Chicago, IL: Heinemann Library, 2001.

Snedden, Robert. Materials. Des Plaines, IL: Heinemann Library, 1999.

"Visual Elements: Group I—The Alkali Metals" (Web site). <http://www.chemsoc.org/viselements/pages/data/intro_groupi_data.html> (May 24, 2001).

ALKALINE
EARTH METALS

CONCEPT

The six alkaline earth metals—beryllium, magnesium, calcium, strontium, barium, and radium—comprise Group 2 on the periodic table of elements. This puts them beside the alkali metals in Group 1, and as their names suggest, the two families share a number of characteristics, most notably their high reactivity. Also, like the alkali metals, or indeed any other family on the periodic table, not all members of the alkali metal family are created equally in terms of their abundance on Earth or their usefulness to human life. Magnesium and calcium have a number of uses, ranging from building and other structural applications to dietary supplements. In fact, both are significant components in the metabolism of living things—including the human body. Barium and beryllium have numerous specialized applications in areas from jewelry to medicine, while strontium is primarily used in fireworks. Radium, on the other hand, is rarely used outside of laboratories, in large part because its radioactive qualities pose a hazard to human life.

HOW IT WORKS

DEFINING A FAMILY

The expression "families of elements" refers to groups of elements on the periodic table that share certain characteristics. These include (in addition to the alkaline earth metals and the alkali metals) the transition metals, halogens, noble gases, lanthanides, and actinides. (All of these are covered in separate essays within this book.) In addition, there are several larger categories with regard to shared traits that often cross family lines; thus all elements are classified either as metals, metalloids, and nonmetals. (These are also discussed in separate essays, which include reference to "orphans," or elements that do not belong to one of the families mentioned above.)

These groupings, both in terms of family and the broader divisions, relate both to external, observable characteristics, as well as to behaviors on the part of electrons in the elements' atomic structures. For instance, metals, which comprise the vast majority of elements on the periodic table, tend to be shiny, hard, and malleable (that is, they can bend without breaking.) Many of them melt at fairly high temperatures, and virtually all of them vaporize (become gases) at high temperatures. Metals also form ionic bonds, the tightest form of chemical bonding.

ELECTRON CONFIGURATIONS OF THE ALKALINE EARTH METALS. Where families are concerned, there are certain observable properties that led chemists in the past to group the alkaline earth metals together. These properties will be discussed with regard to the alkaline earth metals, but another point should be stressed in relation to the division of elements into families. With the advances in understanding that followed the discovery of the electron in 1897, along with the development of quantum theory in the early twentieth century, chemists developed a more fundamental definition of family in terms of electron configuration.

As noted, the alkaline earth metal family occupies the second group, or column, in the periodic table. All elements in a particular group, regardless of their apparent differences, have a common pattern in the configuration of their

DURING WORLD WAR II, MAGNESIUM WAS HEAVILY USED IN AIRCRAFT COMPONENTS. IN THIS 1941 PHOTO, WORK-ERS POUR MOLTEN MAGNESIUM INTO A CAST AT THE WRIGHT AERONAUTICAL CORPORATION. *(Bettmann/Corbis. Reproduced by permission.)*

valence electrons—the electrons at the "outside" of the atom, involved in chemical bonding. (By contrast, the core electrons, which occupy lower regions of energy within the atom, play no role in the bonding of elements.)

All members of the alkaline earth metal family have a valence electron configuration of s^2. This means that two electrons are involved in chemical bonding, and that these electrons move through an orbital, or range of probabilities, roughly corresponding to a sphere. The s orbital pattern corresponds to the first of several sublevels within a principal energy level.

Whatever the number of the principal energy level which corresponds to the period, or row,

on the periodic table the atom has the same number of sublevels. Thus beryllium, on Period 2, has two principal energy levels, and its valence electrons are in sublevel $2s^2$. At the other end of the group is radium, on Period 7. Though radium is far more complex than beryllium, with seven energy levels instead of two, nonetheless it has the same valence electron configuration, only on a higher energy level: $7s^2$.

HELIUM AND THE ALKALINE EARTH METALS. If one studies the valence electron configurations of elements on the periodic table, one notices an amazing symmetry and order. All members of a group, though their principal energy levels differ, share characteristics in their valence shell patterns. Furthermore, for the eight groups numbered in the North American version of the periodic table, the group number corresponds to the number of valence electrons.

There is only one exception: helium, with a valence electron configuration of $1s^2$, is normally placed in Group 8 with the noble gases. Based on that s^2 configuration, it might seem logical to place helium atop beryllium in the alkaline earth metals family; but there are several reasons why this is not done. First of all, helium is obviously not a metal. More importantly, helium behaves in a manner quite different from that of the alkaline earth metals.

Whereas helium, like the rest of the noble gases, is highly resistant to chemical reactions and bonding, alkaline earth metals are known for their high reactivity—that is, a tendency for bonds between atoms or molecules to be made or broken so that materials are transformed. (A similar relationship exists in Group 1, which includes hydrogen and the alkali metals. All have the same valence configuration, but hydrogen is never included as a member of the alkali metals family.)

CHARACTERISTICS OF THE ALKALINE EARTH METALS

Like the alkali metals, the alkaline earth metals have the properties of a base, as opposed to an acid. The alkaline earth metals are shiny, and most are white or silvery in color. Like their "cousins" in the alkali metal family, they glow with characteristic colors when heated. Calcium glows orange, strontium a very bright red, and barium an apple green. Physically they are soft, though not as soft as the alkali metals, many of which can be cut with a knife.

Yet another similarity the alkaline earth metals have with the alkali metals is the fact that four of the them—magnesium, calcium, strontium, and barium—were either identified or isolated in the first decade of the nineteenth century by English chemist Sir Humphry Davy (1778-1829). Around the same time, Davy also isolated sodium and potassium from the alkali metal family.

REACTIVITY. The alkaline earth metals are less reactive than the alkali metals, but like the alkali metals they are much more reactive than most elements. Again like their "cousins," they react with water to produce hydrogen gas and the metal hydroxide, though their reactions are less pronounced than those of the alkali metals. Magnesium metal in its pure form is combustible, and when exposed to air, it burns with an intense white light, combining with the oxygen to produce magnesium oxide. Likewise calcium, strontium, and barium react with oxygen to form oxides.

Due to their high levels of reactivity, the alkaline earth metals rarely appear by themselves in nature; rather, they are typically found with other elements in compound form, often as carbonates or sulfates. This, again, is another similarity with the alkali metals. But whereas the alkali metals tend to form 1+ cations (positively charged atoms), the alkaline earth metals form 2+ cations—that is, cations with a positive charge of 2.

BOILING AND MELTING POINTS. One way that the alkaline earth metals are distinguished from the alkali metals is with regard to melting and boiling points—those temperatures, respectively, at which a solid metal turns into a liquid, and a liquid metal into a vapor. For the alkali metals, the temperatures of the boiling and melting points decrease with an increase in atomic number. The pattern is not so clear, however, for the alkaline earth metals.

The highest melting and boiling points are for beryllium, which indeed has the lowest atomic number. It melts at 2,348.6°F (1,287°C), and boils at 4,789.8°F (2,471°C). These figures are much higher than for lithium, the alkali metal on the same period as beryllium, which melts at 356.9°F (180.5°C) and boils at 2,457°F (1,347°C).

Magnesium, the second alkali earth metal, melts at 1,202°F (650°C), and boils at 1,994°F (1,090°C)—significantly lower figures than for beryllium. However, the melting and boiling points are higher for calcium, third of the alkaline earth metals, with figures of 1,547.6°F (842°C) and 2,703.2°F (1,484°C) respectively. Melting and boiling temperatures steadily decrease as energy levels rise through strontium, barium, and radium, yet these temperatures are never lower than for magnesium.

ABUNDANCE. Of the alkaline earth metals, calcium is the most abundant. It ranks fifth among elements in Earth's crust, accounting for 3.39% of the elemental mass. It is also fifth most abundant in the human body, with a share

MANY ALKALINE EARTH METALS ARE USED IN THE PRODUCTION OF FIREWORKS. *(Bill Ross/Corbis. Reproduced by permission.)*

of 1.4%. Magnesium, which makes up 1.93% of Earth's crust, is the eighth most abundant element on Earth. It ranks seventh in the human body, accounting for 0.50% of the body's mass.

Barium ranks seventeenth among elements in Earth's crust, though it accounts for only 0.04% of the elemental mass. Neither it nor the other three alkali metals appear in the body in significant quantities: indeed, barium and beryllium are poisonous, and radium is so radioactive that exposure to it can be extremely harmful.

Within Earth's crust, strontium is present in quantities of 360 parts per million (ppm), which in fact is rather abundant compared to a number of elements. In the ocean, its presence is about 8 ppm. By contrast, the abundance of beryllium in Earth's crust is measured in parts per billion (ppb), and is estimated at 1,900 ppb. Vastly more rare is radium, which accounts for just 0.6 parts per trillion of Earth's crust—a fact that made its isolation by French-Polish physicist and chemist Marie Curie (1867-1934) all the more impressive.

REAL-LIFE APPLICATIONS

BERYLLIUM

In the eighteenth century, French mineralogist René Just-Haüy (1743-1822) had observed that both emeralds and the mineral beryl had similar properties. French chemist Louis-Nicolas Vauquelin (1763-1829) in 1798 identified the element they had in common: beryllium (Be), which has an atomic number of 4 and an atomic mass of 9.01 amu. Some three decades passed before German chemist Friedrich Wöhler (1800-1882) and French chemist Antoine Bussy (1794-1882), working independently, succeeded in isolating the element.

Beryllium is found primarily in emeralds and aquamarines, both precious stones that are forms of the beryllium alluminosilicate compound beryl. Though it is toxic to humans, beryllium nonetheless has an application in the health-care industry: because it lets through more x rays than does glass, beryllium is often used in x-ray tubes.

Metal alloys that contain about 2% beryllium tend to be particularly strong, resistant to wear, and stable at high temperatures. Copper-beryllium alloys, for instance, are applied in hand tools for industries that use flammable solvents, since tools made of these alloys do not cause sparks when struck against one another. Alloys of beryllium and nickel are applied for specialized electrical connections, as well as for high-temperature uses.

MAGNESIUM

English botanist and physician Nehemiah Grew (1641-1712) in 1695 discovered magnesium sulfate in the springs near the English town of Epsom, Surrey. This compound, called "Epsom salts" ever since, has long been noted for its medicinal value. Epsom salts are used for treating eclampsia, a condition that causes seizures in pregnant women. The compound is also a powerful laxative, and is sometimes used to rid the body of poisons—such as magnesium's sister element, barium.

For some time, scientists confused the oxide compound magnesia with lime or calcium carbonate, which actually involves another alkaline earth metal. In 1754, Scottish chemist and physi-cist Joseph Black (1728-1799) wrote "Experiments Upon Magnesia, Alba, Quick-Lime, and Some Other Alkaline Substances," an important work in which he distinguished between magnesia and lime. Davy in 1808 declared magnesia the oxide of a new element, which he dubbed magnesium, but some 20 years passed before Bussy succeeded in isolating the element.

Magnesium (Mg) has an atomic number of 12, and an atomic mass of 24.31 amu. It is found primarily in minerals such as dolomite and magnesite, both of which are carbonates; and in carnallite, a chloride. Magnesium silicates include asbestos, soapstone or talc, and mica. Not all forms of asbestos contain magnesium, but the fact that many do only serves to show the ways that chemical reactions can change the properties an element possesses in isolation.

AN IMPORTANT COMPONENT OF HEALTH. Whereas magnesium is flammable, asbestos was once used in large quantities as a flame retardant. And whereas asbestos has been largely removed from public buildings throughout the United States due to reports linking asbestos exposure with cancer, magnesium is an important component in the health of living organisms. It plays a critical role in chlorophyll, the green pigment in plants that captures energy from sunlight, and for this reason, it is also used in fertilizers.

In the human body, magnesium ions (charged atoms) aid in the digestive process, and many people take mineral supplements containing magnesium, sometimes in combination with calcium. There is also its use as a laxative, already mentioned. Epsom salts, as befits their base or alkaline quality, are exceedingly bitter—the kind of substance a person only ingests under conditions of the most dire necessity. On the other hand, milk of magnesia is a laxative with a far less unpleasant taste.

MAGNESIUM GOES TO WAR. It is a hallmark of magnesium's chemical versatility that the same element, so important in preserving life, has also been widely used in warfare. Just before World War I, Germany was a leading manufacturer of magnesium, thanks in large part to a method of electrolysis developed by German chemist R. W. Bunsen (1811-1899). When the United States went to war against Germany, American companies began producing magnesium in large quantities.

Bunsen had discovered that powdered magnesium burns with a brilliant white flame, and in the war, magnesium was used in flares, tracer bullets, and incendiary bombs, which ignite and burn upon impact. The bright light produced by burning magnesium has also led to a number of peacetime applications—for instance, in fireworks, and for flashes used in photography.

Magnesium saw service in another world war. By the time Nazi tanks rolled into Poland in 1939, the German military-industrial complex had begun using the metal for building aircraft and other forms of military equipment. America once again put its own war-production machine into operation, dramatically increasing magnesium output to a peak of nearly 184,000 tons (166,924,800 kg) in 1943.

STRUCTURAL APPLICATIONS. Magnesium's principal use in World War I was for its incendiary qualities, but in World War II it was primarily used as a structural metal. It is lightweight, but stronger per unit of mass than any other common structural metal. As a metal for building machines and other equipment, magnesium ranks in popularity only behind iron and aluminum (which is about 50% more dense than magnesium).

The automobile industry is one area of manufacturing particularly interested in magnesium's structural qualities. On both sides of the Atlantic, automakers are using or testing vehicle parts made of alloys of magnesium and other metals, primarily aluminum. Magnesium is easily cast into complex structures, which could mean a reduction in the number of parts needed for building a car—and hence a streamlining of the assembly process.

Among the types of sports equipment employing magnesium alloys are baseball catchers' masks, skis, racecars, and even horseshoes. Various brands of ladders, portable tools, electronic equipment, binoculars, cameras, furniture, and luggage also use parts made of this lightweight, durable metal.

CALCIUM

Davy isolated calcium (Ca) by means of electrolysis in 1808. The element, whose name is derived from the Latin *calx,* or "lime," has an atomic number of 20, and an atomic mass of 40.08. The principal sources of calcium are limestone and dolomite, both of which are carbonates, as well as the sulfate gypsum.

In the form of limestone and gypsum, calcium has been used as a building material since ancient times, and continues to find application in that area. Lime is combined with clay to make cement, and cement is combined with sand and water to make mortar. In addition, when mixed with sand, gravel, and water, cement makes concrete. Marble—once used to build palaces and today applied primarily for decorative touches—also contains calcium.

The steel, glass, paper, and metallurgical industries use slaked lime (calcium hydroxide) and quicklime, or calcium oxide. It helps remove impurities from steel, and pollutants from smokestacks, while calcium carbonate in paper provides smoothness and opacity to the finished product. When calcium carbide (CaC_2) is added to water, it produces the highly flammable gas acetylene (C_2H_2), used in welding torches. In various compounds, calcium is used as a bleach; a material in the production of fertilizers; and as a substitute for salt as a melting agent on icy roads.

The food, cosmetic, and pharmaceutical industries use calcium in antacids, toothpaste, chewing gum, and vitamins. To an even greater extent than magnesium, calcium is important to living things, and is present in leaves, bone, teeth, shells, and coral. In the human body, it helps in the clotting of blood, the contraction of muscles, and the regulation of the heartbeat. Found in green vegetables and dairy products, calcium (along with calcium supplements) is recommended for the prevention of osteoporosis. The latter, a condition involving a loss of bone density, affects elderly women in particular, and causes bones to become brittle and break easily.

STRONTIUM

Irish chemist and physician Adair Crawford (1748-1795) and Scottish chemist and surgeon William Cumberland Cruikshank (1790-1800) in 1790 discovered what Crawford called "a new species of earth" near Strontian in Scotland. A year later, English chemist Thomas Charles Hope (1766-1844) began studying the ore found by Crawford and Cruikshank, which they had dubbed strontia.

In reports produced during 1792 and 1793, Hope explained that strontia could be distinguished from lime or calcium hydroxide on the

IN THIS 1960 PHOTO, A MOTHER TESTS HER SON'S MILK FOR SIGNS OF RADIOACTIVITY, THE RESULT OF NUCLEAR WEAPONS TESTING DURING THE 1950S THAT INVOLVED THE RADIOACTIVE ISOTOPE STRONTIUM-90. THE ISOTOPE FELL TO EARTH IN A FINE POWDER, WHERE IT COATED THE GRASS, WAS INGESTED BY COWS, AND EVENTUALLY WOUND UP IN THE MILK THEY PRODUCED. *(Bettmann/Corbis. Reproduced by permission.)*

one hand, and baryta or barium hydroxide on the other, by virtue of its response to flame tests. Whereas calcium produced a red flame and barium a green one, strontia glowed a brilliant red easily distinguished from the darker red of calcium.

Once again, it was Davy who isolated the new element, using electrolysis, in 1808. Subsequently dubbed strontium (Sr), its atomic number is 38, and its atomic mass 87.62. Silvery white, it oxidizes rapidly in air, forming a pale yellow oxide crust on any freshly cut surface.

Though it has properties similar to those of calcium, the comparative rarity of strontium and the expense involved in extracting it offer no economic incentives for using it in place of its much more abundant sister element. Nonetheless, strontium does have a few uses, primarily because of its brilliant crimson flame. Therefore it is applied in the making of fireworks, signal flares, and tracer bullets—that is, rounds that emit a light as they fly through the air.

One of the more controversial "applications" of strontium involved the radioactive isotope strontium-90, a by-product of nuclear weapons testing in the atmosphere from the late 1940s onward. The isotope fell to earth in a fine powder, coated the grass, was ingested by cows, and eventually wound up in the milk they produced. Because of its similarities to calcium, the isotope became incorporated into the teeth and gums of children who drank the milk, posing health concerns that helped bring an end to atmospheric testing in the early 1960s.

BARIUM

Aspects of barium's history are similar to those of other alkaline earth metals. During the eighteenth century, chemists were convinced that barium oxide and calcium oxide constituted the same substance, but in 1774, Swedish chemist Carl Wilhelm Scheele (1742-1786) demonstrated that barium oxide was a distinct compound. Davy isolated the element, as he did two other alkaline earth metals, by means of electrolysis, in 1808.

Barium (Ba) has an atomic mass of 137.27 and an atomic number of 56. It appears primarily in ores of barite, a sulfate, and witherite, a carbonate. Barium sulfate is used as a white pigment in paints, while barium carbonate is applied in the production of optical glass, ceramics, glazed

ALKALINE EARTH METALS: Group 2 on the periodic table of elements, with valence electron configurations of ns^2. The six alkaline earth metals, all of which are highly reactive chemically, are beryllium, magnesium, calcium, strontium, barium, and radium.

ALKALI METALS: The elements in Group 1 of the periodic table of elements, with the exception of hydrogen. The alkali metals all have one valence electron in the s^1 orbital, and are highly reactive.

CATION: The positive ion that results when an atom loses one or more electrons. All of the alkaline earth metals tend to form 2+ cations (pronounced KAT-ie-unz).

ELECTROLYSIS: The use of an electric current to cause a chemical reaction.

ION: An atom or group of atoms that has lost or gained one or more electrons, and thus has a net electric charge.

ISOTOPES: Atoms that have an equal number of protons, and hence are of the same element, but differ in their number of neutrons. This results in a difference of mass. Isotopes may be either stable or unstable—that is, radioactive. Such is the case with the isotopes of radium, a radioactive member of the alkaline earth metals family.

ORBITAL: A pattern of probabilities regarding the position of an electron for an atom in a particular energy state. The six alkaline earth metals all have valence electrons in an s^2 orbital, which describes a more or less spherical shape.

PERIODS: Rows of the periodic table of elements. These represent successive prin-

---

pottery, and specialty glassware. One of its most important uses is as a drill-bit lubricant—known as a "mud" or slurry—for oil drilling. Like a number of its sister elements, barium (in the form of barium nitrate) is used in fireworks and flares. Motor oil detergents for keeping engines clean use barium oxide and barium hydroxide.

Beryllium is not the only alkaline earth metal used in making x rays, nor is magnesium the only member of the family applied as a laxative. Barium is used in enemas, and barium sulfate is used to coat the inner lining of the intestines to allow a doctor to examine a patient's digestive system. (Though barium is poisonous, in the form of barium sulfate it is safe for ingestion because the compound does not dissolve in water or other bodily fluids.) Prior to receiving x rays, a patient may be instructed to drink a chalky barium sulfate liquid, which absorbs a great deal of the radiation emitted by the x-ray

machine. This adds contrast to the black-and-white x-ray photo, enabling the doctor to make a more informed diagnosis.

RADIUM

Today radium (Ra; atomic number 88; atomic mass 226 amu) has few uses outside of research; nonetheless, the story of its discovery by Marie Curie and her husband Pierre (1859-1906), a French physicist, is a compelling chapter not only in the history of chemistry, but of human endeavor in general. Inspired by the discovery of uranium's radioactive properties by French physicist Henri Becquerel (1852-1908), Marie Curie became intrigued with the subject of radioactivity, on which she wrote her doctoral dissertation. Setting out to find other radioactive elements, she and Pierre refined a large quantity of pitchblende, an ore commonly found in uranium mines. Within a year, they had discovered

KEY TERMS CONTINUED

cipal energy levels in the atoms of the elements involved.

PRINCIPAL ENERGY LEVEL: A value indicating the distance that an electron may move away from the nucleus of an atom. This is designated by a whole-number integer, beginning with 1 and moving upward. The higher the principal energy level, the greater the energy in the atom, and the more complex the pattern of orbitals.

RADIOACTIVITY: A term describing a phenomenon whereby certain materials are subject to a form of decay brought about by the emission of high-energy particles. "Decay" in this sense does not mean "rot"; instead, radioactive isotopes continue changing into other isotopes until they become stable.

REACTIVITY: The tendency for bonds between atoms or molecules to be made or broken in such a way that materials are transformed.

SALT: Generally speaking, a compound that brings together a metal and a nonmetal. More specifically, salts (along with water) are the product of a reaction between an acid and a base.

SHELL: A group of electrons within the same principal energy level.

SUBLEVEL: A region within the principal energy level occupied by electrons in an atom. Whatever the number n of the principal energy level, there are n sublevels. At each principal energy level, the first sublevel to be filled is the one corresponding to the s orbital pattern—where the alkaline earth metals all have their valence electrons.

VALENCE ELECTRONS: Electrons that occupy the highest energy levels in an atom, and which are involved in chemical bonding.

the element polonium, but were convinced that another radioactive ingredient was present—though in much smaller amounts—in pitchblende.

The Curies spent most of their savings to purchase a ton of ore, and began working to extract enough of the hypothesized Element 88 for a usable sample—0.35 oz (1 g). Laboring virtually without ceasing for four years, the Curies—by then weary and in financial difficulties—finally produced the necessary quantity of radium. Their fortunes were about to improve: in 1903 they shared the Nobel Prize in physics with Becquerel, and in 1911, Marie received a second Nobel, this one in chemistry, for her discoveries of polonium and radium. She is the only individual in history to win Nobels in two different scientific categories.

Because the Curies failed to patent their process, however, they received no profits from the many "radium centers" that soon sprung up, touting the newly discovered element as a cure for cancer. In fact, as it turned out, the hazards associated with this highly radioactive substance outweighed any benefits. Thus radium, which at one point was used in luminous paint and on watch dials, was phased out of use. Marie Curie's death from leukemia in 1934 resulted from her prolonged exposure to radiation from radium and other elements.

WHERE TO LEARN MORE

"Alkaline Earth Metals." ChemicalElements.com (Web site). <http://www.chemicalelements.com/groups/alkaline.html>> (May 25, 2001).

"The Alkaline Earth Metals" (Web site). <http://www.nidlink.com/~jfromm/elements/alkaline.htm> (May 25, 2001).

Ebbing, Darrell D.; R. A. D. Wentworth; and James P. Birk. Introductory Chemistry. Boston: Houghton Mifflin, 1995.

Kerrod, Robin. *Matter and Materials.* Illustrated by Terry Hadler. Tarrytown, N.Y.: Benchmark Books, 1996.

Mebane, Robert C. and Thomas R. Rybolt. *Metals.* Illustrated by Anni Matsick. New York: Twenty-First Century Books, 1995.

Oxlade, Chris. *Metal.* Chicago, IL: Heinemann Library, 2001.

Snedden, Robert. *Materials.* Des Plaines, IL: Heinemann Library, 1999.

"Visual Elements: Group 1—The Alkaline Earth Metals" (Web site). <http://www.chemsoc.org/viselements/pages/data/intro_groupii_data.html> (May 25, 2001).

TRANSITION METALS

CONCEPT

By far the largest family of elements is the one known as the transition metals, sometimes called transition elements. These occupy the "dip" in the periodic table between the "tall" sets of columns or groups on either side. Consisting of 10 columns and four rows or periods, the transition metals are usually numbered at 40. With the inclusion of the two rows of transition metals in the lanthanide and actinide series respectively, however, they account for 68 elements—considerably more than half of the periodic table. The transition metals include some of the most widely known and commonly used elements, such as iron—which, along with fellow transition metals nickel and cobalt, is one of only three elements known to produce a magnetic field. Likewise, zinc, copper, and mercury are household words, while cadmium, tungsten, chromium, manganese, and titanium are at least familiar. Other transition metals, such as rhenium or hafnium, are virtually unknown to anyone who is not scientifically trained. The transition metals include the very newest elements, created in laboratories, and some of the oldest, known since the early days of civilization. Among these is gold, which, along with platinum and silver, is one of several precious metals in this varied family.

HOW IT WORKS

TWO NUMBERING SYSTEMS FOR THE PERIODIC TABLE

The periodic table of the elements, developed in 1869 by Russian chemist Dmitri Ivanovitch Mendeleev (1834-1907), is discussed in several places within this volume. Within the table, elements are arranged in columns, or groups, as well as in periods, or rows.

As discussed in the Periodic Table of Elements essay, two basic versions of the chart, differing primarily in their method of number groups, are in use today. The North American system numbers only eight groups—corresponding to the representative or main-group elements—leaving the 10 columns representing the transition metals unnumbered. On the other hand, the IUPAC system, approved by the International Union of Pure and Applied Chemistry (IUPAC), numbers all 18 columns. Both versions of the periodic table show seven periods.

In general, this book applies the North American system, in part because it is the version used in most American classrooms. Additionally, the North American system relates group number to the number of valence electrons in the outer shell of the atom for the element represented. Hence the number of valence electrons (a subject discussed below as it relates to the transition metals) for Group 8 is also eight. Yet where the transition metals are concerned, the North American system is less convenient than the IUPAC system.

ATTEMPTS TO ADJUST THE NORTH AMERICAN SYSTEM. In addressing the transition metals, there is no easy way to apply the North American system's correspondence between the number of valence electrons and the group number. Some versions of the North American system do, however, make an attempt. These equate the number of valence electrons in the transition metals to a group number, and distinguish these by adding a letter "B."

SEVERAL PRECIOUS METALS, INCLUDING GOLD, ARE AMONG THE TRANSITION METALS. *(Charles O'Rear/Corbis. Reproduced by permission.)*

Thus the noble gases are placed in Group 8A or VIIIA because they are a main-group family with eight valence electrons, whereas the transition metal column beginning with iron (Group 8 in the IUPAC system) is designated 8B or VIIIB. But this adjustment to the North American system only makes things more cumbersome, because Group VIIIB also includes two other columns—ones with nine and 10 valence electrons respectively.

VALUE OF THE IUPAC SYSTEM. Obviously, then, the North American system—while it is useful in other ways, and as noted, is applied elsewhere in this book—is not as practical for discussing the transition metals. Of course, one could simply dispense with the term "group" altogether, and simply refer to the columns on the periodic table as columns. To do so, however, is to treat the table simply as a chart, rather than as a marvelously ordered means of organizing the elements.

It makes much more sense to apply the IUPAC system and to treat the transition metals as belonging to groups 3 through 12. This is particularly useful in this context, because those group numbers do correspond to the numbers of valence electrons. Scandium, in group 3 (henceforth all group numbers refer to the IUPAC version), has three valence electrons; likewise zinc, in Group 12, has 12. Neither the IUPAC nor the North American system provide group numbers for the lanthanides and actinides, known collectively as the inner transition metals.

THE TRANSITION METALS BY IUPAC GROUP NUMBER. Here, then, is the list of the transition metals by group number, as designated in the IUPAC system. Under each group heading are listed the four elements in that group, along with the atomic number, chemical symbol, and atomic mass figures for each, in atomic mass units.

Note that, because the fourth member in each group is radioactive and sometimes exists for only a few minutes or even seconds, mass figures are given merely for the most stable isotope. In addition, the last elements in groups 8, 9, and 10, as of 2001, had not received official names. For reasons that will be explained below, lanthanum and actinium, though included here, are discussed in other essays.

Group 1

- 21. Scandium (Sc): 44.9559
- 39. Yttrium (Y): 88.9059
- 57. Lanthanum (La): 138.9055
- 89. Actinium (Ac): 227.0278

Group 2

- 22. Titanium (Ti): 47.9
- 40. Zirconium (Zr): 91.22
- 72. Hafnium (Hf): 178.49
- 104. Rutherfordium (Rf): 261

Group 3

- 23. Vanadium (V): 50.9415
- 41. Niobium (Nb): 92.9064
- 73. Tantalum (Ta): 180.9479
- 105. Dubnium (Db): 262

Group 4

- 24. Chromium (Cr): 51.996
- 42. Molybdenum (Mo): 95.95
- 74. Tungsten (W): 183.85
- 106. Seaborgium (Sg): 263

Group 5

- 25. Manganese (Mn): 54.938
- 43. Technetium (Tc): 98
- 75. Rhenium (Re): 186.207
- 107. Bohrium (Bh): 262

Group 6

- 26. Iron (Fe): 55.847
- 44. Ruthenium (Ru): 101.07
- 76. Osmium (Os): 190.2
- 108. Hassium (Hs): 265

Group 7

- 27. Cobalt (Co): 58.9332
- 45. Rhodium (Rh): 102.9055
- 77. Iridium (Ir): 192.22
- 109. Meitnerium (Mt): 266

Group 8

- 28. Nickel (Ni): 58.7
- 46. Palladium (Pd): 106.4
- 78. Platinum (Pt): 195.09
- 110. Ununnilium (Uun): 271

Group 9

- 29. Copper (Cu): 63.546
- 47. Silver (Ag): 107.868
- 79. Gold (Au): 196.9665
- 111. Unununium (Uuu): 272

Group 10

- 30. Zinc (Zn): 65.38
- 48. Cadmium (Cd): 112.41
- 80. Mercury (Hg): 200.59
- 112. Ununbium (Uub): 277

AN IRON ORE MINE IN SOUTH AFRICA. IRON IS THE FOURTH MOST ABUNDANT ELEMENT ON EARTH, ACCOUNTING FOR 4.71% OF THE ELEMENTAL MASS IN THE PLANET'S CRUST. *(Charles O'Rear/Corbis. Reproduced by permission.)*

ELECTRON CONFIGURATIONS AND THE PERIODIC TABLE

We can now begin to explain why transition metals are distinguished from other elements, and why the inner transition metals are further separated from the transition metals. This distinction relates not to period (row), but to group (column)—which, in turn, is defined by the configuration of valence electrons, or the electrons that are involved in chemical bonding. Valence electrons occupy the highest energy level, or shell, of the atom—which might be thought of as the orbit farthest from the nucleus. However, the term "orbit" is misleading when applied to the ways that an electron moves.

Electrons do not move around the nucleus of an atom in regular orbits, like planets around the Sun; rather, their paths can only be loosely defined in terms of orbitals, a pattern of probabilities indicating the regions that an electron may occupy. The shape or pattern of orbitals is determined by the principal energy level of the

THE AMERICAN FIVE-CENT COIN, CALLED A "NICKEL," IS ACTUALLY AN ALLOY OF NICKEL AND COPPER. *(Layne Kennedy/Corbis. Reproduced by permission.)*

atom, which indicates a distance that an electron may move away from the nucleus.

Principal energy level defines period: elements in Period 4, for instance, all have their valence electrons at principal energy level 4. The higher the number, the further the electron is from the nucleus, and hence, the greater the energy in the atom. Each principal energy level is divided into sublevels corresponding to the number n of the principal energy level: thus, principal energy level 4 has four sublevels, principal energy level 5 has five, and so on.

ORBITAL PATTERNS. The four basic types of orbital patterns are designated as *s, p, d,* and *f.* The *s* shape might be described as spherical, which means that in an *s* orbital, the total electron cloud will probably end up being more or less like a sphere.

The *p* shape is like a figure eight around the nucleus; the *d* like two figure eights meeting at the nucleus; and the *f* orbital pattern is so complex that most basic chemistry textbooks do not even attempt to explain it. In any case, these orbital patterns do not indicate the exact path of an electron. Think of it, instead, in this way: if you could take millions of photographs of the electron during a period of a few seconds, the

resulting blur of images in a *p* orbital, for instance, would somewhat describe the shape of a figure eight.

Since the highest energy levels of the transition metals begin at principal energy level 4, we will dispense with a discussion of the first three energy levels. The reader is encouraged to consult the Electrons essay, as well as the essay on Families of Elements, for a more detailed discussion of the ways in which these energy levels are filled for elements on periods 1 through 3.

ORBITAL FILLING AND THE PERIODIC TABLE

If all elements behaved as they "should," the pattern of orbital filling would be as follows. In a given principal energy level, first the two slots available to electrons on the *s* sublevel are filled, then the six slots on the *p* sublevel, then the 10 slots on the *d* sublevel. The *f* sublevel, which is fourth to be filled, comes into play only at principal energy level 4, and it would be filled before elements began adding valence electrons to the *s* sublevel on the next principal energy level.

That might be the way things "should" happen, but it is not the way that they do happen. The list that follows shows the pattern by which orbitals are filled from sublevel 3p onward. Following each sublevel, in parentheses, is the number of "slots" available for electrons in that shell. Note that in several places, the pattern of filling defies the ideal order described in the preceding paragraph.

Orbital Filling by Principal Energy Level and Sublevel from 3p Onward:

- 3p (6)
- 4s (2)
- 3d (10)
- 4p (6)
- 5s (2)
- 4d (10)
- 5p (6)
- 6s (2)
- 4f (14)
- 5d (10)
- 6p (6)
- 7s (2)
- 5f (14)
- 6d (10)

The 44 representative elements follow a regular pattern of orbital filling through the first 18 elements. By atomic number 18, argon, 3p has been filled, and at that point element 19 (potassium) would be expected to begin filling row 3d. However, instead it begins filling 4s, to which calcium (20) adds the second and last valence electron. It is here that we come to the transition metals, distinguished by the fact that they fill the d orbitals.

Note that none of the representative elements up to this point, or indeed any that follow, fill the d orbital. (The d orbital could not come into play before Period 3 anyway, because at least three sublevels—corresponding to principal energy level 3—are required in order for there to be a d orbital.) In any case, when the representative elements fill the p orbitals of a given energy level, they skip the d orbital and go on to the next principal energy level, filling the s orbitals. Filling of the d orbital on the preceding energy level only occurs with the transition metals: in other words, on period 4, transition metals fill the 3d sublevel, and so on as the period numbers increase.

DISTINGUISHING THE TRANSITION METALS

Given the fact that it is actually the representative elements that skip the d sublevels, and the transition metals that go back and fill them, one might wonder if the names "representative" and "transition" (implying an interruption) should be reversed. But it is the transition metals that are the exception to the rule, for two reasons. First of all, as we have seen, they are the only elements that fill the d orbitals.

Secondly, they are the only elements whose outer-shell electrons are on two different principal energy levels. The orbital filling patterns of transition metals can be identified thus: $ns(n-1)d$, where n is the number of the period on which the element is located. For instance, zinc, on period 4, has an outer shell of $4s^2 3d^{10}$. In other words, it has two electrons on principal energy level 4, sublevel 4s—as it "should." But it also has 10 electrons on principal energy level 3, sublevel 3d. In effect, then, the transition metals "go back" and fill in the d orbitals of the preceding period.

There are further complications in the patterns of orbital filling for the transition metals, which will only be discussed in passing here as a further explanation as to why these elements are not "representative." Most of them have two s valence electrons, but many have one, and palladium has zero. Nor do they add their d electrons in regular patterns. Moving across period 4, for instance, the number of electrons in the d orbital "should" increase in increments of 1, from 1 to 10. Instead the pattern goes like this: 1, 2, 3, 5, 5, 6, 7, 8, 10, 10.

LANTHANIDES AND ACTINIDES. The lanthanides and actinides are set apart even further from the transition metals. In most versions of the periodic table, lanthanum (57) is followed by hafnium (72) in the transition metals section of the chart. Similarly, actinium (89) is followed by rutherfordium (104). The "missing" metals—lanthanides and actinides respectively—are listed at the bottom of the chart. There are reasons for this, as well as for the names of these groups.

After the 6s orbital fills with the representative element barium (56), lanthanum does what a transition metal does—it begins filling the 5d orbital. But after lanthanum, something strange happens: cerium (58) quits filling 5d, and moves to fill the 4f orbital. The filling of that orbital continues throughout the entire lanthanide series, all the way to lutetium (71). Thus lanthanides can be defined as those metals that fill the 4f orbital; however, because lanthanum exhibits similar properties, it is usually included with the lanthanides.

A similar pattern occurs for the actinides. The 7s orbital fills with radium (88), after which actinium (89) begins filling the 6d orbital. Next comes thorium, first of the actinides, which begins the filling of the 5f orbital. This is completed with element 103, lawrencium. Actinides can thus be defined as those metals that fill the 5f orbital; but again, because actinium exhibits similar properties, it is usually included with the actinides.

REAL-LIFE APPLICATIONS

SURVEY OF THE TRANSITION METALS

The fact that the transition elements are all metals means that they are lustrous or shiny in appearance, and malleable, meaning that they can be molded into different shapes without breaking. They are excellent conductors of heat and electricity, and tend to form positive ions by losing electrons.

Generally speaking, metals are hard, though a few of the transition metals—as well as members of other metal families—are so soft they can be cut with a knife. Like almost all metals, they tend to have fairly high melting points, and extremely high boiling points.

Many of the transition metals, particularly those on periods 4, 5, and 6, form useful alloys—mixtures containing more than one metal—with one another, and with other elements. Because of their differences in electron configuration, however, they do not always combine in the same ways, even within an element. Iron, for instance, sometimes releases two electrons in chemical bonding, and at other times three.

ABUNDANCE OF THE TRANSITION METALS. Iron is the fourth most abundant element on Earth, accounting for 4.71% of the elemental mass in the planet's crust. Titanium ranks 10th, with 0.58%, and manganese 13th, with 0.09%. Several other transition metals are comparatively abundant: even gold is much more abundant than many other elements on the periodic table. However, given the fact that only 18 elements account for 99.51% of

Earth's crust, the percentages for elements outside of the top 18 tend to very small.

In the human body, iron is the 12th most abundant element, constituting 0.004% of the body's mass. Zinc follows it, at 13th place, accounting for 0.003%. Again, these percentages may not seem particularly high, but in view of the fact that three elements—oxygen, carbon, and hydrogen—account for 93% of human elemental body mass, there is not much room for the other 10 most common elements in the body. Transition metals such as copper are present in trace quantities within the body as well.

DIVIDING THE TRANSITION METALS INTO GROUPS. There is no easy way to group the transition metals, though certain of these elements are traditionally categorized together. These do not constitute "families" as such, but they do provide useful ways to break down the otherwise rather daunting 40-element lineup of the transition metals.

In two cases, there is at least a relation between group number on the periodic table and the categories loosely assigned to a collection of transition metals. Thus the "coinage metals"—copper, silver, and gold—all occupy Group 9 on the periodic table. These have traditionally been associated with one another because their resistance to oxidation, combined with their malleability and beauty, has made them useful materials for fashioning coins.

Likewise the members of the "zinc group"—zinc, cadmium, and mercury—occupy Group 10 on the periodic table. These, too, have often been associated as a miniature unit due to common properties. Members of the "platinum group"—platinum, iridium, osmium, palladium, rhodium, and ruthenium—occupy a rectangle on the table, corresponding to periods 5 and 6, and groups 6 through 8. What actually makes them a "group," however, is the fact that they tend to appear together in nature.

Iron, nickel, and cobalt, found alongside one another on Period 4, may be grouped together because they are all magnetic to some degree or another. This is far from the only notable characteristic about such metals, but provides a convenient means of further dividing the transition metals into smaller sections.

To the left of iron on the periodic table is a rectangle corresponding to periods 4 through 6, groups 4 through 7. These 11 elements—titanium, zirconium, hafnium, vanadium, niobium, tantalum, chromium, molybdenum, tungsten, manganese, and rhenium—are referred to here as "alloy metals." This is not a traditional designation, but it is nonetheless useful for describing these metals, most of which form important alloys with iron and other elements.

One element was left out of the "rectangle" described in the preceding paragraph. This is technetium, which apparently does not occur in nature. It is lumped in with a final category, "rare and artificial elements."

It should be stressed that there is nothing hard and fast about these categories. The "alloy metals" are not the only ones that form alloys; nickel is used in coins, though it is not called a coinage metal; and platinum could be listed with gold and silver as "precious metals." Nonetheless, the categories used here seem to provide the most workable means of approaching the many transition metals.

COINAGE METALS

GOLD. Gold almost needs no introduction: virtually everyone knows of its value, and history is full of stories about people who killed or died for this precious metal. Part of its value springs from its rarity in comparison to, say iron: gold is present on Earth's crust at a level of about 5 parts per billion (ppb). Yet as noted earlier, it is more abundant than some metals. Furthermore, due to the fact that it is highly unreactive (reactivity refers to the tendency for bonds between atoms or molecules to be made or broken in such a way that materials are transformed), it tends to be easily separated from other elements.

This helps to explain the fact that gold may well have been the first element ever discovered. No ancient metallurgist needed a laboratory in which to separate gold; indeed, because it so often keeps to itself, it is called a "noble" metal—meaning, in this context, "set apart." Another characteristic of gold that made it valuable was its great malleability. In fact, gold is the most malleable of all metals: A single troy ounce (31.1 g) can be hammered into a sheet just 0.00025 in (0.00064 cm) thick, covering 68 ft^2 (6.3 m^2).

Gold is one of the few metals that is not silver, gray, or white, and its beautifully distinctive color caught the eyes of metalsmiths and royalty from the beginning of civilization. Records from India dating back to 5000 B.C. suggest a familiar-

ity with gold, and jewelry found in Egyptian tombs indicates the use of sophisticated techniques among the goldsmiths of Egypt as early as 2600 B.C. Likewise the Bible mentions gold in several passages. The Romans called it *aurum* ("shining dawn"), which explains its chemical symbol, Au.

Early chemistry had its roots among the medieval alchemists, who sought in vain to turn plain metals into gold. Likewise the history of European conquest in the New World is filled with bitter tales of conquistadors, lured by gold, who wiped out entire nations of native peoples.

At one time, gold was used in coins, and nations gauged the value of their currency in terms of the gold reserves they possessed. Few coins today—with special exceptions, such as the South African Krugerrand—are gold. Likewise the United States, for instance, went off the gold standard (that is, ceased to tie its currency to gold reserves) in the 1930s; nonetheless, the federal government maintains a vast supply of gold bullion at Fort Knox in Kentucky.

Gold is as popular as ever for jewelry and other decorative objects, of course, but for the most part, it is too soft to have many other commercial purposes. One of the few applications for gold, a good conductor of electricity, is in some electronic components. Also, the radioactive gold-198 isotope is sometimes implanted in tissues as a means of treating forms of cancer.

SILVER. Like gold, silver has been a part of human life from earliest history. Usually it is considered less valuable, though some societies have actually placed a higher value on silver because it is harder and more durable than gold. In the seventh century B.C., the Lydian civilization of Asia Minor (now Turkey) created the first coins using silver, and in the sixth century B.C., the Chinese began making silver coins. Succeeding dynasties in China continued to mint these coins, round with square holes in them, until the early twentieth century.

The Romans called silver *argentum,* and therefore today its chemical symbol is Ag. Its uses are much more varied than those of gold, both because of its durability and the fact that it is less expensive. Alloyed with copper, which adds strength to it, it makes sterling silver, used in coins, silverware, and jewelry. Silver nitrate compounds are used in silver plating, applied in mirrors and tableware. (Most mirrors today, however, use aluminum.)

Like gold, silver was once widely used for coinage, and silver coins are still more common than their gold counterparts. In 1963, however, the United States withdrew its silver certifications, paper currency exchangeable for silver. Today, silver coins are generally issued in special situations, such as to commemorate an event.

A large portion of the world's silver supply is used by photographers for developing pictures. In addition, because it is an excellent conductor of heat and electricity, silver has applications in the electronics industry; however, its expense has led many manufacturers to use copper or aluminum instead. Silver is also present, along with zinc and cadmium, in cadmium batteries. Like gold, though to a much lesser extent, it is still an important jewelry-making component.

COPPER. Most people think of pennies as containing copper, but in fact the penny is the only American coin that contains no copper alloys. Because the amount of copper necessary to make a penny today costs more than $0.01, a penny is actually made of zinc with a thin copper coating. Yet copper has long been a commonly used coinage metal, and long before that, humans used it for other purposes.

Seven thousand years ago, the peoples of the Tigris-Euphrates river valleys, in what is now Iraq, were mining and using copper, and later civilizations combined copper with zinc to make bronze. Indeed, the history of prehistoric and ancient humans' technological development is often divided according to the tools they made, the latter two of which came from transition metals: the Stone Age, the Bronze Age (c. 3300-1200 B.C.), and the Iron Age.

Thus, by the time the Romans dubbed copper *cuprum* (from whence its chemical symbol, Cu, comes), copper in various forms had long been in use. Today copper is purified by a form of electrolysis, in which impurities such as iron, nickel, arsenic, and zinc are removed. Other "impurities" often present in copper ore include gold, silver, and platinum, and the separation of these from the copper pays for the large amounts of electricity used in the electrolytic process.

Like the other coinage metals, copper is an extremely efficient conductor of heat and electricity, and because it is much less expensive than the other two, pure copper is widely used for electrical wiring. Because of its ability to conduct heat, copper is also applied in materials used for making heaters, as well as for cookware. Due to the high conductivity of copper, a heated copper pan has a uniform temperature, but copper pots must be coated with tin because too much copper in food is toxic.

Copper is also like its two close relatives in that it resists corrosion, and this makes it ideal for plumbing. Its use in making coins resulted from its anti-corrosive qualities, combined with its beauty: like gold, copper has a distinctive color. This aesthetic quality led to the use of copper in decorative applications as well: many old buildings used copper roofs, and the Statue of Liberty is covered in 300 thick copper plates.

Why, then, is the famous statue not copper-colored? Because copper does eventually corrode when exposed to air for long periods of time. Over time, it develops a thin layer of black copper oxide, and as the years pass, carbon dioxide in the air leads to the formation of copper carbonate, which imparts a greenish color.

The human body is about 0.0004% copper, though as noted, larger quantities of copper can be toxic. Copper is found in foods such as shellfish, nuts, raisins, and dried beans. Whereas human blood has hemoglobin, a molecule with an iron atom at the center, the blood of lobsters and other large crustaceans contains hemocyanin, in which copper performs a similar function.

The Zinc Group

ZINC. Together with copper, zinc appeared in another alloy that, like bronze, helped define the ancient world: brass. (The latter is mentioned in the Bible, for instance in the Book of Daniel, when King Nebuchadnezzar dreams of a statue containing brass and other substances, symbolizing various empires.) Used at least from the first millennium B.C. onward, brass appeared in coins and ornaments throughout Asia Minor. Though it is said that the Chinese purified zinc in about A.D. 1000, the Swiss alchemist Paracelsus (1493-1541) is usually credited with first describing zinc as a metal.

Bluish-white, with a lustrous sheen, zinc is found primarily in the ore sulfide sphalerite. The largest natural deposits of zinc are in Australia and the United States, and after mining, the metal is subjected to a purification and reduction process involving carbon. Zinc is used in galvanized steel, developed in the eighteenth century by Italian physicist Luigi Galvani (1737-1798).

Just as a penny is not really copper but zinc, "tin" roofs are usually made of galvanized steel. Highly resistant to corrosion, galvanized steel finds application in everything from industrial equipment to garbage cans. Zinc oxide is applied in textiles, batteries, paints, and rubber products, while luminous zinc sulfide appears in television screens, movie screens, clock dials, and fluorescent light bulbs.

Zinc phosphide is used as a rodent poison. Like several other transition metals, zinc is a part of many living things, yet it can be toxic in large quantities or specific compounds. For a human being, inhaling zinc oxide causes involuntary shaking. On the other hand, humans and many animals require zinc in their diets for the digestion of proteins. Furthermore, it is believed that zinc contributes to the healing of wounds and to the storage of insulin in the pancreas.

CADMIUM. In 1817, German chemist Friedrich Strohmeyer (1776-1835) was working as an inspector of pharmacies for the German state of Hanover. While making his rounds, he discovered that one pharmacy had a sample of zinc carbonate labeled as zinc oxide, and while inspecting the chemical in his laboratory, he discovered something unusual. If indeed it were zinc carbonate, it should turn into zinc oxide when heated, and since both compounds were white, there should be no difference in color. Instead, the mysterious compound turned a yellowish-orange.

Strohmeyer continued to analyze the sample, and eventually realized that he had discovered a new element, which he named after the old Greek term for zinc carbonate, *kadmeia*. Indeed, cadmium typically appears in nature along with zinc or zinc compounds. Silvery white and lustrous or shiny, cadmium is soft enough to be cut with a knife, but chemically it behaves much like zinc: hence the idea of a "zinc group."

Today cadmium is used in batteries, and for electroplating of other metals to protect them against corrosion. Because the cost of cadmium is high due to the difficulty of separating it from zinc, cadmium electroplating is applied only in specialized situations. Cadmium also appears in the control rods of nuclear power plants, where

its ready absorption of neutrons aids in controlling the rate at which nuclear fission occurs.

Cadmium is highly toxic, and is believed to be the cause behind the outbreak of itai-itai ("ouch-ouch") disease in Japan in 1955. People ingested rice oil contaminated with cadmium, and experienced a number of painful side effects associated with cadmium poisoning: nausea, vomiting, choking, diarrhea, abdominal pain, headaches, and difficulty breathing.

MERCURY. One of only two elements— along with bromine—that appears in liquid form at room temperature, mercury is both toxic and highly useful. The Romans called it *hydragyrum* ("liquid silver"), from whence comes its chemical symbol, Hg. Today, however, it is known by the name of the Romans' god Mercury, the nimble and speedy messenger of the gods. Mercury comes primarily from a red ore called cinnabar, and since it often appears in shiny globules that form outcroppings from the cinnabar, it was relatively easy to discover.

Several things are distinctive about mercury, including its bright silvery color. But nothing distinguishes it as much as its physical properties— not only its liquidity, but the fact that it rolls rapidly, like the fleet-footed god after which it is named. Its surface tension (the quality that causes it to bead) is six times greater than that of water, and for this reason, mercury never wets the surfaces with which it comes in contact.

Mercury, of course, is widely used in thermometers, an application for which it is extremely well-suited. In particular, it expands at a uniform rate when heated, and thus a mercury thermometer (unlike earlier instruments, which used water, wine, or alcohol) can be easily calibrated. (Note that due to the toxicity of the element, mercury thermometers in schools are being replaced by other types of thermometers.) At temperatures close to absolute zero, mercury loses its resistance to the flow of electric current, and therefore it presents a promising area of research with regard to superconductivity.

Even with elements present in the human body, such as zinc, there is a danger of toxicity with exposure to large quantities. Mercury, on the other hand, does not occur naturally in the human body, and is therefore extremely dangerous. Because it is difficult for the body to eliminate, even small quantities can accumulate and exert their poisonous effects, resulting in disorders of the nervous system. In fact, the phrase "mad as a hatter" refers to the symptoms of mercury poisoning suffered by hatmakers of the nineteenth century, who used a mercury compound in making hats of beaver and fur.

MAGNETIC METALS

IRON. In its purest form, iron is relatively soft and slightly magnetic, but when hardened, it becomes much more so. As with several of the elements discovered long ago, iron has a chemical symbol (Fe) reflecting an ancient name, the Latin *ferrum*. But long before the Romans' ancestors arrived in Italy, the Hittites of Asia Minor were purifying iron ore by heating it with charcoal over a hot flame.

The Hittites jealously guarded their secret, which gave them a technological advantage over enemies such as the Egyptians. But when Hittite civilization crumbled in the wake of an invasion by the mysterious "Sea Peoples" in about 1200 B.C., knowledge of ore smelting spread throughout the ancient world. Thus began the Iron Age, in which ancient technology matured greatly. (It should be noted that the Bantu peoples, in what is now Nigeria and neighboring countries, developed their own ironmaking techniques a few centuries later, apparently without benefit of contact with any civilization exposed to Hittite techniques.)

Ever since ancient times, iron has been a vital part of human existence, and with the growth of industrialization during the eighteenth century, demand for iron only increased. But heating iron ore with charcoal required the cutting of trees, and by the latter part of that century, England's forests had been so badly denuded that British ironmakers sought a new means of smelting the ore. It was then that British inventor Abraham Darby (1678?-1717) developed a method for making coke from soft coal, which was abundant throughout England. Adoption of Darby's technique sped up the rate of industrialization in England, and thus advanced the entire Industrial Revolution soon to sweep the Western world.

Symbolic of industrialism was iron in its many forms: pig iron, ore smelted in a coke-burning blast furnace; cast iron, a variety of mixtures containing carbon and/or silicon; wrought iron, which contains small amounts of various other elements, such as nickel, cobalt, copper,

chromium, and molybdenum; and most of all steel. Steel is a mixture of iron with manganese and chromium, purified with a blast of hot air in the Bessemer converter, named after English inventor Henry Bessemer (1813-1898). Modern techniques of steelmaking improved on Bessemer's design by using pure oxygen rather than merely hot air.

The ways in which iron is used are almost too obvious (and too numerous) to mention. If iron and steel suddenly ceased to exist, there could be no skyscrapers, no wide-span bridges, no ocean liners or trains or heavy machinery or automobile frames. Furthermore, alloys of steel with other transition metals, such as tungsten and niobium, possess exceptionally great strength, and find application in everything from hand tools to nuclear reactors. Then, of course, there are magnets and electromagnets, which can only be made of iron and/or one of the other magnetic elements, cobalt and nickel.

In the human body, iron is a key part of hemoglobin, the molecule in blood that transports oxygen from the lungs to the cells. If a person fails to get sufficient quantities of iron—present in foods such as red meat and spinach—the result is anemia, characterized by a loss of skin color, weakness, fainting, and heart palpitations. Plants, too, need iron, and without the appropriate amounts are likely to lose their color, weaken, and die.

COBALT. Isolated in about 1735 by Swedish chemist Georg Brandt (1694-1768), cobalt was the first metal discovered since prehistoric, or at least ancient, times. The name comes from *Kobald,* German for "underground gnome," and this reflects much about the early history of cobalt. In legend, the Kobalden were mischievous sprites who caused trouble for miners, and in real life, ores containing the element that came to be known as cobalt likewise caused trouble to men working in mines. Not only did these ores contain arsenic, which made miners ill, but because cobalt had no apparent value, it only interfered with their work of extracting other minerals.

Yet cobalt had been in use by artisans long before Brandt's isolated the element. The color of certain cobalt compounds is a brilliant, shocking blue, and this made it popular for the coloring of pottery, glass, and tile. The element, which makes up less than 0.002% of Earth's crust, is found today primarily in ores extracted from mines in

Canada, Zaire, and Morocco. One of the most important uses of cobalt is in a highly magnetic alloy known as alnico, which also contains iron, nickel, and aluminum. Combined with tungsten and chromium, cobalt makes stellite, a very hard alloy used in drill bits. Cobalt is also applied in jet engines and turbines.

NICKEL. Moderately magnetic in its pure form, nickel had an early history much like that of cobalt. English workers mining copper were often dismayed to find a metal that looked like copper, but was not, and they called it "Old Nick's copper"—meaning that it was a trick played on them by Old Nick, or the devil. The Germans gave it a similar name: *Kupfernickel,* or "imp copper."

Though nickel was not identified as a separate metal by Swedish mineralogist Axel Fredrik Cronstedt (1722-1765) until the eighteenth century, alloys of copper, silver, and nickel had been used as coins even in ancient Egypt. Today, nickel is applied, not surprisingly, in the American five-cent piece—that is, the "nickel"—made from an alloy of nickel and copper. Its anti-corrosive nature also provides a number of other applications for nickel: alloyed with steel, for instance, it makes a protective layer for other metals.

THE PLATINUM GROUP

PLATINUM. First identified by an Italian physician visiting the New World in the mid-sixteenth century, platinum—now recognized as a precious metal—was once considered a nuisance in the same way that nickel and cadmium were. Miners, annoyed with the fact that it got in the way when they were looking for gold, called it *platina,* or "little silver." One of the reasons why platinum did not immediately catch the world's fancy is because it is difficult to extract, and typically appears with the other metals of the "platinum group": iridium, osmium, palladium, rhodium, and ruthenium.

Only in 1803 did English physician and chemist William Hyde Wollaston (1766-1828) develop a means of extracting platinum, and when he did, he discovered that the metal could be hammered into all kinds of shapes. Platinum proved such a success that it made Wollaston financially independent, and he retired from his medical practice at age 34 to pursue scientific research. Today, platinum is used in everything from thermometers to parts for rocket engines,

both of which take advantage of its ability to withstand high temperatures.

IRIDIUM AND OSMIUM. In the same year that Wollaston isolated platinum, three French chemists applied a method of extracting it with a mixture of nitric and hydrochloric acids. The process left a black powder that they were convinced was a new element, and in 1804 English chemist Smithson Tennant (1761-1815) gave it the name iridium. Iris was the Greek goddess of the rainbow, and the name reflected the element's brilliant, multi-colored sheen. As with platinum, iridium is used today in a number of ways, particularly for equipment such as laser crystals that must withstand high temperatures. (In addition, one particular platinum-iridium bar provides the standard measure of a kilogram.)

Tennant discovered a second element in 1804, also from the residue left over from the acid process for extracting platinum. This one had a distinctive smell when heated, so he named it osmium after *osme,* Greek for "odor." In 1898, Austrian chemist Karl Auer, Baron von Welsbach (1858-1929), developed a light bulb using osmium as a filament, the material that is heated. Though osmium proved too expensive for commercial use, Auer's creation paved the way for the use of another transition metal, tungsten, in making long-lasting filaments. Osmium, which is very hard and resistant to wear, is also used in electrical devices, fountain-pen tips, and phonograph needles.

PALLADIUM, RHODIUM, AND RUTHENIUM. By the time Tennant isolated osmium, Wollaston had found yet another element that emerged from the refining of platinum with nitric and hydrochloric acids. This was palladium, named after a recently discovered asteroid, Pallas. Soft and malleable, palladium resists tarnishing, which makes it valuable to the jewelry industry. Combined with yellow gold, it makes the alloy known as white gold. Because palladium has the unusual ability of absorbing up to 900 times its volume in hydrogen, it provides a useful means of extracting that element.

Also in 1805, Wollaston discovered rhodium, which he named because it had a slightly red color. The density of rhodium is lower, and the melting point higher, than for most of the platinum group elements, and it is often used as an alloy to harden platinum. As with many elements in this group, it has a number of high-temperature applications. When electroplated, it is highly reflective, and this makes it useful as a material for optical instruments.

The element that came to be known as ruthenium had been detected several times before 1844, when Russian chemist Carl Ernest Claus (1796-1864) finally identified it as a distinct element. Thus it was the only platinum group element in whose isolation neither Tennant nor Wollaston played a leading role. Since it had been found in Russia, the element was called ruthenium, in honor of that country's ancient name, Ruthenia.

ALLOY METALS

The elements described here as "alloy metals" will be treated briefly, but the amount of space given to them here does not reflect their importance to civilization. Several of these are relatively abundant, and many—titanium, tungsten, manganese, chromium and others—are vital parts of daily life. However, due to space considerations, the treatment of these metals that follows is a regrettably abbreviated one.

As suggested by the informal name given to them here, they are often applied in alloys, typically with other transition metals such as iron, or with other "alloy metals." They are also used in a variety of compounds for a wide variety of applications.

TITANIUM, ZIRCONIUM, AND HAFNIUM

English clergyman and amateur scientist William Gregor (1761-1817) in 1791 noticed what he called "a new metallic substance" in a mineral called menchanite, which he found in the Menachan region of Cornwall, England. Four years later, German chemist Martin Heinrich Klaproth (1743-1817) studied menchanite and found that it contained a new element, which he named titanium after the Titans of Greek mythology. Only in 1910, however, was the element isolated. Due to its high strength-to-weight ratio, titanium is applied today in alloys used for making airplanes and spacecraft. It is also used in white paints and in a variety of other combinations, mostly in alloys containing other transition metals, as well as in compounds with nonmetals such as oxygen.

Six years before he identified titanium, in 1789, Klaproth had found another element, which he named zirconium after the mineral zircon in which it was found. Zircon had been in use for many centuries, and another zirconium-containing mineral, jacinth, is mentioned in the Bible. As with titanium, there was a long period from identification to isolation, which did not occur until 1914. Zirconium alloys are used today in a number of applications, including superconducting magnets and as a construction metal in nuclear power plants. Compounds including zirconium also have a variety of uses, ranging from furnace linings to poison ivy lotion.

Because it is almost always found with zirconium, hafnium—named after the ancient name of Copenhagen, Denmark—proved extremely difficult to isolate. Only in 1923 was it isolated, using x-ray analysis, and today it is applied primarily in light bulb filaments and electrodes. Sometimes it is accidentally "applied," when zirconium is actually the desired metal, because the two are so difficult to separate.

VANADIUM, TANTALUM, AND NIOBIUM. Vanadium, named after the Norse goddess Vanadis or Freyja, was first identified in 1801, but only isolated in 1867. It is often applied in alloys with steel, and sometimes used for bonding steel and titanium.

Tantalum and niobium are likewise named after figures from mythology—in this case, Greek. These two elements appear together so often that their identification as separate substances was a long process, involving numerous scientists and heated debates. For this reason, alloys involving tantalum—valued for its high melting point—usually include niobium as well.

CHROMIUM, TUNGSTEN, AND MANGANESE, AND OTHER METALS. Because it displays a wide variety of colors, chromium was named after the Greek word for "color." It is used in chrome and tinted glass, and as a strengthening agent for stainless steel. Tungsten likewise strengthens steel, in large part because it has the highest melting point of any metal: 6,170°F (3,410°C). Its name comes from a Swedish word meaning "heavy stone," and its chemical symbol (W) refers to the ore named wolframite, in which it was first discovered. Its high resistance to heat gives tungsten alloys applications in areas ranging from light bulb filaments to furnaces to missiles.

KEY TERMS

ACTINIDES: Those transition metals that fill the 5*f* orbital. Because actinium—which does not fill the 5*f* orbital—exhibits characteristics similar to those of the actinides, it is usually considered part of the actinides family.

ALLOY: A mixture containing more than one metal.

ELECTROLYSIS: The use of an electric current to cause a chemical reaction.

ELECTRON CLOUD: A term used to describe the pattern formed by orbitals.

GROUPS: Columns on the periodic table of elements. These are ordered according to the numbers of valence electrons in the outer shells of the atoms for the elements represented.

INNER TRANSITION METALS: The lanthanides and actinides, both of which fill the *f* orbitals. For this reason, they are usually treated separately.

ION: An atom or group of atoms that has lost or gained one or more electrons, and thus has a net electric charge.

ISOTOPES: Atoms that have an equal number of protons, and hence are of the same element, but differ in their number of neutrons. This results in a difference of mass. Isotopes may be either stable or unstable. The latter type, known as radioisotopes, are radioactive.

IUPAC SYSTEM: A version of the periodic table of elements, authorized by the International Union of Pure and Applied Chemistry (IUPAC), which numbers all groups on the table from 1-18. Thus in the IUPAC system, in use primarily outside of North America, both the representative or main-group elements and the transition metals are numbered.

LANTHANIDES: The transition metals that fill the 4*f* orbital. Because lanthanum—which does not fill the 4*f* orbital—exhibits characteristics similar to those of the lanthanides, it is usually considered part of the lanthanide family.

NORTH AMERICAN SYSTEM: A version of the periodic table of elements that only numbers groups of elements in which the number of valence electrons equals the group number. Hence the transition metals are usually not numbered in this system. Some North American charts, however, do provide group numbers for transition metals by including an "A" after the group numbers for representative or main-group elements, and "B" after those of transition metals.

ORBITAL: A pattern of probabilities indicating the regions that may be occupied by an electron. The higher the principal energy level, the more complex the pattern of orbitals.

PERIODIC TABLE OF ELEMENTS: A chart that shows the elements arranged in order of atomic number, along with chemical symbol and the average atomic mass (in atomic mass units) for that particular element. Elements are arranged in groups or columns, as well as in rows or periods, according to specific aspects of their valence electron configurations.

PERIODS: Rows on the periodic table of elements. These represent successive principal energy levels for the valence electrons in the atoms of the elements involved. Elements in the transition metal family occupy periods 4, 5, 6, or 7.

KEY TERMS CONTINUED

PRINCIPAL ENERGY LEVEL: A value indicating the distance that an electron may move away from the nucleus of an atom. This is designated by a whole-number integer, beginning with 1 and moving upward. The higher the principal energy level, the greater the energy in the atom, and the more complex the pattern of orbitals. Elements in the transition metal family have principal energy levels of 4, 5, 6, or 7.

RADIOACTIVITY: A term describing a phenomenon whereby certain isotopes known as radioisotopes are subject to a form of decay brought about by the emission of high-energy particles. "Decay" does not mean that the isotope "rots"; rather, it decays to form another isotope until eventually (though this may take a long time), it becomes stable.

REACTIVITY: The tendency for bonds between atoms or molecules to be made or broken in such a way that materials are transformed.

REPRESENTATIVE OR MAIN-GROUP ELEMENTS: The 44 elements in Groups 1 through 8 on the periodic table of elements in the North American system, for which the number of valence electrons equals the group number. (The only exception is helium.) In the IUPAC system, these

elements are assigned group numbers 1, 2, and 13 through 18. (In these last six columns on the IUPAC chart, there is no relation between the number of valence electrons and group number.)

SHELL: The orbital pattern of the valence electrons at the outside of an atom.

SUBLEVEL: A region within the principal energy level occupied by electrons in an atom. Whatever the number n of the principal energy level, there are n sublevels. At each principal energy level, the first sublevel to be filled is the one corresponding to the s orbital pattern, followed by the p and then the d pattern.

TRANSITION METALS: A group of 40 elements, which are not assigned a group number in the version of the periodic table of elements known as the North American system. These are the only elements that fill the d orbital. In addition, the transition metals—unlike the representative or main-group elements—have their valence electrons on two different principal energy levels. Though the lanthanides and actinides are considered inner transition metals, they are usually treated separately.

VALENCE ELECTRONS: Electrons that occupy the highest principal energy level in an atom. These are the electrons involved in chemical bonding.

Likewise manganese, first identified in the mid-eighteenth century, is extraordinarily strong. Its major use is in steelmaking. In addition, compounds such as manganese oxide are used in fertilizers. Other alloy metals include rhenium (identified only in 1925) and molybdenum. These two are often applied in alloys together, and with tungsten, for a variety of industrial purposes.

RARE AND ARTIFICIAL ELEMENTS

RARE EARTH-LIKE ELEMENTS. Though they are not part of the lanthanide series, scandium and yttrium share many properties with those elements, once known as the "rare earths." (In this context, "rarity" refers not so much to scarcity as to difficulty of extraction.) Furthermore, like most of the lanthanides, these two elements were first discovered in Scandinavia.

Their origin is reflected in the name given by Swedish chemist Lars Nilson (1840-1899) to scandium, which he discovered in 1876. As for yttrium, it came from the complex mineral known as ytterite, from whence emerged many of the lanthanides. Yttrium is used in alloys to impart strength, and has a wide if specialized array of applications. Scandium has no important uses.

ARTIFICIAL ELEMENTS. Technetium, with an atomic number of 43, is unusual in that it is one of the few elements with an atomic number less than 92 that does not occur in nature. This is reflected in its name, from the Greek *technetos,* meaning "artificial." The strange thing is that technetium, produced in a laboratory in 1936, does occur naturally in the universe—but it seems to be found only in the spectral lines from older stars, and not in those of younger stars such as our Sun.

The other elements in the transition metals family are referred to as "transuranium elements," meaning that they have atomic numbers higher than that of uranium (92). These have all been created artificially; all are highly radioactive; few survive for more than a few minutes (at most); and few have applications in ordinary life.

WHERE TO LEARN MORE

The Copper Page (Web site). <http://www.copper.org> (May 26, 2001).

Gold Institute (Web site). <http://www.goldinstitute.org> (May 26, 2001).

Kerrod, Robin. *Matter and Materials.* Illustrated by Terry Hadler. Tarrytown, N.Y.: Benchmark Books, 1996.

Mebane, Robert C. and Thomas R. Rybolt. *Metals.* Illustrated by Anni Matsick. New York: Twenty-First Century Books, 1995.

Oxlade, Chris. *Metal.* Chicago, IL: Heinemann Library, 2001.

"*The Pictorial Periodic Table*" (Web site). <http://chemlab.pc.maricopa.edu/periodic/periodic.html> (May 22, 2001).

Stwertka, Albert. *A Guide to the Elements.* New York: Oxford University Press, 1998.

"*Transition Metals*" *ChemicalElements.com* (Web site). <http://www.chemicalelements.com/groups/transition.html> (May 26, 2001).

"The Transition Metals." *University of Colorado Department of Physics* (Web site). <http://www.colorado.edu/physics/2000/periodic_table/transition_elements.html> (May 26, 2001).

WebElements (Web site). <http://www.webelements.com> (May 22, 2001).

Zincworld (Web site). <http://www.zincworld.org> (May 22, 2001).

ACTINIDES

CONCEPT

The actinides (sometimes called actinoids) occupy the "bottom line" of the periodic table—a row of elements normally separated from the others, placed at the foot of the chart along with the lanthanides. Both of these families exhibit unusual atomic characteristics, properties that set them apart from the normal sequence on the periodic table. But there is more that distinguishes the actinides, a group of 14 elements along with the transition metal actinium. Only four of them occur in nature, while the other 10 have been produced in laboratories. These 10 are classified, along with the nine elements to the right of actinium on Period 7 of the periodic table, as transuranium (beyond uranium) elements. Few of these elements have important applications in daily life; on the other hand, some of the lower-number transuranium elements do have specialized uses. Likewise several of the naturally occurring actinides are used in areas ranging from medical imaging to powering spacecraft. Then there is uranium, "star" of the actinide series: for centuries it seemed virtually useless; then, in a matter of years, it became the most talked-about element on Earth.

HOW IT WORKS

THE TRANSITION METALS

Why are actinides and lanthanides set apart from the periodic table? This can best be explained by reference to the transition metals and their characteristics. Actinides and lanthanides are referred to as inner transition metals, because, although they belong to this larger family, they are usually considered separately—rather like grown children who have married and started families of their own.

The qualities that distinguish the transition metals from the representative or main-group elements on the periodic table are explained in depth within the Transition Metals essay. The reader is encouraged to consult that essay, as well as the one on Families of Elements, which further places the transition metals within the larger context of the periodic table. Here these specifics will be discussed only briefly.

ORBITAL PATTERNS OF THE TRANSITION METALS. The transition metals are distinguished by their configuration of valence electrons, or the outer-shell electrons involved in chemical bonding. Together with the core electrons, which are at lower energy levels, valence electrons move in areas of probability referred to as orbitals. The pattern of orbitals is determined by the principal energy level of the atom, which indicates a distance that an electron may move away from the nucleus.

Each principal energy level is divided into sublevels corresponding to the number n of the principal energy level. The actinides, which would be on Period 7 if they were included on the periodic table with the other transition metals, have seven principal energy levels. (Note that period number and principal energy level number are the same.) In the seventh principal energy level, there are seven possible sublevels.

The higher the energy level, the larger the number of possible orbital patterns, and the more complex the patterns. Orbital patterns loosely define the overall shape of the electron cloud, but this does not necessarily define the paths along which the electrons move. Rather, it means that if you could take millions of photographs of the electron during a period of a few seconds, the resulting blur of images would describe more or less the shape of a specified orbital.

The four basic types of orbital patterns are discussed in the Transition Metals essay, and will not be presented in any detail here. It is important only to know that, unlike the representative elements, transition metals fill the sublevel corresponding to the *d* orbitals. In addition, they are the only elements that have valence electrons on two different principal energy levels.

LANTHANIDES AND ACTINIDES. The lanthanides and actinides are further set apart even from the transition metals, due to the fact that these elements also fill the highly complex *f* orbitals. Thus these two families are listed by themselves. In most versions of the periodic table, lanthanum (57) is followed by hafnium (72) in the transition metals section of the chart; similarly, actinium (89) is followed by rutherfordium (104). The "missing" metals—lanthanides and actinides, respectively—are shown at the bottom of the chart.

The lanthanides can be defined as those metals that fill the 4*f* orbital. However, because lanthanum (which does not fill the 4*f* orbital) exhibits similar properties, it is usually included with the lanthanides. Likewise the actinides can be defined as those metals that fill the 5*f* orbital; but again, because actinium exhibits similar properties, it is usually included with the actinides.

ISOTOPES

One of the distinguishing factors in the actinide family is its great number of radioactive isotopes. Two atoms may have the same number of protons, and thus be of the same element, yet differ in their number of neutrons—neutrally charged patterns alongside the protons at the nucleus. Such atoms are called isotopes, atoms of the same element having different masses.

Isotopes are represented symbolically in one of several ways. For instance, there is this format: $\frac{m}{a}S$, where S is the chemical symbol of the element, a is the atomic number (the number of protons in its nucleus), and m the mass number—the sum of protons and neutrons. For the

isotope known as uranium-238, for instance, this is shown as $\frac{238}{92}U$.

Because the atomic number of any element is established, however, isotopes are usually represented simply with the mass number thus: ^{238}U. They may also be designated with a subscript notation indicating the number of neutrons, so that this information can be obtained at a glance without it being necessary to do the arithmetic. For the uranium isotope shown here, this is written as $\frac{238}{92}U_{146}$.

RADIOACTIVITY

The term radioactivity describes a phenomenon whereby certain materials are subject to a form of decay brought about by the emission of high-energy particles, or radiation.

Types of particles emitted in radiation include:

- Alpha particles, or helium nuclei;
- Beta particles—either an electron or a subatomic particle called a positron;
- Gamma rays or other very high-energy electromagnetic waves.

Isotopes are either stable or unstable, with the unstable variety, known as radioisotopes, being subject to radioactive decay. In this context, "decay" does not mean "rot"; rather, a radioisotope decays by turning into another isotope. By continuing to emit particles, the isotope of one element may even turn into the isotope of another element.

Eventually the radioisotope becomes a stable isotope, one that is not subject to radioactive decay. This is a process that may take seconds, minutes, hours, days, years—and sometimes millions or even billions of years. The rate of decay is gauged by the half-life of a radioisotope sample: in other words, the amount of time it takes for half the nuclei (plural of nucleus) in the sample to become stable.

Actinides decay by a process that begins with what is known as K-capture, in which an electron of a radioactive atom is captured by the nucleus and taken into it. This is followed by the splitting, or fission, of the atom's nucleus. This fission produces enormous amounts of energy, as well as the release of two or more neutrons, which may in turn bring about further K-capture. This is called a chain reaction.

GLENN T. SEABORG HOLDS A SAMPLE CONTAINING ARTIFICIAL ELEMENTS BETWEEN 94 AND 102 ON THE PERIOD-
IC TABLE. *(Roger Ressmeyer/Corbis. Reproduced by permission.)*

REAL-LIFE APPLICATIONS

THE FIRST THREE NATURALLY OCCURRING ACTINIDES

In the discussion of the actinides that follows, atomic number and chemical symbol will follow the first mention of an element. Atomic mass figures are available on any periodic table, and these will not be mentioned in most cases. The atomic mass figures for actinide elements are very high, as fits their high atomic number, but for most of these, figures are usually for the most stable isotope, which may exist for only a matter of seconds.

Though it gives its name to the group as a whole, actinium (Ac, 89) is not a particularly significant element. Discovered in 1902 by German chemist Friedrich Otto Giesel (1852-1927), it is found in uranium ores. Actinium is 150 times more radioactive than radium, a highly radioactive alkaline earth metal isolated around the same time by French-Polish physicist and chemist Marie Curie (1867-1934) and her husband Pierre (1859-1906).

THORIUM. More significant than actinium is thorium (Th, 90), first detected in 1815 by the renowned Swedish chemist Jons Berzelius (1779-1848). Berzelius promptly named the element after the Norse god Thor, but eventually concluded that what he had believed to be a new element was actually the compound yttrium phosphate. In 1829, however, he examined another mineral and indeed found the element he believed he had discovered 14 years earlier.

It 1898, Marie Curie and an English chemist named Gerhard Schmidt, working independently, announced that thorium was radioactive. Today it is believed that the enormous amounts of energy released by the radioactive decay of subterranean thorium and uranium plays a significant part in Earth's high internal temperature. The energy stored in the planet's thorium reserves may well be greater than all the energy available from conventional fossil and nuclear fuels combined.

Thorium appears on Earth in an abundance of 15 parts per million (ppm), many times greater than the abundance of uranium. With its high energy levels, thorium has enormous potential as a nuclear fuel. When struck by neutrons, thorium-232 converts to uranium-233, one of the few known fissionable isotopes—that is, isotopes that can be split to start nuclear reactions.

It is perhaps ironic that this element, with its potential for use in some of the most high-tech applications imaginable, is widely applied in a very low-tech fashion. In portable gas lanterns for camping and other situations without electric power, the mantle often contains oxides of thorium and cerium, which, when heated, emit a brilliant white light. Thorium is also used in the manufacture of high-quality glass, and as a catalyst in various industrial processes.

PROTACTINIUM. Russian chemist Dmitri Ivanovitch Mendeleev (1834-1907), father of the periodic table, used the table's arrangement of elements as a means of predicting the discovery of new substances: wherever he found a "hole" in the table, Mendeleev could say with assurance that a new element would eventually be found to fill it. In 1871, Mendeleev predicted the discovery of "eka-tantalum," an element that filled the space below the transition metal tantalum. (At this point in history, just two years after Mendeleev created the periodic table, the lanthanides and actinides had not been separated from the rest of the elements on the chart.)

Forty years after Mendeleev foretold its existence, two German chemists found what they thought might be Element 91. It had a half-life of only 1.175 minutes, and, for this reason, they named it "brevium." Then in 1918, Austrian physicist Lise Meitner (1878-1968)—who, along with Curie and French physicist Marguerite Perey (1909-1975) was one of several women involved in the discovery of radioactive elements—was working with German chemist Otto Hahn (1879-1968) when the two discovered another isotope of Element 91. This one had a much, much longer half-life: $3.25 \cdot 10^4$ years, or about five times as long as the entire span of human civilization.

Originally named protoactinium, the name of Element 91 was changed to protactinium (Pa), whose longest-lived isotope has an atomic mass of 231. Shiny and malleable, protactinium has a melting point of 2,861.6°F (1,572°C). It is highly toxic, and so rare that no commercial uses have been found for it. Indeed, protactinium could only be produced from the decay products of uranium and radium, and thus it is one of the few elements with an atomic number less than 93 that cannot be said to occur in nature.

URANIUM'S EARLY HISTORY. Chemistry books, in fact, differ as to the number of naturally occurring elements. Some say 88, which is the most correct figure, because protactinium, along with technetium (43) and two others, cannot be said to appear naturally on Earth. Other books say 92, a less accurate figure that nonetheless reflects an indisputable fact: above uranium (U, 92) on the periodic table, there are no elements that generally occur in nature. (However, a few do occur as radioactive by-products of uranium.)

But uranium is much more than just the last truly natural element, though for about a century it apparently had no greater importance. When German chemist Martin Heinrich Klaproth (1743-1817) discovered it in 1789, he named it after another recent discovery: the planet Uranus. During the next 107 years, uranium had a very quiet existence, befitting a rather dull-looking material.

Though it is silvery white when freshly cut, uranium soon develops a thin coating of black uranium oxide, which turns it a flat gray. Yet glassmakers did at least manage to find a use for it—as a coating for decorative glass, to which it imparted a hazy, fluorescent yellowish green hue. Little did they know that they were using one of the most potentially dangerous substances on Earth.

THE DESTRUCTIVE POWER OF URANIUM. In 1895, German physicist Wilhelm Röntgen (1845-1923) noticed that photographic plates held near a Crookes tube—a device for analyzing electromagnetic radiation—became fogged. He dubbed the rays that had caused this x rays. A year later, in 1896, French physicist Henri Becquerel (1852-1908) left some photographic plates in a drawer with a sample of uranium, and discovered that the uranium likewise caused a fogging of the photographic plates. This meant that uranium was radioactive.

With the development of nuclear fission by Hahn and German chemist Fritz Strassman in 1938, uranium suddenly became all-important because of its ability to undergo nuclear fission, accompanied by the release of huge amounts of energy. During World War II, in what was known as the Manhattan Project, a team of scientists in Los Alamos, New Mexico, developed the atomic

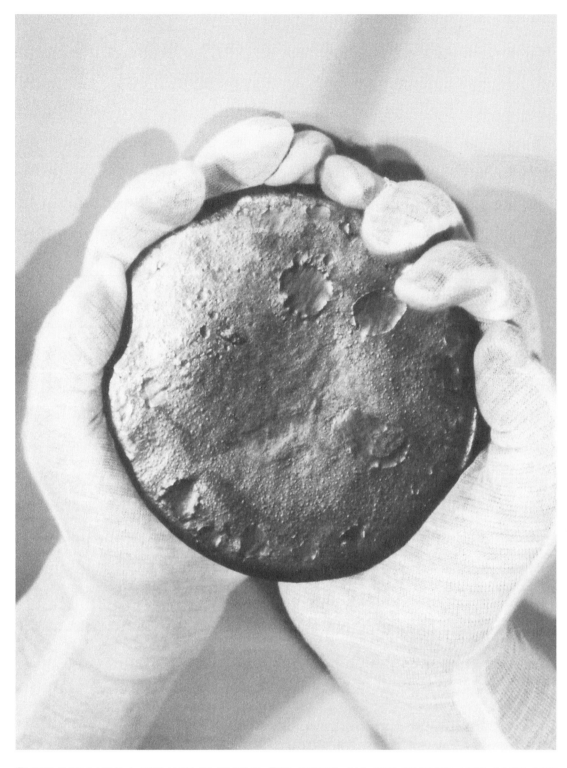

GLOVED HANDS HOLD A GRAY LUMP OF URANIUM. THIS MATERIAL HAS BEEN REMOLDED AFTER HAVING BEEN REMOVED FROM A TITAN II MISSILE, PART OF THE DISARMAMENT AFTER THE END OF THE COLD WAR. *(Martin Marietta; Roger Ressmeyer/Corbis. Reproduced by permission.)*

bomb. The first of the two atomic bombs dropped on Japan in 1945 contained uranium, while the second contained the transuranium element plutonium.

OTHER USES FOR URANIUM. Though nuclear weapons have fortunately not been used against human beings since 1945, uranium has remained an important component of nuclear energy—both in the development of bombs and in the peaceful application of nuclear power. It has other uses as well, due to the fact that it is extremely dense.

Indeed, uranium has a density close to that of gold and platinum, but is much cheaper, because it is more abundant on Earth. In addition, various isotopes of uranium are a by-product of nuclear power, which separates these isotopes from the highly fissionable ^{235}U isotope. Thus, quantities of uranium are available for use in situations where a great deal of mass is required in a small space: in counterweights for aircraft control systems, for instance, or as ballast for missile reentry vehicles.

Because ^{238}U has a very long half-life—4.47 · 10^9 years, or approximately the age of Earth—it is used to estimate the age of rocks and other geological features. Uranium-238 is the "parent" of a series of "daughter" isotopes that geologists find in uranium ores. Uranium-235 also produces "daughter" isotopes, including isotopes of radium, radon, and other radioactive series. Eventually, uranium isotopes turn into lead, but this can take a very long time: even ^{235}U, which lasts for a much shorter period than ^{238}U, has a half-life of about 700 million years.

The radiation associated with various isotopes of uranium, as well as other radioactive materials, is extremely harmful. It can cause all manner of diseases and birth defects, and is potentially fatal. The tiny amounts of radiation produced by uranium in old pieces of decorative glass is probably not enough to cause any real harm, but the radioactive fallout from Hiroshima resulted in birth defects among the Japanese population during the late 1940s.

TRANSURANIUM ELEMENTS

Transuranium elements are those elements with atomic numbers higher than that of uranium. None of these occur in nature, except as isotopes that develop in trace amounts in uranium ore. The first such element was neptunium (Np, 93), created in 1940 by American physicist Edwin Mattison McMillan (1907-1991) and American physical chemist Philip Hauge Abelson.

The development of the cyclotron by American physicist Ernest Lawrence (1901-1958) at the University of California at Berkeley in the 1930s made possible the artificial creation of new elements. A cyclotron speeds up protons or ions (charged atoms) and shoots them at atoms of uranium or other elements with the aim of adding positive charges to the nucleus. In the first two decades after the use of the cyclotron to create neptunium, scientists were able to develop eight more elements, all the way up to mendelevium (Md, 101), named in honor of the man who created the periodic table.

Most of these efforts occurred at Berkeley under the leadership of American nuclear chemist Glenn T. Seaborg (1912-1999) and American physicist Albert Ghiorso. The pace of development in transuranium elements slowed after about 1955, however, primarily due to the need for ever more powerful cyclotrons and ion-accelerating machines. In addition to Berkeley, there are two other centers for studying these high-energy elements: the Joint Institute for Nuclear Research in Dubna, Russia, and the Gesellschaft für Schwerionenforschung (GSI) in Darmstadt, Germany.

THE TRANSURANIUM ACTINIDES

PLUTONIUM. Just as neptunium had been named for the next planet beyond Uranus, a second transuranium element, discovered by Seaborg and two colleagues in 1940, was named after Pluto. Among the isotopes of plutonium (Pu, 94) is plutonium-239, one of the few fissionable isotopes other than uranium-233 and uranium-235. For that reason, it was applied in the second bomb dropped on Japan.

In addition to its application in nuclear weapons, plutonium is used in nuclear power reactors, and in thermoelectric generators, which convert the heat energy it releases into electricity. Plutonium is also used as a power source in artificial heart pacemakers. Huge amounts of the element are produced each year as a by-product of nuclear power reactors.

BERKELEY DISCOVERIES OF THE 1950S. Because the lanthanide element above it on the periodic table was named europium after Europe, americium (Am, 95) was named after America. Discovered by Seaborg, Ghiorso, and two others in 1944, it was first produced in a nuclear reaction involving plutonium-239. Americium radiation is used in measuring the thickness of glass during production; in addition, the isotope americium-241 is used as an ionization source in smoke detectors, and

KEY TERMS

ACTINIDES: Those elements that fill the 5f orbital. Because actinium—which does not fill the 5f orbital—exhibits characteristics similar to those of the actinides, it is usually considered part of the actinides family. Of the other 14 actinides, usually shown at the bottom of the periodic table, only the first three occur in nature.

ATOMIC NUMBER: The number of protons in the nucleus of an atom. Since this number is different for each element, elements are listed on the periodic table of elements in order of atomic number.

ELECTRON CLOUD: A term used to describe the pattern formed by orbitals.

HALF-LIFE: The length of time it takes a substance to diminish to one-half its initial amount. For a sample of radioisotopes, the half-life is the amount of time it takes for half of the nuclei to become stable isotopes. Half-life can be a few seconds; on the other hand, for uranium-238, it is a matter of several billion years.

INNER TRANSITION METALS: The lanthanides and actinides, which are unique in that they fill the f orbitals. For this reason, they are usually treated separately.

ISOTOPES: Atoms that have an equal number of protons, and hence are of the same element, but differ in their number of neutrons. This results in a difference of mass. Isotopes may be either stable or unstable. The unstable type, known as radioisotopes, are radioactive.

LANTHANIDES: The transition metals that fill the 4f orbital.

MASS NUMBER: The sum of protons and neutrons in the atom's nucleus. The designation ^{238}U, for uranium-238, means that this particular isotope of uranium has a mass number of 238. Since uranium has an atomic number of 92, this means that uranium-238 has 146 neutrons in its nucleus.

NEUTRON: A subatomic particle that has no electric charge. Neutrons are found at the nucleus of an atom, alongside protons.

NUCLEUS: The center of an atom, a region where protons and neutrons are located, and around which electrons spin. The plural of "nucleus" is nuclei.

ORBITAL: A region of probabilities regarding the position of an electron for an atom in a particular energy state. The higher the principal energy level, the more complex the pattern of orbitals.

in portable devices for taking gamma-ray photographs.

Above curium (Cm, 96) on the periodic table is the lanthanide gadolinium, named after Finnish chemist Johan Gadolin (1760-1852). Therefore, the discoverers of Element 96 also decided to name it after a person, Marie Curie. As with some of the other relatively low-number transuranium elements, this one is not entirely artificial: its most stable isotope, some geologists believe, may have been present in rocks many millions of years ago, but these isotopes have long since decayed. Because curium generates great amounts of energy as it decays, it is used for providing compact sources of power in remote locations on Earth and in space vehicles.

KEY TERMS CONTINUED

PRINCIPAL ENERGY LEVEL: A value indicating the distance that an electron may move away from the nucleus of an atom. This is designated by a whole-number integer, beginning with 1 and moving upward. The higher the principal energy level, the greater the energy in the atom, and the more complex the pattern of orbitals. Elements in the transition metal family have principal energy levels of 4, 5, 6, or 7.

RADIATION: In a general sense, radiation can refer to anything that travels in a stream, whether that stream be composed of subatomic particles or electromagnetic waves. In a more specific sense, the term relates to the radiation from radioactive materials, which can be harmful to human beings.

RADIOACTIVITY: A term describing a phenomenon whereby certain isotopes, known as radioisotopes, are subject to a form of decay brought about by the emission of high-energy particles. "Decay" does not mean that the isotope "rots"; rather, it decays to form another isotope until eventually (though this may take a long time) it becomes stable.

RADIOISOTOPE: An isotope subject to the decay associated with radioactivity. A radioisotope is thus an unstable isotope.

SUBLEVEL: A region within the principal energy level occupied by electrons in an atom. Whatever the number n of the principal energy level, there are n sublevels. Actinides are distinguished by the fact that their valence electrons are in a sublevel corresponding to the 5f orbital.

TRANSITION METALS: A group of 40 elements (counting lanthanum and actinium), which are the only elements that fill the d orbital. In addition, the transition metals have their valence electrons on two different principal energy levels. Though the lanthanides and actinides are considered inner transition metals, they are usually considered separately.

TRANSURANIUM ELEMENTS: Elements with an atomic number higher than that of uranium (92). These have all have been produced artificially. The transuranium elements include 11 actinides, as well as 9 transition metals.

VALENCE ELECTRONS: Electrons that occupy the highest principal energy level in an atom. These are the electrons involved in chemical bonding.

When Seaborg, Ghiorso, and others created Element 97, berkelium (Bk), they again took the naming of lanthanides as their cue. Just as terbium, directly above it on the periodic table, had been named for the Swedish town of Ytterby, where so many lanthanides were discovered, they named the new element after the American city where so many transuranium elements had been developed. Berkelium has no known applications outside of research. The Berkeley team likewise named californium (Cf, 98) after the state where Berkeley is located. Researchers today are studying the use of californium radiation for treatment of tumors involved in various forms of cancer.

THE REMAINING TRANSURANIUM ACTINIDES. The remaining transuranium actinides were all named after famous people: einsteinium (Es, 99) for Albert Einstein (1879-1955); fermium (Fm, 100) after Italian-American physicist Enrico Fermi (1901-1954); mendelevium after Mendeleev; nobelium (No, 102) after Swedish inventor and philanthropist Alfred Nobel (1833-1896); and lawrencium (Lr, 103) after Ernest Lawrence.

Both einsteinium and fermium were by-products of nuclear testing at Bikini Atoll in the south Pacific in 1952. For this reason, their existence was kept a secret for two years. Neither element has a known application. The same is true of mendelevium, produced by Seaborg, Ghiorso, and others with a cyclotron at Berkeley in 1955, as well as the other two transuranium actinides.

BEYOND THE TRANSURANIUM ACTINIDES

As noted earlier, there are nine additional transuranium elements, which properly belong to the transition metals. The first of these is rutherfordium, discovered in 1964 by the Dubna team and named after Ernest Rutherford (1871-1937), the British physicist who discovered the nucleus. The Dubna team named dubnium, discovered in 1967, after their city, just as berkelium had been named after the Berkeley team's city. Both groups developed versions of Element 106 in 1974, and both agreed to name it seaborgium after Seaborg, but this resulted in a controversy that was not settled for some time.

The name of bohrium, created at Dubna in 1976, honors Danish physicist Niels Bohr (1885-1962), who developed much of the model of electron energy levels discussed earlier in this essay. Hassium, produced at the GSI in 1984, is named for the German state of Hess. Two years earlier, the GSI team also created the last named element on the periodic table, meitnerium (109), named after Meitner. Beyond meitnerium are three elements, as yet unnamed, created at the GSI in the mid-1990s.

WHERE TO LEARN MORE

Cooper, Dan. *Enrico Fermi and the Revolutions in Modern Physics.* New York: Oxford University Press, 1999.

"Exploring the Table of Isotopes" (Web site). <http://ie.lbl.gov/education/isotopes.htm> (May 15, 2001).

Kidd, J. S. and Renee A. Kidd. *Quarks and Sparks: The Story of Nuclear Power.* New York: Facts on File, 1999.

Knapp, Brian J. *Elements.* Illustrated by David Woodroffe and David Hardy. Danbury, CT: Grolier Educational, 1996.

"A Periodic Table of the Elements" Los Alamos National Laboratory (Web site). <http://pearl1.lanl.gov/periodic/> (May 22, 2001).

"The Pictorial Periodic Table" (Web site). <http://chem-lab.pc.maricopa.edu/periodic/periodic.html> (May 22, 2001).

Sherrow, Victoria. *The Making of the Atom Bomb.* San Diego, CA: Lucent Books, 2000.

"Some Physics of Uranium." The Uranium Institute (Web site). <http://www.uilondon.org/education_resources/physics_of_uranium/in dex1.htm> (May 27, 2001).

Stwertka, Albert. *A Guide to the Elements.* New York: Oxford University Press, 1998.

Uranium Information Center (Web site). <http://www.uic.com.au/> (May 27, 2001).

LANTHANIDES

CONCEPT

Along the bottom of the periodic table of elements, separated from the main body of the chart, are two rows, the first of which represents the lanthanides. Composed of lanthanum and the 14 elements of the lanthanide series, the lanthanides were once called the "rare earth" metals. In fact, they are not particularly rare: many of them appear in as much abundance as more familiar elements such as mercury. They are, however, difficult to extract, a characteristic that defines them as much as their silvery color; sometimes high levels of reactivity; and sensitivity to contamination. Though some lanthanides have limited uses, members of this group are found in everything from cigarette lighters to TV screens, and from colored glass to control rods in nuclear reactors.

HOW IT WORKS

DEFINING THE LANTHANIDES

The lanthanide series consists of the 14 elements, with atomic numbers 58 through 71, that follow lanthanum on the periodic table of elements. These 14, along with the actinides—atomic numbers 90 through 103—are set aside from the periodic table due to similarities in properties that define each group.

Specifically, the lanthanides and actinides are the only elements that fill the f-orbitals. The lanthanides and actinides are actually "branches" of the larger family known as transition metals. The latter appear in groups 3 through 12 on the IUPAC version of the periodic table, though they are not numbered on the North American version.

The lanthanide series is usually combined with lanthanum, which has an atomic number of 57, under the general heading of lanthanides. As their name indicates, members of the lanthanide series share certain characteristics with lanthanum; hence the collective term "lanthanides." These 15 elements, along with their chemical symbols, are:

- Lanthanum (La)
- Cerium (Ce)
- Praseodymium (Pr)
- Neodymium (Nd)
- Promethium (Pm)
- Samarium (Sm)
- Europium (Eu)
- Gadolinium (Gd)
- Terbium (Tb)
- Dysprosium (Dy)
- Holmium (Ho)
- Erbium (Er)
- Thulium (Tm)
- Ytterbium (Yb)
- Lutetium (Lu)

Most of these are discussed individually in this essay.

PROPERTIES OF LANTHANIDES. Bright and silvery in appearance, many of the lanthanides—though they are metals—are so soft they can be cut with a knife. Lanthanum, cerium, praseodymium, neodymium, and europium are highly reactive. When exposed to oxygen, they form an oxide coating. (An oxide is a compound formed by metal with an oxygen.) To prevent this result, which tarnishes the

MISCH METAL, WHICH INCLUDES CERIUM, IS OFTEN USED IN CIGARETTE LIGHTERS. *(Massimo Listri/Corbis. Reproduced by permission.)*

metal, these five lanthanides are kept stored in mineral oil.

The reactive tendencies of the other lanthanides vary: for instance, gadolinium and lutetium do not oxidize until they have been exposed to air for a very long time. Nonetheless, lanthanides tend to be rather "temperamental" as a class. If contaminated with other metals, such as calcium, they corrode easily, and if contaminated with nonmetals, such as nitrogen or oxygen, they become brittle. Contamination also alters their boiling points, which range from 1,506.2°F (819°C) for ytterbium to 3,025.4°F (1,663°C) for lutetium.

Lanthanides react rapidly with hot water, or more slowly with cold water, to form hydrogen gas. As noted earlier, they also are quite reactive with oxygen, and they experience combustion readily in air. When a lanthanide reacts with another element to form a compound, it usually loses three of its outer electrons to form what are called tripositive ions, or atoms with an electric charge of +3. This is the most stable ion for lanthanides, which sometimes develop less stable +2 or +4 ions. Lanthanides tend to form ionic compounds, or compounds containing either positive or negative ions, with other substances—in particular, fluorine.

ARE THEY REALLY "RARE"?

Though they were once known as the rare earth metals, lanthanides were so termed because, as we shall see, they are difficult to extract from compounds containing other substances— including other lanthanides. As for rarity, the scarcest of the lanthanides, thulium, is more abundant than either arsenic or mercury, and certainly no one thinks of those as rare substances. In terms of parts per million (ppm), thulium has a presence in Earth's crust equivalent to 0.2 ppm. The most plentiful of the lanthanides, cerium, has an abundance of 46 ppm, greater than that of tin.

If, on the other hand, rarity is understood not in terms of scarcity, but with regard to difficulty in obtaining an element in its pure form, then indeed the lanthanides are rare. Because their properties are so similar, and because they are inclined to congregate in the same substances, the original isolation and identification of the lanthanides was an arduous task that took well over a century. The progress followed a common pattern.

First, a chemist identified a new lanthanide; then a few years later, another scientist came along and extracted another lanthanide from the sample that the first chemist had believed to be a single element. In this way, the lanthanides emerged over time, each from the one before it, rather like Russian matryoshka or "nesting" dolls.

EXTRACTING LANTHANIDES. Though most of the lanthanides were first isolated in Scandinavia, today they are found in considerably warmer latitudes: Brazil, India, Australia, South Africa, and the United States. The principal source of lanthanides is monazite, a heavy, dark sand from which about 50% of the lanthanide mass available to science and industry has been extracted.

In order to separate lanthanides from other elements, they are actually combined with other substances—substances having a low solubility, or tendency to dissolve. Oxalates and fluorides are low-solubility substances favored for this purpose. Once they are separated from non-lanthanide elements, ion exchange is used to separate one lanthanide element from another.

There is a pronounced decrease in the radii of lanthanide atoms as they increase in atomic number: in other words, the higher the atomic number, the smaller the radius. This decrease, known as the lanthanide contraction, aids in the process of separation by ion exchange. The lanthanides are mixed in an ionic solution, then passed down a long column containing a resin. Various lanthanide ions bond more or less tightly, depending on their relative size, with the resin.

After this step, the lanthanides are washed out of the ion exchange column and into various solutions. One by one, they become fully separated, and are then mixed with acid and heated to form an oxide. The oxide is then converted to a fluoride or chloride, which can then be reduced to metallic form with the aid of calcium.

REAL-LIFE APPLICATIONS

THE HISTORICAL APPROACH

In studying the lanthanides, one can simply move along the periodic table, from lanthanum all the way to lutetium. However, in light of the difficulties involved in extracting the lanthanides, one from another, an approach along historical lines aids in understanding the unique place each lanthanide occupies in the overall family.

SAMARIUM IS USED IN NUCLEAR POWER PLANT CONTROL RODS, SUCH AS THE ONE SHOWN HERE. (*Tim Wright/Corbis. Reproduced by permission.*)

The terms "lanthanide series" or even "lanthanides" did not emerge for some time—in other words, scientists did not immediately know that they were dealing with a whole group of metals. As is often the case with scientific discovery, the isolation of lanthanides followed an irregular pattern, and they did not emerge in order of atomic number.

Cerium was in fact discovered long before lanthanum itself, in the latter half of the eighteenth century. There followed, a few decades later, the discovery of a mineral called ytterite, named after the town of Ytterby, Sweden, near which it was found in 1787. During the next century, most of the remaining lanthanides were extracted from ytterite, and the man most responsible for this was Swedish chemist Carl Gustav Mosander (1797-1858).

Because Mosander had more to do with the identification of the lanthanides than any one individual, the middle portion of this historical overview is devoted to his findings. The recognition and isolation of lanthanides did not stop with Mosander, however; therefore another group of minerals is discussed in the context of the latter period of lanthanide discovery.

EARLY LANTHANIDES

CERIUM. In 1751, Swedish chemist Axel Crönstedt (1722-1765) described what he thought was a new form of tungsten, which he had found at the Bastnäs Mine near Riddarhyttan, Sweden. Later, German chemist Martin Heinrich Klaproth (1743-1817) and Swedish chemist Wilhelm Hisinger (1766-1852) independently analyzed the material Crönstedt had discovered, and both concluded that this must be a new element. It was named cerium in honor of Ceres, an asteroid between Mars and Jupiter discovered in 1801. Not until 1875 was cerium actually extracted from an ore.

Among the applications for cerium is an alloy called misch metal, prepared by fusing the chlorides of cerium, lanthanum, neodymium, and praseodymium. The resulting alloy ignites at or below room temperature, and is often used as the "flint" in a cigarette lighter, because it sparks when friction from a metal wheel is applied.

Cerium is also used in jet engine parts, as a catalyst in making ammonia, and as an anti-knock agent in gasoline—that is, a chemical that reduces the "knocking" sounds sometimes produced in an engine by inferior grades of fuel. In cerium (IV) oxide, or CeO_2, it is used to extract the color from formerly colored glass, and is also applied in enamel and ceramic coatings.

GADOLINIUM. In 1794, seven years after the discovery of ytterite, Finnish chemist Johan Gadolin (1760-1852) concluded that ytterite contained a new element, which was later named gadolinite in his honor. A very similar name would be applied to an element extracted from ytterite, and the years between Gadolin's discovery and the identification of this element spanned the period of the most fruitful activity in lanthanide identification.

During the next century, all the other lanthanides were discovered within the composition of gadolinite; then, in 1880, Swiss chemist Jean-Charles Galissard de Marignac (1817-1894) found yet another element hiding in it. French chemist Paul Emile Lecoq de Boisbaudran (1838-1912) rediscovered the same element six years later, and proposed that it be called gadolinium.

Silvery in color, but with a sometimes yellowish cast, gadolinium has a high tendency to oxidize in dry air. Because it is highly efficient for capturing neutrons, it could be useful in nuclear power reactors. However, two of its seven isotopes are in such low abundance that it has had little nuclear application. Used in phosphors for color television sets, among other things, gadolinium shows some promise for ultra high-tech applications: at very low temperatures it becomes highly magnetic, and may function as a superconductor.

MOSANDER'S LANTHANIDES

LANTHANUM. Between 1839 and 1848, Mosander was consumed with extracting various lanthanides from ytterite, which by then had come to be known as gadolinite. When he first succeeded in extracting an element, he named it lanthana, meaning "hidden." The material, eventually referred to as lanthanum, was not prepared in pure form until 1923.

Like a number of other lanthanides, lanthanum is very soft—so soft it can be cut with a knife—and silvery-white in color. Among the most reactive of the lanthanides, it decomposes rapidly in hot water, but more slowly in cold water. Lanthanum also reacts readily with oxygen, and corrodes quickly in moist air.

As with cerium, lanthanum is used in misch metal. Because lanthanum compounds bring about special optical qualities in glass, it also used for the manufacture of specialized lenses. In addition, compounds of lanthanum with fluorine or oxygen are used in making carbon-arc lamps for the motion picture industry.

SAMARIUM. While analyzing an oxide formed from lanthanide in 1841, Mosander decided that he had a new element on his hands, which he called didymium. Four decades later, Boisbaudran took another look at didymium, and concluded that it was not an element; rather, it contained an element, which he named samarium after the mineral samarskite, in which it is found. Still later, Marignac was studying samarskite when he discovered what came to be known as gadolinium. But the story did not end there: even later, in 1901, French chemist Eugène-Anatole Demarçay (1852-1903) found yet another element, europium, in samarskite.

Samarium is applied today in nuclear power plant control rods, in carbon-arc lamps, and in optical masers and lasers. In alloys with cobalt, it is used in manufacturing the most permanent electromagnets available. Samarium is also utilized in the manufacture of optical glass, and as a catalyst in the production of ethyl alcohol.

ERBIUM AND TERBIUM. To return to Mosander, he was examining ytterite in 1843 when he identified three different "earths," all of which he also named after Ytterby: yttria, erbia, and terbia. Erbium was the first to be extracted. A pure sample of its oxide was prepared in 1905 by French chemist Georges Urbain (1872-1938) and American chemist Charles James (1880-1928), but the pure metal itself was only extracted in 1934.

Soft and malleable, with a lustrous silvery color, erbium produces salts (which are usually combinations of a metal with a nonmetal) that are pink and rose, making it useful as a tinting agent. One of its oxides is utilized, for instance, to tint glass and porcelain with a pinkish cast. It is also applied, to a limited extent, in the nuclear power industry.

Mosander also identified another element, terbium, in ytterite in 1839, and Marignac isolated it in a purer form nearly half a century later, in 1886. To repeat a common theme, it is silvery-gray and soft enough to be cut with a knife. When hit by an electron beam, a compound containing terbium emits a greenish color, and thus it is used as a phosphor in color television sets.

Later Isolation of Lanthanides

YTTERBIUM, HOLMIUM, AND THULIUM. For many years after Mosander, there was little progress in the discovery of lanthanides, and when it came, it was in the form of a third element, named after the town where so many of the lanthanides were discovered. In 1878, while analyzing what Mosander had called erbia, Marignac realized that it contained one or possibly two elements.

A year later, Swedish chemist Lars Frederik Nilson (1840-1899) concluded that it did indeed contain two elements, which were named ytterbium and scandium. (Scandium, with an atomic number of 21, is not part of the lanthanide series.) Urbain is sometimes credited for discovering ytterbium: in 1907, he showed that the materials Nilson had studied were actually a mixture of two oxides. In any case, Urbain said that

KEY TERMS

ALLOY: A mixture of two or more metals.

ATOMIC NUMBER: The number of protons in the nucleus of an atom. Since this number is different for each element, elements are listed on the periodic table of elements in order of atomic number.

ION: An atom or atoms that has lost or gained one or more electrons, and thus has a net electrical charge.

LANTHANIDE CONTRACTION: A progressive decrease in the radius of lanthanide atoms as they increase in atomic number.

LANTHANIDE SERIES: A group of 14 elements, with atomic numbers 58 through 71, that follow lanthanum on the periodic table of elements.

LANTHANIDES: The lanthanide series, along with lanthanum.

OXIDE: A compound formed by the chemical bonding of a metal with oxygen.

PERIODIC TABLE OF ELEMENTS: A chart showing the elements arranged in order of atomic number, grouping them according to common characteristics.

RARE EARTH METALS: An old name for the lanthanides, reflecting the difficulty of separating them from compounds containing other lanthanides or other substances.

TRANSITION METALS: Groups 3 through 12 on the IUPAC or European version of the periodic table of elements. The lanthanides and actinides, which appear at the bottom of the periodic table, are "branches" of this family.

the credit should be given to Marignac, who is the most important figure in the history of lanthanides other than Mosander. As for ytterbium, it is highly malleable, like other lanthanides, but does not have any significant applications in industry.

Swedish chemist Per Teodor Cleve (1840-1905) found in 1879 that erbia contained two more elements, which he named holmium and thulium. Thulium refers to the ancient name for Scandinavia, Thule. Rarest of all the lanthanides, thulium is highly malleable—and also highly expensive. Hence it has few commercial applications.

DYSPROSIUM. Named for the Greek word *dysprositos*, or "hard to get at," dysprosium was discovered by Boisbaudran. Separating ytterite in 1886, he found gallium (atomic number 31—not a lanthanide); samarium (discussed above); and dysprosium. Yet again, a mineral extracted from ytterite had been named after a previously discovered element, and, yet again, it turned out to contain several elements. The substance in question this time was holmium, which, as Boisbaudran discovered, was actually a complex mixture of terbium, erbium, holmium, and the element he had identified as dysprosium. A pure sample was not obtained until 1950.

Because dysprosium has a high affinity for neutrons, it is sometimes used in control rods for nuclear reactors, "soaking up" neutrons rather as a sponge soaks up water. Soft, with a lustrous silver color like other lanthanides, dysprosium is also applied in lasers, but otherwise it has few uses.

EUROPIUM AND LUTETIUM. Whereas many other lanthanides are named for regions in northern Europe, the name for europium refers to the European continent as a whole, and that of lutetium is a reference to the old Roman name for Paris. As mentioned earlier,

Demarçay found europium in samarskite, a discovery he made in 1901. Actually, Boisbaudran had noticed what appeared to be a new element about a decade previously, but he did not pursue it, and thus the credit goes to his countryman.

Most reactive of the lanthanides, europium responds both to cold water and to air. In addition, it is capable of catching fire spontaneously. Among the most efficient elements for the capture of neutrons, it is applied in the control systems of nuclear reactors. In addition, its compounds are utilized in the manufacture of phosphors for TV sets: one such compound, for instance, emits a reddish glow. Yet another europium compound is added to the glue on postage stamps, making possible the electronic scanning of stamps.

Urbain, who discovered lutetium, named it after his hometown. James also identified a form of the lanthanide, but did not announce his discovery until much later. Except for some uses at a catalyst in the production of petroleum, lutetium has few industrial applications.

WHERE TO LEARN MORE

Cotton, Simon. *Lanthanides and Actinides.* New York: Oxford University Press, 1991.

Heiserman, David L. *Exploring Chemical Elements and Their Compounds.* Blue Ridge Summit, PA: Tab Books, 1992.

"*Luminescent Lanthanides*" (Web site). <http://orgwww.chem.uva.nl/lanthanides/> (May 16, 2001).

Snedden, Robert. *Materials.* Des Plaines, IL: Heinemann Library, 1999.

Oxlade, Chris. *Metal.* Chicago: Heinemann Library, 2001.

Stwertka, Albert. *A Guide to the Elements.* New York: Oxford University Press, 1996.

Whyman, Kathryn. *Metals and Alloys.* Illustrated by Louise Nevett and Simon Bishop. New York: Gloucester Press, 1988.

NONMETALS AND METALLOIDS

NONMETALS

CONCEPT

Nonmetals, as their name implies, are elements that display properties quite different from those of metals. Generally, they are poor conductors of heat and electricity, and they are not ductile: in other words, they cannot be easily reshaped. Included in this broad grouping are the six noble gases, the five halogens, and eight "orphan" elements. Two of these eight—hydrogen and carbon—are so important that separate essays are devoted to them. Two more, addressed in this essay, are absolutely essential to human life: oxygen and nitrogen. Hydrogen and helium, a nonmetal of the noble gas family, together account for about 99% of the mass of the universe, while Earth and the human body are composed primarily of oxygen, with important components of carbon, nitrogen, and hydrogen. Indeed, much of life—human, animal, and plant—can be summed up with these four elements, which together make Earth different from any other known planet. Among the other "orphan" nonmetals are phosphorus and sulfur, the "brimstone" of the Bible, as well as boron and selenium.

HOW IT WORKS

DEFINING THE NONMETALS

The majority of elements on the periodic table are metals: solids (along with one liquid, mercury) which are lustrous or shiny in appearance. Metals are ductile, or malleable, meaning that they can be molded into different shapes without breaking. They are excellent conductors of heat and electricity, and tend to form positive ions by losing electrons. The vast majority of elements—87 in all—are metals, and these occupy the left, center, and part of the right-hand side of the periodic table.

With the exception of hydrogen, placed in Group 1 above the alkali metals, the nonmetals fill a triangle-shaped space in the upper right corner of the periodic table. All gaseous elements are nonmetals, a group that also includes some solids, as well as bromine, the only element other than mercury that is liquid at room temperature. In general, the nonmetals are characterized by properties opposite those of metals. They are poor conductors of heat and electricity (though carbon, a good electrical conductor, is an exception), and they tend to form negative ions by gaining electrons. They are not particularly malleable, and whereas most metals are shiny, nonmetals may be dull-colored, black (in the case of carbon in some forms or allotropes), or invisible, as is the case with most of the gaseous nonmetals.

GROUPING THE NONMETALS

Whereas the metals can be broken down into five families, along with seven "orphans"—elements that do not fit into a family grouping—the nonmetals are arranged in two families, as well as eight "orphans." Hydrogen, because of its great abundance in the universe and its importance to chemical studies, is addressed in a separate essay within this book. The same is true of carbon: though not nearly as abundant as hydrogen, carbon is a common element of all living things, and therefore it is discussed in the essay on Carbons and Organic Chemistry.

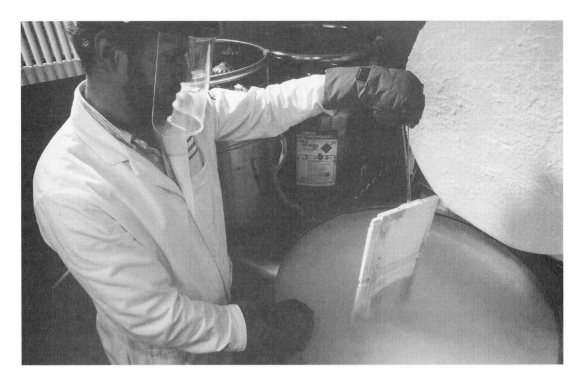

LIQUID NITROGEN ACCOUNTS FOR ABOUT ONE-THIRD OF ALL COMMERCIAL USES OF NITROGEN. HERE, A MEDICAL WORKER PUTS A SAMPLE OF BONE MARROW INTO A TANK OF LIQUID NITROGEN FOR PRESERVATION. (*Leif Skoogfors/ Corbis. Reproduced by permission.*)

The other six "orphan" nonmetals, which will be examined in detail later in this essay, are listed below by atomic number:

- 5. Boron (B)
- 7. Nitrogen (N)
- 8. Oxygen (O)
- 15. Phosphorus (P)
- 16. Sulfur (S)
- 34. Selenium (Se)

THE NOBLE GASES. The noble gases, discussed in detail within an essay devoted to that subject, are listed below by atomic number:

- 2. Helium (He)
- 10. Neon (Ne)
- 18. Argon (Ar)
- 36. Krypton (Kr)
- 54. Xenon (Xe)
- 86. Radon (Rn)

Occupying Group 8 of the North American periodic table, the noble gases—with the exception of helium—have valence electron configurations of ns^2np^6. This means that they have two valence electrons (the electrons involved in chemical bonding) on the orbital designated as s, and six more on the p orbital. As for n, this des-ignates the energy level, a number that corresponds to the period or row that an element occupies on the periodic table.

Most elements bond according to what is known as the octet rule, forming a shell composed of eight electrons. Since the noble gases already have eight electrons on their shell, they tend not to bond with other elements: hence the name "noble," which in this context means "set apart." Helium, on the other hand, has a valence shell of s^2; however, it too is characterized by a lack of reactivity, and therefore is included in the noble gas family. Noble gases are all monatomic, meaning that they exist as individual atoms, rather than in molecules. To put it another way, their molecules are single atoms. (By contrast, atoms of oxygen, for instance, usually combine to form a diatomic molecule, designated O_2.)

Due to their apparent lack of reactivity, the noble gases—also known as the rare gases—were once called the inert (inactive) gases. Indeed, helium, neon, and argon have not been found to combine with other elements to form compounds. However, in 1962, English chemist Neil Bartlett (1932-) succeeded in preparing a compound of xenon with platinum and fluorine ($XePtF_6$), thus overturning the idea that the

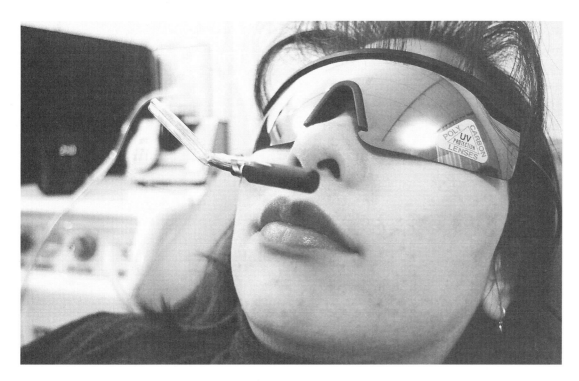

AT AN "OXYGEN BAR," YOU CAN RELAX BY BREATHING OXYGEN. HERE, A WOMAN IS BREATHING FLAVORED OXYGEN AT A BAR IN JAPAN. *(AFP/Corbis. Reproduced by permission.)*

noble gases were entirely "inert." Since that time, numerous compounds of xenon with other elements, most notably oxygen and fluorine, have been developed. Fluorine has also been used to form simple compounds with krypton and radon.

THE HALOGENS. Next to the noble gases, in Group 7, are the halogens, a family discussed in a separate essay. These are listed below, along with atomic number and chemical symbol. Note that astatine is sometimes included with the metalloids, elements that display characteristics both of metals and nonmetals.

- 9. Fluorine (F)
- 17. Chlorine (Cl)
- 35. Bromine (Br)
- 53. Iodine (I)
- 85. Astatine (At)

The halogens all have valence electron configurations of ns^2np^5: in other words, at any energy level n, they have two valence electrons in the s orbital, and 5 in the p orbital. In terms of phase of matter, the halogens are the most varied family on the periodic table. Fluorine and chlorine are gases, iodine is a solid, and bromine (as noted earlier) is one of only two elements existing at room temperature as a liquid. As for asta-

tine, it is a solid too, but so highly radioactive it is hard to determine much about its properties. (The only other nonmetal that has a large number of radioactive isotopes is the noble gas radon, considered highly dangerous.)

Though they are "next door" to the noble gases, the halogens could not be more different. Whereas their neighbors to the right are the least reactive elements on the periodic table, the halogens are the most reactive. Indeed, fluorine has the highest possible value of electronegativity, the relative ability of an atom to attract valence electrons. It is therefore one of the only elements that will bond with a noble gas.

All of the halogens tend to form salts, compounds—formed, along with water, from the reaction of an acid and base—that bring together a metal and a nonmetal. Due to this tendency, the first of the family to be isolated—chlorine, in 1811—was originally named "halogen," a combination of the Greek words halos, or salt, and gennan, "to form or generate." In their pure form, halogens are diatomic, and in contact with other elements, they form ionic bonds, which are the strongest form of chemical bond. In the process of bonding, they become negatively charged ions, or anions.

ABUNDANCE OF THE NONMETALS

IN THE UNIVERSE. As is stated many times throughout this book, humans may be created equal, but the elements are not. Though 88 elements exist in nature, this certainly does not mean that each occupies a 1/88 share of the total pie. Just two elements—the nonmetals hydrogen and helium, which occupy the first two positions on the periodic table—account for 99.9% of the matter in the entire universe, yet the percentage of matter they make up on Earth is very small.

The reason for this disparity is that, whereas our own planet (as far as we know) is unique in sustaining life—requiring oxygen, carbon, nitrogen, and other elements—the vast majority of the universe is made up of stars, composed primarily of hydrogen and helium.

ON EARTH AND IN THE ATMOSPHERE. Nonmetals account for a large portion of Earth's total known elemental mass—that is, the composition of the planet's crust, waters, and atmosphere. Listed below are figures ranking nonmetals within the overall picture of elements on Earth. The numbers following each element's name indicate the percentage of each within the planet's total known elemental mass.

- 1. Oxygen (49.2%)
- 9. Hydrogen (0.87%)
- 11. Chlorine (0.19%)
- 12. Phosphorus (0.11%)
- 14. Carbon (0.08%)
- 15. Sulfur (0.06%)
- 17. Nitrogen (0.03%)
- 18. Fluorine (0.03%)

Some of the figures above may seem rather small, but in fact only 18 elements account for all but 0.49% of the planet's known elemental mass, the remainder being composed of numerous other elements in small quantities. In Earth's atmosphere, the composition is all nonmetallic, as one might expect:

- 1. Nitrogen (78%)
- 2. Oxygen (21%)
- 3. Argon (0.93%)
- 4. Various trace gases, including water vapor, carbon dioxide, and ozone or O_3 (0.07%)

IN THE HUMAN BODY. As noted earlier, carbon is a component in all living things, and it constitutes the second-most abundant element in the human body. Together with oxygen and hydrogen, it accounts for 93% of the body's mass. Listed below are the most abundant nonmetals in the human body, by ranking.

- 1. Oxygen (65%)
- 2. Carbon (18%)
- 3. Hydrogen (10%)
- 4. Nitrogen (3%)
- 6. Phosphorus (1.0%)
- 9. Sulfur (0.26%)
- 11. Chlorine (0.14%)

REAL-LIFE APPLICATIONS

BORON

The first "orphan" nonmetal, by atomic number, is boron, named after the Arabic word buraq or the Persian burah. (It is thus unusual in being one of the only elements whose name is not based on a word from a European language, or—for the later elements—the name of a person or place.) It was discovered in 1808 by English chemist Sir Humphry Davy (1778-1829), a man responsible for identifying or isolating numerous elements; French physicist and chemist Joseph Gay-Lussac (1778-1850), known in part for the gas law named after him; and French chemist Louis Jacques Thénard (1777-1857). These scientists used the reaction of boric acid (H_3BO_3) with potassium to isolate the element.

Sometimes classified as a metalloid because it is a semiconductor of electricity, boron is applied in filaments used in fiber optics research. Few of its other applications, however, have much to do with electrical conductivity. Boric or boracic acid is used as a mild antiseptic, and in North America, it is applied for the control of cockroaches, silverfish, and other pests. A compound known as borax is a water softener in washing powders, while other compounds are used to produce enamels for the coating of refrigerators and other appliances. Compounds involving boron are also present in pyrotechnic flares, because they emit a distinctive green color, and in the igniters of rockets.

NITROGEN

Though most people think of air as consisting primarily of oxygen, in fact—as noted above—the greater part of the air we breathe is made up of nitrogen. Scottish chemist David Rutherford (1749-1819) is usually given credit for discovering the element: in 1772, he identified nitrogen as the element that remained when oxygen was removed from air. Several other scientists around the same time made a similar discovery.

Because of its heavy presence in air, nitrogen is obtained primarily by cooling air to temperatures below the boiling points of its major components. Nitrogen boils (that is, turns into a gas) at a lower temperature than oxygen: -320.44°F (-195.8°C), as opposed to -297.4°F (-183°C). When air is cooled to -328°F (-200°C) and then allowed to warm slowly, the nitrogen boils first, and therefore evaporates first. The nitrogen gas is captured, cooled, and liquefied once more.

Nitrogen can also be obtained from compounds such as potassium nitrate or saltpeter, found primarily in India; or sodium nitrate (Chilean saltpeter), which comes from the desert regions of Chile. To isolate nitrogen, various processes are undertaken in a laboratory—for instance, heating barium azide or sodium azide, both of which contain nitrogen.

REACTIONS WITH OTHER ELEMENTS. Existing as diatomic molecules, nitrogen forms very strong triple bonds, and as a result tends to be fairly unreactive at low temperatures. Thus, for instance, when a substance burns, it reacts with the oxygen in the air, but not with the nitrogen. At high temperatures, however, nitrogen combines with other elements, reacting with metals to form nitrides; with hydrogen to form ammonia; with O_2 to form nitrites; and with O_3 to form nitrates.

Nitrogen and oxygen, in particular, react at high temperatures to form numerous compounds: nitric oxide (NO), nitrous oxide (N_2O), nitrogen dioxide (NO_2), dinitrogen trioxide (N_2O_3), and dinitrogen pentoxide (N_2O_5). In reaction with halogens, nitrogen forms unstable, explosive compounds such as nitrogen trifluoride (NF_3), nitrogen trichloride (NCl_3), and nitrogen triiodide (NI_3). One thing is for sure: never mix ammonia with bleach, which involves the halogen chlorine. Together, these two produce chloramines, gases with poisonous vapors.

USES FOR NITROGEN. One of the most striking scenes in a memorable film, "Terminator 2: Judgment Day" (1991), occurs near the end of the movie, when the villainous T-1000 robot steps into a pool of liquid nitrogen

and cracks to pieces. As noted, nitrogen must be at extremely low temperatures to assume liquid form. Liquid nitrogen, which accounts for about one-third of all commercial uses of the element, is applied for quick-freezing foods, and for preserving foods in transit. Liquid nitrogen also makes it possible to process materials, such as some forms of rubber, that are too pliable for machining at room temperature. These materials are first cooled in liquid nitrogen, and then become more rigid.

In processing iron or steel, which forms undesirable oxides if exposed to oxygen, a blanket of nitrogen is applied to prevent this reaction. The same principle is applied in making computer chips and even in processing foods, since these items too are detrimentally affected by oxidation. Because it is far less combustible than air (magnesium is one of the few elements that burns nitrogen in combustion), nitrogen is also used to clean tanks that have carried petroleum or other combustible materials.

As noted, nitrogen combines with hydrogen to form ammonia, used in fertilizers and cleaning materials. Ammonium nitrate, applied primarily as a fertilizer, is also a dangerous explosive, as shown with horrifying effect in the bombing of the Alfred P. Murrah Federal Building in Oklahoma City on April 19, 1995—a tragedy that took 168 lives. Nor is ammonium nitrate the only nitrogen-based explosive. Nitric acid is used in making trinitrotoluene (TNT), nitroglycerin, and dynamite, as well as gunpowder and smokeless powder.

NITROGEN, THE ENVIRONMENT, AND HEALTH. The nitrogen cycle is the process whereby nitrogen passes from the atmosphere into living things, and ultimately back into the atmosphere. Both through the action of lightning in the sky, and of bacteria in the soil, nitrogen is converted to nitrates and nitrites—compounds of nitrogen and oxygen. These are then absorbed by plants to form plant proteins, which convert to animal proteins in the bodies of animals who eat the plants. When an animal dies, the proteins are returned to the soil, and denitrifying bacteria break down these compounds, returning elemental nitrogen to the atmosphere.

Not all nitrogen in the atmosphere is healthful, however. Oxides of nitrogen, formed in the high temperatures of internal combustion engines, pass into the air as nitric oxide. This compound reacts readily with oxygen in the air to form nitrogen dioxide, a toxic reddish-brown gas that adds to the tan color of smog over major cities.

Another health concern is posed by sodium nitrate and sodium nitrite, added to bacon, sausage, hot dogs, ham, bologna and other food products to inhibit the growth of harmful microorganisms. Many researchers believe that nitrites impair the ability of a young child's blood to carry oxygen. Furthermore, nitrites often combine with amines, a form of organic compound, to create a variety of toxins known as nitrosoamines. Because of concerns about these dangers, scientists and health activists have called for a ban on the use of nitrites and nitrates as food additives.

OXYGEN

Discovered independently by Swedish chemist Carl W. Scheele (1742-1786) and English chemist Joseph Priestley (1733-1804) in the period 1773-1774, oxygen was named by a third scientist, French chemist Antoine Lavoisier (1743-1794). Believing (incorrectly) that all acids contained the newly discovered element, Lavoisier called it oxygen, which comes from a French word meaning "acid-former."

Like many elements, oxygen has been in use since the beginning of time. But this is quite different from saying, for instance, that iron has been in use since the early stages of human history. A person can live without iron (except for the necessary quantities in the human body), and can survive for weeks or even months without food. One can live for a few days without water (the most famous and plentiful of all oxygen-containing compounds); but one cannot survive for more than a few minutes without the oxygen in air.

Oxygen appears in three allotropes (different versions of the same element, distinguished by molecular structure): monatomic oxygen (O); diatomic oxygen or O_2; and triatomic oxygen (O_3), better known as ozone. The diatomic form dominates the natural world, but in the upper atmosphere, ozone forms a protective layer that keeps the Sun's harmful ultraviolet radiation from reaching Earth. Concerns that chlorofluorocarbons (CFCs) may be depleting the ozone layer by converting these triatomic molecules to

O_2 has led to a reduction in the output of CFCs by industrialized nations.

OBTAINING OXYGEN. Oxygen, of course, is literally "in the air," mixed with larger quantities of nitrogen. Higher up in the atmosphere, it occurs as a free element. Through electrolysis,, it can be obtained from water; however, this process is prohibitively expensive for most commercial applications.

Oxygen-containing compounds are also sources of oxygen for commercial use, but generally oxygen is obtained by the fractional distillation of liquid air, described above with regard to nitrogen. After the nitrogen has been separated, argon and neon (which also have lower boiling points than oxygen) also boil off, leaving behind an impure form of oxygen. This is further purified by a process of cooling, liquefying, and evaporation, which eliminates traces of noble gases such as krypton and xenon.

Many millions of years ago, when Earth was first formed, there was no oxygen on the planet. The growth of oxygen on Earth coincided with the development of organisms that, as they evolved, increasingly needed oxygen. The present concentration of oxygen in the atmosphere, oceans, and the rocks of Earth's crust was reached about 580 million years ago, and is sustained today by biological activity. When plants undergo photosynthesis, carbon dioxide and water react in the presence of chlorophyll to produce carbohydrates and oxygen.

OXIDES AND OTHER COMPOUNDS. Though the bond in diatomic oxygen is strong, once it is broken, monatomic oxygen reacts readily with other elements to form a seemingly limitless range of compounds: oxides, silicates, carbonates, phosphates, sulfates, and other more complex substances.

The process known as oxidation results in the formation of numerous oxides. Sometimes oxygen and another element form several oxides, as for example in the case of nitrogen, whose five oxides are listed above. Water is an oxide; so too are carbon dioxide and carbon monoxide. When animals and plants die, the organic materials that make them up react with oxygen in the air, resulting in a complex form of oxidation known as decay—or, in common language, "rotting."

Oxygen, reacting with compounds such as hydrocarbons, produces carbon dioxide and water vapor at high temperatures. If the oxidation process is extremely rapid, and takes place at high temperatures, it is usually identified as combustion. In addition, oxygen reacts with iron and other metals to form oxides. Many of these oxides, commonly known as rust, are undesirable.

Every year, millions upon millions of dollars are spent on painting metal structures, or for other precautions to protect against the formation of metallic oxides. On the other hand, metallic oxides may be produced deliberately for applications in materials such as mortar color, to enhance the appearance of a brick building.

USES OF OXYGEN. Aside from the obvious application of oxygen for breathing, there are four major fields that make use of this element: medicine, metallurgy, rocketry, and the field of chemistry concerned with chemical synthesis. The medical application is closest to how we normally use oxygen in our daily lives. In oxygen therapy, a patient having difficulty breathing is given doses of pure, or nearly pure, oxygen. This is used during surgical procedures, and to treat patients who have had heart attacks, as well as those suffering from various infectious or respiratory diseases.

The use of oxygen in metallurgy involves refining coke, which is almost pure carbon, to make carbon monoxide. Carbon monoxide, in turn, reduces iron oxides to pure metallic iron. Oxygen is also used in blast furnaces to convert pig iron to steel by removing excess carbon, silicon, and metallic impurities. In addition, oxygen is applied in torches for welding and cutting. In the form of liquefied oxygen, or LOX, oxygen is used in rockets and missiles. The space shuttle, for instance, carries a huge internal tank containing oxygen and hydrogen, which, when they react, give the vehicle enormous thrust.

In chemical synthesis—the preparation of compounds (especially organic ones) from easily available chemicals—commercial chemists use oxygen, for instance, to loosen the bonds in hydrocarbons. If this is done too quickly, it results in combustion; but at a controlled rate, the chemical synthesis of hydrocarbons can generate products such as acetylene, ethylene, and propylene.

Oxygen can be used to produce synthetic fuels, as well as for water purification and sewage treatment. Airplanes carry oxygen supplies in case of depressurization at high altitudes; in addition, divers carry tanks in which oxygen is mixed with helium, rather than nitrogen, to prevent the dangerous condition known as "the bends."

As early as the 1960s, smog-ridden cities such as Tokyo and Mexico City were equipped with coin-operated oxygen booths—a sort of "phone booth for the lungs." After inserting the appropriate amount of money, a person received a dose of oxygen for inhaling. This idea, spawned by necessity, is the likely inspiration for a rather bizarre fad that took hold in the trendier cities of North America during the mid-1990s: oxygen bars. Popular in Los Angeles, New York, and Toronto, these are establishments in which patrons pay up to a dollar per minute to inhale pure or flavored oxygen. Enthusiasts have touted the health benefits of this practice, but some physicians have warned of oxygen toxicity and other dangers.

THE OTHER "ORPHAN" NONMETALS

PHOSPHORUS. Some elements, such as iron or gold, were known from ancient or even prehistoric times—meaning that the identity of the discoverer is unknown. Phosphorus was the first element whose discoverer is known: German chemist Hennig Brand (c. 1630-c. 1692), who identified it in 1674.

Highly reactive with oxygen, phosphorus is used in the production of safety matches, smoke bombs, and other incendiary devices. It is also important in fertilizers, and in various industrial applications. Phosphorus forms a number of important compounds, most notably phosphates, on which animals and plants depend.

Phosphorus pollution, created by the use of household detergents containing phosphates, raised environmental concerns in the 1960s and 1970s. It was feared that high phosphate levels in rivers and creeks would lead to runaway, detrimental growth of plants and algae near bodies of water, a condition known as eutrophication. These concerns led to a ban on the use of phosphates in detergents.

SULFUR. On its own, sulfur has no smell, but in combination with other elements, it often acquires a foul odor, which has given it an unpleasant reputation. The element's smell, combined with its combustibility, led to the associa-

KEY TERMS

ALLOTROPES: Different versions of the same element, distinguished by molecular structure.

DIATOMIC: A term describing an element that exists as molecules composed of two atoms. This is in contrast to monatomic elements.

ELECTROLYSIS: The use of an electric current to cause a chemical reaction.

HALOGENS: Group 7 of the periodic table of elements, with valence electron configurations of ns^2np^5. In contrast to the noble gases, the halogens are known for high levels of reactivity.

ION: An atom or group of atoms that has lost or gained one or more electrons, and thus has a net electric charge.

ISOTOPES: Atoms that have an equal number of protons, and hence are of the same element, but differ in their number of neutrons. This results in a difference of mass. Isotopes may be either stable or unstable. The unstable type, known as radioisotopes, are radioactive.

METALS: Elements that are lustrous or shiny in appearance; malleable, meaning that they can be molded into different shapes without breaking; and excellent conductors of heat and electricity. Metals, which constitute the vast majority of all elements, tend to form positive ions by losing electrons.

MONATOMIC: A term describing an element that exists as single atoms. This in contrast to diatomic elements.

NOBLE GASES: Group 8 of the periodic table of elements, all of whom (with the exception of helium) have valence electron configurations of ns^2np^6. The noble gases are noted for their extreme lack of reactivity—in other words, they tend not to react to, or bond with, other elements.

NONMETALS: Elements that have a dull appearance; are not malleable; are poor conductors of heat and electricity; and tend to gain electrons to form negative ions. They are thus opposite of metals in most regards, as befits their name. In addition to hydrogen, in Group 1 of the periodic table, the other 18 nonmetals occupy the upper right-hand side of the chart. They include the noble gases, halogens, and seven "orphan" elements: boron, carbon, nitrogen, oxygen, phosphorus, sulfur, and selenium.

ORBITAL: A pattern of probabilities regarding the position of an electron for an atom in a particular energy state. The higher the principal energy level, the more complex the pattern of orbitals.

RADIOACTIVITY: A term describing a phenomenon whereby certain isotopes known as radioisotopes are subject to a form of decay brought about by the emission of high-energy particles. "Decay" does not mean that the isotope "rots"; rather, it decays to form another isotope until eventually (though this may take a long time), it becomes stable.

REACTIVITY: The tendency for bonds between atoms or molecules to be made or broken in such a way that materials are transformed.

SHELL: The orbital pattern of the valence electrons at the outside of an atom.

VALENCE ELECTRONS: Electrons that occupy the highest principal energy level in an atom. These are the electrons involved in chemical bonding.

tion of "brimstone"—the ancient name for sulfur—with the fires of hell.

Because it is not usually combined with other elements in nature, the discovery of sulfur was relatively easy. Pure, or nearly pure, sulfur is mined on the Gulf Coast of the United States, as well as in Poland and Sicily. Sulfur compounds also appear in a number of ores, such as gypsum (calcium sulfate), or magnesium sulfate, better known as Epsom salts.

Applications of the aforementioned compounds are discussed in the Alkaline Earth Metals essay. In addition, sulfates are used as agricultural insecticides, and for killing algae in water supplies. Potassium aluminum sulfate, a gelatinous solid that sinks to the bottom when dropped into water, is also used in water purification. Sulfur is sometimes applied in pure form as a fungicide, or in matches, fireworks, and gunpowder. More often, however, it is found in compounds such as the sulfates or the sulfides—including sulfuric acid and the evil-smelling gas known as hydrogen sulfide.

Rotten eggs and intestinal gas are two examples of hydrogen sulfide, which, though poisonous, usually poses little danger, because the smell keeps people away. Another sulfur compound is mercaptan, an ingredient in the skunk's distinctive aroma. Tiny quantities of mercaptan are added to natural gas (which has no odor) so that dangerous gas leaks can be detected by smell.

SELENIUM. When Swedish chemist Jons Berzelius (1779-1848) first discovered selenium in 1817, in deposits at the bottom of a tank in a sulfuric acid factory, he thought it was tellurium, a metalloid discovered in 1800. A few months later, he reconsidered the evidence, and realized he had found a new element. Because tellurium, which lies just below selenium on the periodic table, had been named for the Earth (*tellus* in Latin), he named his new discovery after the Greek word for the Moon, *selene*.

Found primarily in impurities from sulfide ores, selenium is often obtained commercially as a by-product of the refining of copper by electrolysis. It occurs in at least three allotropic forms, variously black and red in color. Plants and animals, including humans, need small amounts of selenium to survive, but larger quantities can be toxic. This was demonstrated in the late 1970s, when waterfowl in the area of Kesterson Reservoir in northern California began turning up with birth defects. The cause was later traced to the dumping of selenium from agricultural wastes and industrial plants.

Because selenium is photovoltaic (able to convert light directly into electricity) and photoconductive (meaning that its resistance to the flow of electric current decreases in the presence of light), it has applications in photocells, exposure meters, and solar cells. It is also used for the conversion of alternating current to direct current, and is applied as a semiconductor in electronic and solid-state appliances. Photocopiers use selenium in toners, and compounds containing selenium are used to tint glass red, orange, or pink.

WHERE TO LEARN MORE

Beatty, Richard. *Phosphorus.* New York: Benchmark Books, 2001.

Blashfield, Jean F. *Nitrogen.* Austin, TX: Raintree Steck-Vaughn, 1999.

Fitzgerald, Karen. *The Story of Oxygen.* New York: F. Watts, 1996.

"*Group Trends: Selected Nonmetals*" (Web site). <http://wine1.sb.fsu.edu/chm1045/notes/Periodic/NMTrends/Period08.htm> (May 29, 2001).

Knapp, Brian J. *Elements.* Illustrated by David Woodroffe and David Hardy. Danbury, CT: Grolier Educational, 1996.

"*Nonmetals*" Rutgers University Department of Chemistry (Web site). <http://chemistry.rutgers.edu/genchem/nonmet.html> (May 29, 2001).

"*Nonmetals, Semimetals, and Their Compounds*" (Web site). <http://ull.chemistry.uakron.edu/genchem/21/title.html> (May 29, 2001).

WebElements (Web site). <http://www.webelements.com> (May 22, 2001).

METALLOIDS

CONCEPT

The term "metalloid" may sound like a reference to a heavy-metal music fan, but in fact it describes a small collection of elements on the right-hand side of the periodic table. Forming a diagonal between boron and astatine, which lies four rows down and four columns to the right of boron, the metalloids are six elements that display qualities of both metals and nonmetals. (Some classifications include boron and astatine as well, but in this book, they are treated as a nonmetal and a halogen respectively.) Of these six—silicon, germanium, arsenic, antimony, tellurium, and polonium—only a few are household names. People know that arsenic is poison, and they have some general sense that silicon is important in surgical implants. But most people do not know that silicon, also the principal material in sand and glass, is the second-most plentiful element on Earth. Without silicon, from which computer chips are made, our computer-based society simply could not exist.

HOW IT WORKS

FAMILIES AND "ORPHANS"

Most elements fit into some sort of "family" grouping: the alkali metals, the alkaline earth metals, transition metals, lanthanides, actinides, halogens, and noble gases. These seven families, five metallic and two nonmetallic, account for 91 of 112 elements on the periodic table.

The "orphan" elements, or those not readily classifiable within a family, occupy groups 3 through 6 on the North American version of the periodic table, which only numbers the "tall" columns.

Twenty elements, occupying a rectangle that stretches across four groups or columns, and five periods or rows, are "orphans." (This is in addition to hydrogen, on Period 1, Group 1.) These are best classified, not by family, but by cross-family characteristics that unite large groups of elements. These cross-family groupings can be likened to nations: on one level, a person identifies with his or her family ("I'm a Smith"); but on another level, a person has a national identity ("I'm an American.")

Likewise gold, for instance, is both a member of the transition metals family, and of the larger metals grouping. The "orphan" elements, because they have no family, are best identified with characteristics that unite a large body of elements. (For this reason, the "orphan" metals are discussed in the Metals essay, and the "orphan" nonmetals in the Nonmetals essay.)

BETWEEN METALS AND NONMETALS

Of the twenty "orphan" elements in groups 3 through 6, seven are metals and seven nonmetals. Between them runs a diagonal, comprising the six metalloids. (As noted earlier, boron is sometimes considered a metalloid, but in this book, it is discussed in the Nonmetals essay, while astatine—also sometimes grouped with the metalloids—is treated in the Halogens essay.) What is a metalloid? The best way to answer that question is by evaluating the differences between a metal and a nonmetal.

On the periodic table, metals fill the left, center, and part of the right-hand side of the

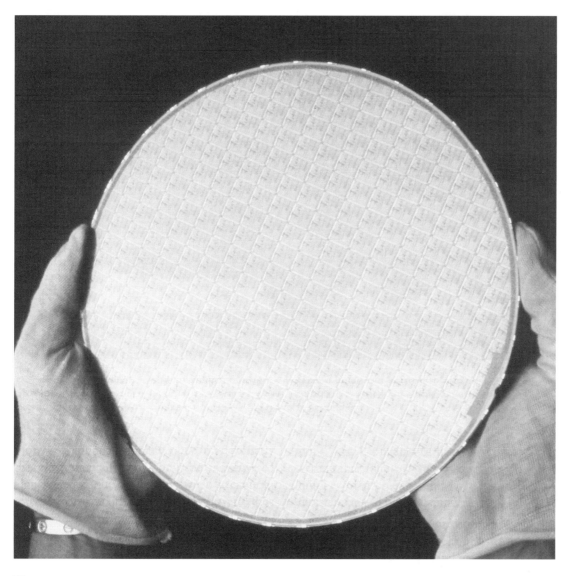

WITHOUT SILICON, FROM WHICH COMPUTER CHIPS ARE MADE, OUR COMPUTER-BASED SOCIETY SIMPLY COULD NOT EXIST. HERE, A WORKER HOLDS A SILICON WAFER WITH INTEGRATED CIRCUITS. *(Charles O'Rear/Corbis. Reproduced by permission.)*

chart, as well as the two rows—corresponding to the lanthanides and actinides—placed separately at the bottom. Thus it should not come as a surprise that most elements (87, in fact) are metals. Metals are lustrous or shiny in appearance, and ductile, meaning that they can be molded into different shapes without breaking. They are excellent conductors of heat and electricity, and tend to form positive ions by losing electrons.

Nonmetals, which occupy the upper right-hand side of the periodic table, include the noble gases, halogens, and the seven "orphan" elements alluded to above. With the addition of hydrogen (which, as noted, is covered in a separate essay), they comprise 19 elements. Whereas all metals

are solids, except for mercury—a liquid at room temperature—the nonmetals are a collection of gases, solids, and one other liquid, bromine. (Not surprisingly, all gaseous elements are nonmetals.) As their name suggests, nonmetals are opposite to metals in most regards: dull in appearance, they are not particularly ductile or malleable. With the exception of carbon, they are poor conductors of heat and electricity, and tend to gain electrons to form negative ions.

SURVEY OF THE METALLOIDS

The metalloids can thus be defined as those elements which exhibit characteristics of both metals and nonmetals. They are all solids, but not

POLONIUM WAS FIRST ISOLATED BY MARIE AND PIERRE CURIE. MARIE CURIE NAMED THE ELEMENT IN HONOR OF HER HOMELAND, POLAND.

lustrous, and conduct heat and electricity moderately well.

The six metalloids treated here are listed below, with the atomic number and chemical symbol of each.

- 14. Silicon (Si)
- 32. Germanium (Ge)
- 33. Arsenic (As)
- 51. Antimony (Sb)
- 52. Tellurium (Te)
- 84. Polonium (Po)

Silicon is the second most abundant element on Earth, comprising 25.7% of the planet's known elemental mass. Together with oxygen, it accounts for nearly three-quarters of the known total. (The planet's core is probably composed largely of iron, with significant deposits of uranium and other metals between the crust and the core, but geologists can only make educated guesses as to the elemental composition beneath the crust.)

THE METALLOIDS IN HISTORY. By far the most important of the metalloids, silicon has been in use as a compound for centuries, but was only isolated in the early nineteenth century. Arsenic and antimony were likewise known from ancient times, and both attracted the attention of alchemists—medieval mystics, in many regards the forerunners of chemists, who believed it was possible to change plain metals into gold.

Traces of the alchemical mindset remained a part of European science as late as the mid-eighteenth century, a tendency exemplified by the fact that tellurium, discovered around that time, was believed to be "unripe gold." On the other hand, the discovery of germanium more than a century later—following predictions of its existence based on the periodic table—reflected a much greater degree of scientific rigor. Finally, the isolation of polonium, known for its radioactivity, was evidence of an even more advanced stage of development in chemistry and science as a whole.

REAL-LIFE APPLICATIONS

SILICON

Swedish chemist Jons Berzelius (1779-1848) discovered silicon in 1823, but humans had been using the element—in the form of compounds known as silicates—for thousands of years. Indeed, silicon may have been one of the first elements formed, many millions of years before life appeared on this planet. Geologists believe that the Earth was once composed primarily of molten iron, oxygen, silicon, and aluminum, still the predominant elements in the planet's crust. Because iron has a greater atomic mass, it settled toward the center, while the more lightweight elements rose to the surface.

Silicon is found in everything from the Sun and other stars, as well as meteorites, to plants and animal bones. Many hundreds of minerals on Earth contain silicon and oxygen in various forms: for instance, silicon and oxygen form silica (SiO_2), commonly known as sand. This sand is mixed with lime and soda (sodium carbonate), as well other substances, to make glass. The purest varieties of silica form quartz, known for its clear crystals, while impure quartz forms crystals of semiprecious gems such as amethyst, opal, and agate.

COMPOUNDS: SILICATES AND SILICONES. Silicon is directly below carbon on the periodic table, and the elements in that group (which includes germanium, selenium, and lead) are sometimes called the "carbon family." Like carbon, silicon forms a huge array of compounds, because it has four electrons to share in chemical bonding, and thus can form long strings of atoms. Because silicon atoms are much larger than carbon atoms, however, not as many other atoms can get close to silicon. Thus the number of possible compounds involving silicon—though still quite impressive—is less than the many millions of carbon-based compounds

Often silicon is combined with oxygen in long chains that use the oxygen atoms as separators. Because oxygen has a valence of two, meaning that it can bond to two other atoms at a time, it bonds easily between two other silicon atoms. The silicon, meanwhile, has two other electrons to use in bonding, so it typically attaches to two more, non-bridging, oxygen atoms. Compounds of this type are known as silicates, and most of the rocks and clay on Earth—with the exception of lime, which is calcium oxide—are silicates made up of combinations of silicon, oxygen, and various metals such as aluminum, iron, sodium, and potassium.

Silicones are another variety of compound involving silicon atoms strung together by bridging oxygen atoms. Instead of attaching to two other, non-bridging oxygen atoms, as in a silicate, the silicon atoms in a silicone attach to organic groups—that is, molecules containing carbon. Because silicones often resemble organic compounds such as oils, greases, and rubber, silicone oils are frequently used in place of organic petroleum as a lubricant because they can withstand greater variations in temperature. Silicone rubbers are used in everything from bouncing balls to space vehicles. And because the body tolerates the introduction of silicone implants better than it does organic ones, silicones are used in surgical implants as well.

THE MANY OTHER USES OF SILICON. Silicones are also present in electrical insulators, rust preventives, fabric softeners, hair sprays, hand creams, furniture and automobile polishes, paints, adhesives (including bathtub sealers), and chewing gum. Yet even this list does not exhaust the many applications of elemental silicon, as opposed to silicone.

Due to its semi-metallic qualities, silicon is used as a semiconductor of electricity. Computer chips are tiny slices of ultra-pure silicon, etched with as many as half a million microscopic and intricately connected electronic circuits. These chips manipulate voltages using binary codes, for which 1 means "voltage on" and 0 means "voltage off." By means of these pulses, silicon chips perform multitudes of calculations in seconds—calculations that would take humans hours or months or even years.

A porous form of silica, known as silica gel, absorbs water vapor from the air, and is often packed alongside moisture-sensitive products such as electronics components, in order to keep them dry. Silicon carbine, an extremely hard crystalline material manufactured by fusing sand with coke (almost pure carbon) at high temperatures, has applications as an abrasive.

ARSENIC

Silicon is unquestionably the "star" of the metalloids, but several other elements in this broad grouping have been known since ancient times. Among these is arsenic, well-known from many a detective novel. Highly poisonous, it is often a fixture of such stories: in a typical plot, a greedy nephew drops some arsenic into the tea of an elderly aunt who has left money to him in her will, and it is up to the detective/hero/heroine to uncover the details behind this dastardly scheme. The use of arsenic as a prop in such tales had already become something of a cliché by 1941, when "Arsenic and Old Lace," a play by Joseph Kesselring that satirized an old-fashioned "whodunit," made its debut on Broadway.

Known since ancient times, arsenic is named after the Greek word *arsenikon*, meaning "yellow orpiment." (Orpiment is arsenic trisulfide, or As_2S_3.) But the Greeks were not the only peoples in antiquity who knew of arsenic and its toxic properties: the Egyptians and Chinese were likewise familiar with it. Alchemists took an interest in arsenic, and one of them, Albertus Magnus (1193-1280)—a German physician who contributed significantly to the rebirth of interest in science during the late Middle Ages—probably isolated the element from arsenic trisulfide in about 1250.

Arsenic is more than just a prop in a detective story. (And in fact it is not a very effective poison for a murderer who hopes to get away with his crime: traces of arsenic remain in the body of a murder victim—even in the person's hair—for years after death.) Today, arsenic is used for purposes such as bronzing, and for hardening and improving the sphericity (roundness) of lead shot in shotgun shells.

In solid-state devices such as transistors, arsenic or antimony is used for the purposes of "doping." This may sound like another version of the whodunit-style poisoning described above, but actually doping involves the alteration of chemical properties by deliberately introducing impurities. Through the doping of silicon, its electrical conductivity is improved.

ANTIMONY

In ancient times, antimony was known by the Latin name *stibium*; hence its chemical symbol of Sb. The word, which can be deciphered as "a metal that does not occur by itself," has roots much older than the Roman civilization, however. Many centuries before the rise of the Romans, the biblical prophet Jeremiah railed against women's use of "stibic stone" to paint their faces. In fact, stibic stone is antimony (III) sulfide, or Sb_2S_3, a black mineral which has long been used as an eyeliner.

During the Middle Ages, alchemists studied *antimony* and its compounds, but the first scientist to distinguish between the compounds and the element itself was French chemist Jean-Baptiste Buquet, in 1771. During the early modern era, antimony was used in making bells, tools, and type for printing presses, because its addition to an alloy strengthened the other metals. Today, antimony is added to lead in storage batteries to make the lead harder and stronger. It is still used in metal type, as well as bullets and cable sheathing.

Antimony was a component in early varieties of matches, but by 1855, the less volatile phosphorus had replaced it. In altered form, however, antimony still appears in some matches. Oxides and sulfides of antimony are also applied in paints, ceramics, fireworks, pharmaceuticals, and other products. These are also utilized for dyeing cloth and fireproofing materials. In addition, as noted above with regard to arsenic, antimony is applied for the doping of silicon and other materials in semiconductors, particularly those that use infrared detectors.

TELLURIUM

When Hungarian chemist Joseph Ramacsaházy first described tellurium in the mid-eighteenth century, he did so in alchemical terms, referring to it as "unripe gold." This reflected the belief, still lingering from the Middle Ages, that metals "grow" to higher states. At least part of his confusion is understandable, since the element did appear in minerals that also contained gold.

A few years later, Baron Franz Joseph Müller von Reichenstein, an Austrian mining inspector in Transylvania, analyzed the substance, which he concluded was bismuth sulfide. Later he changed his mind, then subjected the material to a series of tests over a period of many years, until finally he sent a sample to the distinguished German chemist, Martin Heinrich Klaproth (1743-1817). Klaproth concluded that it was indeed a new element, and suggested the Latin word tellus, for "earth," as a name.

Grayish-white and lustrous, tellurium looks like a metal, but it is much more brittle than most metallic elements. Furthermore, it is a semiconductor, which further separates it from the typically conductive metals. Its semiconductive properties have led to applications in building electronic components. Added to copper and stainless steel, tellurium improves the machinability of those metals; in addition, tellurium compounds are used for the coloring of glass, porcelain, enamel, and ceramics.

GERMANIUM AND POLONIUM

Germanium. Both named after European countries, germanium and polonium were identified much later than the other metalloids. In 1871, two years after he created his famous periodic table, Russian chemist Dmitri Ivanovitch Mendeleev (1834-1907) predicted the existence of an element he called "eka-silicon." This turned out to be germanium, which sits directly beneath silicon on the periodic table, and which was discovered in 1886 by German chemist Clemens Alexander Winkler (1838-1904).

Germanium is principally used as a doping agent in making transistor elements. It is also used as a phosphor in fluorescent lamps. In addition, germanium and germanium oxide, transparent to the infrared portion of the electromagnetic spectrum, are present in infrared spectroscopes and infrared detectors.

POLONIUM. While the applications of germanium are limited compared to those of, say, silicon, polonium is useful primarily as a source

KEY TERMS

ION: An atom or group of atoms that has lost or gained one or more electrons, and thus has a net electric charge.

ISOTOPES: Atoms that have an equal number of protons, and hence are of the same element, but differ in their number of neutrons. This results in a difference of mass. Isotopes may be either stable or unstable. The latter type, known as radioisotopes, are radioactive.

METALLOIDS: Elements which exhibit characteristics of both metals and nonmetals. Metalloids are all solids, but are not lustrous or shiny, and they conduct heat and electricity moderately well. The six metalloids occupy a diagonal region between the metals and nonmetals on the right side of the periodic table. Sometimes boron and astatine are included with the metalloids, but in this volume, they are treated as an "orphan" nonmetal and a halogen, respectively.

METALS: Elements that are lustrous, or shiny in appearance, and malleable, meaning that they can be molded into different shapes without breaking. They are excellent conductors of heat and electricity, and tend to form positive ions by losing electrons.

NONMETALS: Elements that have a dull appearance; are not malleable; are poor conductors of heat and electricity; and tend to gain electrons to form negative ions. They are thus the opposite of metals in most regards, as befits their name.

NORTH AMERICAN SYSTEM: A version of the periodic table of elements that only numbers groups of elements in which the number of valence electrons equals the group number—that is, the two "tall" columns to the left of the transition metals, as well as the six "tall" columns to the right.

"ORPHAN": An element that does not belong to any clearly defined family of elements. The metalloids are all "orphans."

RADIOACTIVITY: A term describing a phenomenon whereby certain isotopes known as radioisotopes are subject to a form of decay brought about by the emission of high-energy particles. "Decay" does not mean that the isotope "rots"; rather, it decays to form another isotope until eventually (though this may take a long time) it becomes stable.

of neutrons in atomic laboratories—hardly something with which a person comes in contact during the course of an ordinary day. Yet the isolation of polonium by Polish-French physicist and chemist Marie Curie (1867-1934), along with her husband Pierre (1859-1906), a French physicist, was a prelude to one of the greatest stories in the history of chemistry.

Setting out to find radioactive elements other than uranium, the Curies refined a large quantity of pitchblende, an ore commonly found in uranium mines. Within a year, they had discovered polonium, which Marie named after her homeland. Polonium is about 100 times as radioactive as uranium, but the Curies were certain that yet another element lay in the pitchblende, and they devoted the next four years to the exhausting process of isolating it. The story of radium, its isolation, and the great price Marie Curie paid for working with radioactive materials is discussed in the essay on Alkaline Earth Metals.

WHERE TO LEARN MORE

"*The Arsenic Website Project.*" *Harvard University Department of Physics* (Web site). <http://phys4.harvard.edu/~wilson/arsenic_project_main.html> (May 30, 2001).

Knapp, Brian J. *Elements.* Illustrated by David Woodroffe and David Hardy. Danbury, CT: Grolier Educational, 1996.

"*Metalloids.*" *ChemicalElements.com* (Web site). <http://www.chemicalelements.com/groups/metalloids.html> (May 30, 2001).

"*Metalloids*" (Web site). <http://www.wealthhealthandwisdom.com/metalloids.asp> (May 30, 2001).

Pflaum, Rosalynd. *Marie Curie and Her Daughter Irène.* Minneapolis, MN: Lerner Publications, 1993.

Thomas, Jens. *Silicon.* New York: Benchmark Books, 2001.

HALOGENS

CONCEPT

Table salt, bleach, fluoride in toothpaste, chlorine in swimming pools—what do all of these have in common? Add halogen lamps to the list, and the answer becomes more clear: all involve one or more of the halogens, which form Group 7 of the periodic table of elements. Known collectively by a term derived from a Greek word meaning "salt-producing," the halogen family consists of five elements: fluorine, chlorine, bromine, iodine, and astatine. The first four of these are widely used, often in combination; the last, on the other hand, is a highly radioactive and extremely rare substance. The applications of halogens are many and varied, including some that are dangerous, controversial, and deadly.

HOW IT WORKS

THE HALOGENS ON THE PERIODIC TABLE

As noted, the halogens form Group 7 of the periodic table of elements. They are listed below, along with chemical symbol and atomic number:

- Fluorine (F) 9
- Chlorine (Cl): 17
- Bromine (Br): 35
- Iodine (I): 53
- Astatine (At): 85

On the periodic table, as displayed in chemistry labs around the world, the number of columns and rows does not vary, since these configurations are the result of specific and interrelated properties among the elements. There are always 18 columns; however, the way in which these are labeled differs somewhat from place to

place. Many chemists outside the United States refer to these as 18 different groups of elements; however, within the United States, a somewhat different system is used.

In many American versions of the chart, there are only eight groups, sometimes designated with Roman numerals. The 40 transition metals in the center are not designated by group number, nor are the lanthanides and actinides, which are set apart at the bottom of the periodic table. The remaining eight columns are the only ones assigned group numbers. In many ways, this is less useful than the system of 18 group numbers; however, it does have one advantage.

ELECTRON CONFIGURATIONS AND BONDING. In the eight-group system, group number designates the number of valence electrons. The valence electrons, which occupy the highest energy levels of an atom, are the electrons that bond one element to another. These are often referred to as the "outer shell" of an atom, though the actual structure is much more complex. In any case, electron configuration is one of the ways halogens can be defined: all have seven valence electrons.

Because the rows in the periodic table indicate increasing energy levels, energy levels rise as one moves up the list of halogens. Fluorine, on row 2, has a valence-shell configuration of $2s^2 2p^5$; while that of chlorine is $3s^2 3p^5$. Note that only the energy level changes, but not the electron configuration at the highest energy level. The same goes for bromine ($4s^2 4p^5$), iodine ($4s^2 4p^5$), and astatine ($5s^2 5p^5$).

All members of the halogen family have the same valence-shell electron configurations, and thus tend to bond in much the same way. As we

SALT, LIKE THE MOUND SHOWN HERE, IS A SAFE AND COMMON SUBSTANCE THAT CONTAINS CHLORINE. (*Charles O'Rear/Corbis. Reproduced by permission.*)

shall see, they are inclined to form bonds more readily than most other substances, and indeed fluorine is the most reactive of all elements.

Thus it is ironic that they are "next door" to the Group 8 noble gases, the least reactive among the elements. The reason for this, as discussed in the Chemical Bonding essay, is that most elements bond in such a way that they develop a valence shell of eight electrons; the noble gases are already there, so they do not bond, except in some cases—and then principally with fluorine.

CHARACTERISTICS OF THE HALOGENS

In terms of the phase of matter in which they are normally found, the halogens are a varied group. Fluorine and chlorine are gases, iodine is a solid, and bromine is one of only two elements that exists at room temperature as a liquid. As for astatine, it is a solid too, but so highly radioactive that it is hard to know much about its properties.

Despite these differences, the halogens have much in common, and not just with regard to their seven valence electrons. Indeed, they were identified as a group possessing similar characteristics long before chemists had any way of knowing about electrons, let alone electron con-

figurations. One of the first things scientists noticed about these five elements is the fact that they tend to form salts. In everyday terminology, "salt" refers only to a few variations on the same thing—table salt, sea salt, and the like. In chemistry, however, the meaning is much broader: a salt is defined as the result of bonding between an acid and a base.

Many salts are formed by the bonding of a metal and a nonmetal. The halogens are all nonmetals, and tend to form salts with metals, as in the example of sodium chloride (NaCl), a bond between chlorine, a halogen, and the metal sodium. The result, of course, is what people commonly call "salt." Due to its tendency to form salts, the first of the halogens to be isolated—chlorine, in 1811—was originally named "halogen." This is a combination of the Greek words *halos*, or salt, and *gennan*, "to form or generate."

In their pure form, halogens are diatomic, meaning that they exist as molecules with two atoms: F_2, Cl_2, and so on. When bonding with metals, they form ionic bonds, which are the strongest form of chemical bond. In the process, halogens become negatively charged ions, or anions. These are represented by the symbols F-, Cl-, Br-, and I-, as well as the names fluoride, chloride, bromide, and iodide. All of the halogens are highly reactive, and will combine directly with almost all elements.

Due to this high level of reactivity, the halogens are almost never found in pure form; rather, they have to be extracted. Extraction of halogens is doubly problematic, because they are dangerous. Exposure to large quantities can be harmful or fatal, and for this reason halogens have been used as poisons to deter unwanted plants and insects—and, in one of the most horrifying chapters of twentieth century military history, as a weapon in World War I.

REAL-LIFE APPLICATIONS

CHLORINE

Chlorine is a highly poisonous gas, greenish-yellow in color, with a sharp smell that induces choking in humans. Yet, it can combine with other elements to form compounds safe for human consumption. Most notable among these compounds is salt, which has been used as a food preservative since at least 3000 B.C.

Salt, of course, occurs in nature. By contrast, the first chlorine compound made by humans was probably hydrochloric acid, created by dissolving hydrogen chloride gas in water. The first scientist to work with hydrochloric acid was Persian physician and alchemist Rhazes (ar-Razi; c. 864-c. 935), one of the most outstanding scientific minds of the medieval period. Alchemists, who in some ways were the precursors of true chemists, believed that base metals such as iron could be turned into gold. Of course this is not possible, but alchemists in about 1200 did at least succeed in dissolving gold using a mixture of hydrochloric and nitric acids known as *aqua regia*.

The first modern scientist to work with chlorine was Swedish chemist Carl W. Scheele (1742-1786), who also discovered a number of other elements and compounds, including barium, manganese, oxygen, ammonia, and glycerin. However, Scheele, who isolated it in 1774, thought that chlorine was a compound; only in 1811 did English chemist Sir Humphry Davy (1778-1829) identify it as an element. Another chemist had suggested the name "halogen" for the alleged compound, but Davy suggested that it

THIS WORKER IS REMOVING FREON FROM AN OLD REFRIGERATOR. CFCS LIKE FREON MAY BE RESPONSIBLE FOR THE DEPLETION OF THE OZONE LAYER. *(James A. Sugar/Corbis. Reproduced by permission.)*

be called chlorine instead, after the Greek word *chloros*, which indicates a sickly yellow color.

USES OF CHLORINE. The dangers involved with chlorine have made it an effective substance to use against stains, plants, animals—and even human beings. Chlorine gas is highly irritating to the mucous membranes of the nose, mouth, and lungs, and it can be detected in air at a concentration of only 3 parts per million (ppm).

The concentrations of chlorine used against troops on both sides in World War I (beginning in 1915) was, of course, much higher. Thanks to the use of chlorine gas and other antipersonnel agents, one of the most chilling images to emerge from that conflict was of soldiers succumbing to poisonous gas. Yet just as it is harmful to humans, chlorine can be harmful to microbes, thus preserving human life. As early as 1801, it had been used in solutions as a disinfectant; in 1831, its use in hospitals made it effective as a weapon against a cholera epidemic that swept across Europe.

Another well-known use of chlorine is as a bleaching agent. Until 1785, when chlorine was first put to use as a bleach, the only way to get stains and unwanted colors out of textiles or paper was to expose them to sunlight, not always an effective method. By contrast, chlorine, still used as a bleach today, can be highly effective—a good reason not to use regular old-fashioned bleach on anything other than white clothing. (Since the 1980s, makers of bleaches have developed all-color versions to brighten and take out stains from clothing of other colors.)

Calcium hydrocholoride (CaOCl), both a bleaching powder and a disinfectant used in swimming pools, combines both the disinfectant and bleaching properties of chlorine. This and the others discussed here are just some of many, many compounds formed with the highly reactive element chlorine. Particularly notable—and controversial—are compounds involving chlorine and carbon.

CHLORINE AND ORGANIC COMPOUNDS. Chlorine bonds well with organic substances, or those containing carbon. In a number of instances, chlorine becomes part of an organic polymer such as PVC (polyvinyl chloride), used for making synthetic pipe. Chlorine polymers are also applied in making synthetic rubber, or neoprene. Due to its resistance to heat, oxidation, and oils, neoprene is used in a number of automobile parts.

The bonding of chlorine with substances containing carbon has become increasingly controversial because of concerns over health and the environment, and in some cases chlorine-carbon compounds have been outlawed. Such was the fate of DDT, a pesticide soluble in fats and oils rather than in water. When it was discovered that DDT was carcinogenic, or cancer-causing, in humans and animals, its use in the United States was outlawed.

Other, less well-known, chlorine-related insecticides have likewise been banned due to their potential for harm to human life and the environment. Among these are chlorine-containing materials once used for dry cleaning. Also notable is the role of chlorine in chlorofluorocarbons (CFCs), which have been used in refrigerants such as Freon, and in propellants for aerosol sprays. CFCs tend to evaporate easily, and concerns over their effect on Earth's atmosphere have led to the phasing out of their use.

FLUORINE

Fluorine has the distinction of being the most reactive of all the elements, with the highest electronegativity value on the periodic table. Because of this, it proved extremely difficult to isolate. Davy first identified it as an element, but was poisoned while trying unsuccessfully to decompose hydrogen fluoride. Two other chemists were also later poisoned in similar attempts, and one of them died as a result.

French chemist Edmond Fremy (1814-1894) very nearly succeeded in isolating fluorine, and though he failed to do so, he inspired his student Henri Moissan (1852-1907) to continue the project. One of the problems involved in isolating this highly reactive element was the fact that it tends to "attack" any container in which it is placed: most metals, for instance, will burst into flames in the presence of fluorine. Like the others before him, Moissan set about to isolate fluorine from hydrogen fluoride by means of electrolysis—the use of an electric current to cause a chemical reaction—but in doing so, he used a platinum-iridium alloy that resisted attacks by fluorine. In 1906, he received the Nobel Prize for his work, and his technique is still used today in modified form.

PROPERTIES AND USES OF FLUORINE. A pale green gas of low density, fluorine can combine with all elements except some of the noble gases. Even water will burn in the presence of this highly reactive substance. Fluorine is also highly toxic, and can cause severe burns on contact, yet it also exists in harmless compounds, primarily in the mineral known as fluorspar, or calcium fluoride. The latter gives off a fluorescent light (fluorescence is the term for a type of light not accompanied by heat), and fluorine was named for the mineral that is one of its principal "hosts".

Beginning in the 1600s, hydrofluoric acid was used for etching glass, and is still used for that purpose today in the manufacture of products such as light bulbs. The oil industry uses it as a catalyst—a substance that speeds along a chemical reaction—to increase the octane number in gasoline. Fluorine is also used in a polymer commonly known as Teflon, which provides a non-stick surface for frying pans and other cooking-related products.

Just as chlorine saw service in World War I, fluorine was enlisted in World War II to create a weapon far more terrifying than poison gas: the atomic bomb. Scientists working on the Manhattan Project, the United States' effort to develop the bombs dropped on Japan in 1945, needed large quantities of the uranium-235 isotope. This they obtained in large part by diffusion of the compound uranium hexafluoride, which consists of molecules containing one uranium atom and six fluorine anions.

FLUORIDATION OF WATER. Long before World War II, health officials in the United States noticed that communities having high concentration of fluoride in their drinking water tended to suffer a much lower incidence of tooth decay. In some areas the concentration of fluoride in the water supply was high enough that it stained people's teeth; still, at the turn of the century—an era when dental hygiene as we know it today was still in its infancy—the prevention of tooth decay was an attractive prospect. Perhaps, officials surmised, it would be possible to introduce smaller concentrations of fluoride into community drinking water, with a resulting improvement in overall dental health.

After World War II, a number of municipalities around the United States undertook the fluoridation of their water supplies, using concentrations as low as 1 ppm. Within a few years, fluoridation became a hotly debated topic, with proponents pointing to the potential health benefits and opponents arguing from the standpoint of issues not directly involved in science. It was an invasion of personal liberty, they said, for governments to force citizens to drink water which had been supplemented with a foreign substance.

During the 1950s, in fact, fluoridation became associated in some circles with Communism—just another manifestation of a government trying to control its citizens. In later years, ironically, antifluoridation efforts became associated with groups on the political left rather than the right. By then, the argument no longer revolved around the issue of government power; instead the concern was for the health risks involved in introducing a substance lethal in large doses.

Fluoride had meanwhile gained application in toothpastes. Colgate took the lead, introducing "stannous fluoride" in 1955. Three years later, the company launched a memorable advertising campaign with commercials in which a little girl showed her mother a "report card" from the den-

tist and announced "Look, Ma! No cavities!" Within a few years, virtually all brands of toothpaste used fluoride; however, the use of fluoride in drinking water remained controversial.

As late as 1993, in fact, the issue of fluoridation remained heated enough to spawn a study by the U.S. National Research Council. The council found some improvement in dental health, but not as large as had been claimed by early proponents of fluoridation. Furthermore, this improvement could be explained by reference to a number of other factors, including fluoride in toothpastes and a generally heightened awareness of dental health among the U.S. populace.

CHLOROFLUOROCARBONS. Another controversial application of fluorine is its use, along with chlorine and carbon, in chlorofluorocarbons. As noted above, CFCs have been used in refrigerants and propellants; another application is as a blowing agent for polyurethane foam. This continued for several decades, but in the 1980s, environmentalists became concerned over depletion of the ozone layer high in Earth's atmosphere.

Unlike ordinary oxygen (O_2), ozone or O_3 is capable of absorbing ultraviolet radiation from the Sun, which would otherwise be harmful to human life. It is believed that CFCs catalyze the conversion of ozone to oxygen, and that this may explain the "ozone hole," which is particularly noticeable over the Antarctic in September and October.

As a result, a number of countries signed an agreement in 1996 to eliminate the manufacture of halocarbons, or substances containing halogens and carbon. Manufacturers in countries that signed this agreement, known as the Montreal Protocol, have developed CFC substitutes, most notably hydrochlorofluorocarbons (HCFCs), CFC-like compounds also containing hydrogen atoms.

The ozone-layer question is far from settled, however. Critics argue that in fact the depletion of the ozone layer over Antarctica is a natural occurrence, which may explain why it only occurs at certain times of year. This may also explain why it happens primarily in Antarctica, far from any place where humans have been using CFCs. (Ozone depletion is far less significant in the Arctic, which is much closer to the population centers of the industrialized world.)

In any case, natural sources, such as volcano eruptions, continue to add halogen compounds to the atmosphere.

Bromine

Bromine is a foul-smelling reddish-brown liquid whose name is derived from a Greek word meaning "stink." With a boiling point much lower than that of water—137.84°F (58.8°C)—it readily transforms into a gas. Like other halogens, its vapors are highly irritating to the eyes and throat. It is found primarily in deposits of brine, a solution of salt and water. Among the most significant brine deposits are in Israel's Dead Sea, as well as in Arkansas and Michigan.

Credit for the isolation of bromine is usually given to French chemist Antoine-Jérôme Balard (1802-1876), though in fact German chemist Carl Löwig (1803-1890) actually isolated it first, in 1825. However, Balard, who published his results a year later, provided a much more detailed explanation of bromine's properties.

The first use of bromine actually predated both men by several millennia. To make their famous purple dyes, the Phoenicians used murex mollusks, which contained bromine. (Like the names of the halogens, the word "Phoenicians" is derived from Greek—in this case, a word meaning "red" or "purple," which referred to their dyes.) Today bromine is also used in dyes, and other modern uses include applications in pesticides, disinfectants, medicines, and flame retardants.

At one time, a compound containing bromine was widely used by the petroleum industry as an additive for gasoline containing lead. Ethylene dibromide reacts with the lead released by gasoline to form lead bromide ($PbBr_2$), referred to as a "scavenger," because it tends to clean the emissions of lead-containing gasoline. However, leaded gasoline was phased out during the late 1970s and early 1980s; as a result, demand for ethylene dibromide dropped considerably.

HALOGEN LAMPS. The name "halogen" is probably familiar to most people because of the term "halogen lamp." Used for automobile headlights, spotlights, and floodlights, the halogen lamp is much more effective than ordinary incandescent light. Incandescent "heat-producing" light was first developed in the 1870s and improved during the early part of the twentieth century with the replacement of carbon by tungsten as the principal material in the filament, the area that is heated.

Tungsten proved much more durable than carbon when heated, but it has a number of problems when combined with the gases in an incandescent bulb. As the light bulb continues to burn for a period of time, the tungsten filament begins to thin and will eventually break. At the same time, tungsten begins to accumulate on the surface of the bulb, dimming its light. However, by adding bromine and other halogens to the bulb's gas filling—thus making a halogen lamp—these problems are alleviated.

As tungsten evaporates from the filament, it combines with the halogen to form a gaseous compound that circulates within the bulb. Instead of depositing on the surface of the bulb, the compound remains a gas until it comes into contact with the filament and breaks down. It is then redeposited on the filament, and the halogen gas is free to combine with newly evaporated tungsten. Though a halogen bulb does eventually break down, it lasts much longer than an ordinary incandescent bulb and burns with a much brighter light. Also, because of the decreased tungsten deposits on the surface, it does not begin to dim as it nears the end of its life.

Iodine

First isolated in 1811 from ashes of seaweed, iodine has a name derived from the Greek word meaning "violet-colored"—a reference to the fact it forms dark purple crystals. During the 1800s, iodine was obtained commercially from mines in Chile, but during the twentieth century wells of brine in Japan, Oklahoma, and Michigan have proven a better source.

Among the best-known properties of iodine is its importance in the human diet. The thyroid gland produces a growth-regulating hormone that contains iodine, and lack of iodine can cause a goiter, a swelling around the neck. Table salt does not naturally contain iodine; however, sodium chloride sold in stores usually contains about 0.01% sodium iodide, added by the manufacturer.

Iodine was once used in the development of photography: during the early days of photographic technology, the daguerreotype process used silver plates sensitized with iodine vapors.

KEY TERMS

ANION: The negative ion that results when an atom gains one or more electrons. An anion (pronounced "AN-ie-un") of an element is never called, for instance, the chlorine anion. Rather, an anion involving a single element is named by adding the suffix -ide to the name of the original element—in this case, "chloride." Other rules apply for more complex anions.

ATOMIC NUMBER: The number of protons in the nucleus of an atom. Since this number is different for each element, elements are listed on the periodic table of elements in order of atomic number.

CHEMICAL SYMBOL: A one-or two-letter abbreviation for the name of an element.

DIATOMIC: A term describing an element that exists as molecules composed of two atoms. All of the halogens are diatomic.

ELECTROLYSIS: The use of an electrical current to cause a chemical reaction.

HALF-LIFE: The length of time it takes a substance to diminish to one-half its initial amount.

HALOGENS: Group 7 of the periodic table of elements, including fluorine, chlorine, bromine, iodine, and astatine. The halogens are diatomic, and tend to form salts; hence their name, which comes from two Greek terms meaning "salt-forming."

ION: An atom that has lost or gained one or more electrons, and thus has a net electric charge.

IONIC BONDING: A form of chemical bonding that results from attractions between ions with opposite electrical charges.

ISOTOPES: Atoms that have an equal number of protons, and hence are of the same element, but differ in their number of neutrons. This results in a difference of mass. Isotopes may be either stable or unstable—that is, radioactive.

PERIODIC TABLE OF ELEMENTS: A chart that shows the elements arranged in order of atomic number. Vertical columns within the periodic table indicate groups or "families" of elements with similar chemical characteristics.

POLYMER: A large molecule containing many small units that hook together.

RADIOACTIVITY: A term describing a phenomenon whereby certain materials are subject to a form of decay brought about by the emission of high-energy particles. "Decay" in this sense does not mean "rot"; instead, radioactive isotopes continue to emit particles, changing into isotopes of other elements, until they become stable.

SALT: A compound formed by the reaction of an acid with a base. Salts are usually formed by the joining of a metal and a nonmetal.

VALENCE ELECTRONS: Electrons that occupy the highest energy levels in an atom, and are involved in chemical bonding. The halogens all have seven valence electrons.

Iodine compounds are used today in chemical analysis and in synthesis of organic compounds.

ASTATINE

Just as fluorine has the distinction of being the most reactive, astatine is the rarest of all the elements. Long after its existence was predicted, chemists still had no luck finding it in nature, and it was only created in 1940 by bombarding bismuth with alpha particles (positively charged helium nuclei). The newly isolated element was given a Greek name meaning "unstable."

Indeed, none of astatine's 20 known isotopes is stable, and the longest-lived has a half-life of only 8.3 hours. This has only added to the difficulties involved in learning about this strange element, and therefore it is difficult to say what applications, if any, astatine may have. The most promising area involves the use of astatine to treat a condition known as hyperthyroidism, related to an overly active thyroid gland.

WHERE TO LEARN MORE

"The Chemistry of the Halogens." Purdue University Department of Chemistry (Web site). <http://chemed.chem.purdue.edu/genchem/topicreview/bp/ch10/group7.html> (May 20, 2001).

"Halogens." Chemical Elements.com (Web site). <http://www.chemicalelements.com/groups/halogens.html> (May 20, 2001).

"Halogens." Corrosion Source (Web site). <http://www.corrosionsource.com/handbook/periodic/halogens.htm> (May 20, 2001).

"Halogens" (Web site). <http://registrar.ies.ncsu.edu/ol_2000/module6/halogen/halogen/htm> (May 20, 2001).

"The Halogens" (Web site). <http://www.nidlink.com/~jfromm/elements/halogen.htm> (May 20, 2001).

Knapp, Brian J. Chlorine, Fluorine, Bromine, and Iodine. Henley-on-Thames, England: Atlantic Europe, 1996.

Oxlade, Chris. Elements and Compounds. Chicago: Heinemann Library, 2001.

Stwertka, Albert. A Guide to the Elements. New York: Oxford University Press, 1998.

"Visual Elements: Group VII—The Halogens" (Web site). <http://www.chemsoc.org/viselements/pages/data/intro_groupvii_data.html> (May 20, 2001).

NOBLE GASES

CONCEPT

Along the extreme right-hand column of the periodic table of elements is a group known as the noble gases: helium, neon, argon, krypton, xenon, and radon. Also known as the rare gases, they once were called inert gases, because scientists believed them incapable of reacting with other elements. Rare though they are, these gases are a part of everyday life, as evidenced by the helium in balloons, the neon in signs—and the harmful radon in some American homes.

HOW IT WORKS

DEFINING THE NOBLE GASES

The periodic table of elements is ordered by the number of protons in the nucleus of an atom for a given element (the atomic number), yet the chart is also arranged in such a way that elements with similar characteristics are grouped together. Such is the case with Group 8, which is sometimes called Group 18, a collection of non-metals known as the noble gases. The six noble gases are helium (He), neon (Ne), argon (Ar), krypton (Kr), xenon (Xe), and radon (Rn). Their atomic numbers are, respectively, 2, 10, 18, 36, 54, and 86.

Several characteristics, aside from their placement on the periodic table, define the noble gases. Obviously, all are gases, meaning that they only form liquids or solids at extremely low temperatures—temperatures that, on Earth at least, are usually only achieved in a laboratory. They are colorless, odorless, and tasteless, as well as monatomic—meaning that they exist as individual atoms, rather than in molecules. (By contrast,

atoms of oxygen—another gas, though not among this group—usually combine to form a molecule, O_2.)

LOW REACTIVITY

There is a reason why noble gas atoms tend not to combine: one of the defining characteristics of the noble gas "family" is their lack of chemical reactivity. Rather than reacting to, or bonding with, other elements, the noble gases tend to remain apart—hence the name "noble," implying someone or something that is set apart from the crowd, as it were. Due to their apparent lack of reactivity, the noble gases—also known as the rare gases—were once known as the inert gases.

Indeed, helium, neon, and argon have not been found to combine with other elements to form compounds. However, in 1962 English chemist Neil Bartlett (1932-) succeeded in preparing a compound of xenon with platinum and fluorine ($XePtF_6$), thus overturning the idea that the noble gases were entirely "inert." Since that time, numerous compounds of xenon with other elements, most notably oxygen and fluorine, have been developed. Fluorine has also been used to form simple compounds with krypton and radon.

Nonetheless, low reactivity—instead of no reactivity, as had formerly been thought—characterizes the rare gases. One of the factors governing the reactivity of an element is its electron configuration, and the electrons of the noble gases are arranged in such a way as to discourage bonding with other elements.

A SCIENTIST USING THE POTASSIUM-ARGON DATING PROCESS TO DETERMINE THE AGE OF MATERIALS. *(Dean Conger/Corbis. Reproduced by permission.)*

REAL-LIFE APPLICATIONS

ISOLATION OF THE NOBLE GASES

HELIUM. Helium is an unusual element in many respects—not least because it is the only element to have first been identified in the Solar System before it was discovered on Earth. This is significant, because the elements on Earth are the same as those found in space: thus, it is more than just an attempt at sounding poetic when scientists say that humans, as well as the world around them, are made from "the stuff of stars."

In 1868, a French astronomer named Pierre Janssen (1824-1907) was in India to observe a total solar eclipse. To aid him in his observations, he used a spectroscope, an instrument for analyzing the spectrum of light emitted by an object. What Janssen's spectroscope showed was surprising: a yellow line in the spectrum, never seen before, which seemed to indicate the presence of a previously undiscovered element. Janssen called it "helium" after the Greek god Helios, or Apollo, whom the ancients associated with the Sun.

Janssen shared his findings with English astronomer Sir Joseph Lockyer (1836-1920), who had a worldwide reputation for his work in analyzing light waves. Lockyer, too, believed that what Janssen had seen was a new element, and a few months later, he observed the same unusual spectral lines. At that time, the spectroscope was still a new invention, and many members of the worldwide scientific community doubted its usefulness, and therefore, in spite of Lockyer's reputation, they questioned the existence of this "new" element. Yet during their lifetimes, Janssen and Lockyer were proven correct.

NEON, ARGON, KRYPTON, AND XENON. They had to wait a quarter century, however. In 1893, English chemist Sir William Ramsay (1852-1916) became intrigued by the presence of a mysterious gas bubble left over when nitrogen from the atmosphere was combined with oxygen. This was a phenomenon that had also been noted by English physicist Henry Cavendish (1731-1810) more than a century before, but Cavendish could offer no explanation. Ramsay, on the other hand, had the benefit of observations made by English physicist John William Strutt, Lord Rayleigh (1842-1919).

Up to that time, scientists believed that air consisted only of oxygen, carbon dioxide, and water vapor. However, Rayleigh had noticed that when nitrogen was extracted from air after a process of removing those other components, it had a slightly higher density than nitrogen prepared from a chemical reaction. In light of his own observations, Ramsay concluded that whereas nitrogen obtained from chemical reactions was pure, the nitrogen extracted from air contained trace amounts of an unknown gas.

Ramsay was wrong in only one respect: hidden with the nitrogen was not one gas, but five. In order to isolate these gases, Ramsay and Rayleigh subjected air to a combination of high pressure and low temperature, allowing the various gases to boil off at different temperatures. One of the gases was helium—the first confirmation that the element existed on Earth—but the other four gases were previously unknown. The Greek roots of the names given to the four gases reflected scientists' wonder at discovering these hard-to-find elements: *neos* (new), *argos* (inactive), *kryptos* (hidden), and *xenon* (stranger).

RADON. Inspired by the studies of Polish-French physicist and chemist Marie Curie (1867-1934) regarding the element radium and the phenomenon of radioactivity (she discovered the element, and coined the latter term), German physicist Friedrich Dorn (1848-1916) became fascinated with radium. Studying the element, he discovered that it emitted a radioactive gas, which he dubbed "radium emanation." Eventually, however, he realized that what was being produced was a new element. This was the first clear proof that one element could become another through the process of radioactive decay.

Ramsay, who along with Rayleigh had received the Nobel Prize in 1904 for his work on the noble gases, was able to map the new element's spectral lines and determine its density and atomic mass. A few years later, in 1918, another scientist named C. Schmidt gave it the name "radon." Due to its behavior and the configuration of its electrons, chemists classified radon among what they continued to call the "inert gases" for another half-century—until Bartlett's preparation of xenon compounds in 1962.

PRESENCE OF THE RARE GASES ON EARTH

IN THE ATMOSPHERE. Though the rare gases are found in minerals and meteorites on Earth, their greatest presence is in the planet's atmosphere. It is believed that they were

LAS VEGAS HAS NUMEROUS EXAMPLES OF SIGNS ILLUMINATED BY NEON. *(Dave Bartruff/Corbis. Reproduced by permission.)*

released into the air long ago as a by-product of decay on the part of radioactive materials in the Earth's crust. Within the atmosphere, argon is the most "abundant"—in comparative terms, given the fact that the "rare gases" are, by definition, rare.

Nitrogen makes up about 78% of Earth's atmosphere and oxygen 21%, meaning that these two elements constitute fully 99% of the air above the Earth. Argon ranks a distant third, with 0.93%. The remaining 0.07% is made up on water vapor, carbon dioxide, ozone (O_3), and traces of the noble gases. These are present in such small quantities that the figures for them are not typically presented as percentages, but rather in terms of parts per million (ppm). The concentrations of neon, helium, krypton, and xenon in the atmosphere are 18, 5, 1, and 0.09 ppm respectively.

IN THE SOIL. Radon in the atmosphere is virtually negligible, which is a fortunate thing, in light of its radioactive qualities. Few Americans, in fact, even knew of its existence until 1988, when the United States Environmental Protection Agency (EPA) released a report estimating that some ten million American homes had potentially harmful radon levels. This

A RESEARCHER WORKS WITH LIQUID HELIUM AS PART OF AN EXPERIMENT ON COSMIC BACKGROUND RADIATION. CLOSE TO ABSOLUTE ZERO, HELIUM TRANSFORMS INTO A HIGHLY UNUSUAL LIQUID THAT HAS NO MEASURABLE RESISTANCE TO FLOW. (*Roger Ressmeyer/Corbis. Reproduced by permission.*)

set off a scare, and during the late 1980s and 1990s, sales of home radon detectors boomed. Meanwhile, the federal government increased concerns with additional reports, advising people to seal their basements and ventilate their homes if radon exceeded certain levels.

A number of scientists have disputed the government's claims, yet some regions of the United States appear to be at relatively high risk due to the presence of radon in the soil. The element seems to be most plentiful in soils containing high concentrations of uranium. If radon is present in a home that has been weather-sealed to improve the efficiency of heating and cooling systems, it is indeed potentially dangerous to the residents.

Chinese scientists in the 1960s made an interesting discovery regarding radon and its application to seismography, or the area of the earth sciences devoted to studying and predicting earthquakes. Radon levels in groundwater, the Chinese reports showed, rise considerably just before an earthquake. Since then, the Chinese have monitored radon concentrations in water, and used this data to predict earthquakes.

EXTRACTING RARE GASES. Radon, in fact, is not the only rare gas that can be obtained as the result of radioactive decay: in 1903, Ramsay and British chemist Frederick Soddy (1877-1956) showed that the breakdown of either uranium or radium results in the production of helium atoms (beta particles). A few years later, English physicist Ernest Rutherford (1871-1937) demonstrated that radiation carrying a positive electrical charge (alpha rays) was actually a stream of helium atoms stripped of an electron.

Many of the noble gases are extracted by liquefying air—that is, by reducing it to temperatures at which it assumes the properties of a liquid rather than a gas. By controlling temperatures in the liquefied air, it is possible to reach the boiling point for a particular noble gas and thereby extract it, much as was done when these gases were first isolated in the 1890s.

THE UNIQUE SITUATION OF HELIUM. Helium is remarkable, in that it only liquefies at a temperature of -457.6°F (-272°C), just above absolute zero. Absolute zero is the temperature at which the motion of atoms or molecules comes to a virtual stop, but the motion of helium atoms never completely ceases. In order to liquefy it, in fact, even at those low temperatures, it must be subjected to pressures many times that exerted by Earth's atmosphere.

Given these facts, it is difficult to extract helium from air. More often, it is obtained from natural gas wells, where it is present in relatively large concentrations—between 1% and 7% of the natural gas. The majority of the Earth's helium supply belongs to the United States, where the greatest abundance of helium-supplying wells are in Texas, Oklahoma, and Kansas. During World War II, the United States took advantage of this supply of relatively inexpensive helium to provide buoyancy for a fleet of airships used for reconnaissance.

There is one place with an abundant supply of helium, but there are no plans for a mining expedition any time soon. That place is the Sun, where the nuclear fusion of hydrogen atoms creates helium. Indeed, helium seems to be the most plentiful element of all, after hydrogen, constituting 23% of the total mass of the universe. Why, then, is it so difficult to obtain on Earth? Most likely because it is so light in comparison to air; it simply floats off into space.

APPLICATIONS FOR THE NOBLE GASES

RADON, ARGON, KRYPTON, AND XENON. Though radon is known primarily for the hazards it poses to human life and well-being, it has useful applications. As noted above, its presence in groundwater appears to provide a possible means of predicting earthquakes. In addition, it is used for detecting leaks, measuring flow rates, and inspecting metal welds.

One interesting use of argon and, in particular, the stable isotope argon-40, is in dating techniques used by geologists, paleontologists, and other scientists studying the distant past. When volcanic rocks are subjected to extremely high temperatures, they release argon, and as the rocks cool, argon-40 accumulates. Because argon-40 is formed by the radioactive decay of a potassium isotope, potassium-40, the amount of argon-40 that forms is proportional to the rate of decay for potassium-40. The latter has a half-life of 1.3 billion years, meaning that it takes 1.3 billion years for half the potassium-40 originally present to be converted to argon-40. Using argon-40, paleontologists have been able to estimate the age of

volcanic layers above and below fossil and artifact remains in east Africa.

Krypton has a number of specialized applications—for instance, it is mixed with argon and used in the manufacture of windows with a high level of thermal efficiency. Used in lasers, it is often mixed with a halogen such as fluorine. In addition, it is also sometimes used in halogen sealed-beam headlights. Many fans of *Superman*, no doubt, were disappointed at some point in their lives to discover that there is no such thing as "kryptonite," the fictional element that caused the Man of Steel to lose his legendary strength. Yet krypton—the real thing—has applications that are literally out of this world. In the development of fuel for space exploration, krypton is in competition with its sister element, xenon. Xenon offers better performance, but costs about ten times more to produce; thus krypton has become more attractive as a fuel for space flight.

In addition to its potential as a space fuel, xenon is used in arc lamps for motion-picture film projection, in high-pressure ultraviolet radiation lamps, and in specialized flashbulbs used by photographers. One particular isotope of xenon is utilized for tracing the movement of sands along a coastline. Xenon is also applied in high-energy physics for detecting nuclear radiation in bubble chambers. Furthermore, neuroscientists are experimenting with the use of xenon in diagnostic procedures to clarify x-ray images of the human brain.

NEON. Neon, of course, is best-known for its application in neon signs, which produce an eye-catching glow when lit up at night. French chemist Georges Claude (1870-1960), intrigued by Ramsay's discovery of neon, conducted experiments that led to the development of the neon light in 1910. That first neon light was simply a glass tube filled with neon gas, which glowed a bright red when charged with electricity.

Claude eventually discovered that mixing other gases with neon produced different colors of light. He also experimented with variations in the shapes of glass tubes to create letters and pictures. By the 1920s, neon light had come into vogue, and it is still popular today. Modern neon lamps are typically made of plastic rather than glass, and the range of colors is much greater than in Claude's day: not only the gas filling, but the coating inside the tube, is varied, resulting in a variety of colors from across the spectrum.

Though the neon sign is its best-known application, neon is used for many other things. Neon glow lamps are often used to indicate on/off settings on electronic instrument panels, and lightweight neon lamps are found on machines ranging from computers to voltage regulators. In fact, the first practical color television, produced in 1928, used a neon tube to produce the red color in the receiver. Green came from mercury, but the blue light in that early color TV came from another noble gas, helium.

HELIUM. Helium, of course, is widely known for its use in balloons—both for large airships and for the balloons that have provided joy and fun to many a small child. Though helium is much more expensive than hydrogen as a means of providing buoyancy to airships, hydrogen is extremely flammable, and after the infamous explosion of the airship *Hindenburg* in 1937, helium became the preferred medium for airships. As noted earlier, the United States military made extensive use of helium-filled airships during the World War II.

The use of helium for buoyancy is one of the most prominent applications of this noble gas, but far from the only one. In fact, not only have people used helium to go up in balloons, but divers use helium for going down beneath the surface of the ocean. In that situation, of course, helium is not used for providing buoyancy, but as a means of protecting against the diving-related condition known as "the bends," which occurs when nitrogen in the blood bubbles as the diver rises to the surface. Helium is mixed with oxygen in diver's air tanks because it does not dissolve in the blood as easily as nitrogen.

Among the most fascinating applications of helium relate to its extraordinarily low freezing point. Helium has played a significant role in the low-temperature science known as cryogenics, and has found application in research concerning superconductivity: the use of very low temperatures to develop materials that conduct electrical power with vastly greater efficiency than ordinary conductors. Close to absolute zero, helium transforms into a highly unusual liquid unlike any known substance, in that it has no measurable resistance to flow. This means that it could carry an electrical current hundreds of times more efficiently than a copper wire.

WHERE TO LEARN MORE

"*The Chemistry of the Rare Gases*" (Web site). <http://chemed.chem.purdue.edu/genchem/topicreview/bp/ch10/raregas.html> (May 13, 2001).

"*Homework: Science: Chemistry: Gases*" *Channelone.com* (Web site). <http://www.channelone.com/fasttrack/science/chemistry/gases.html> (May 12, 2001).

Knapp, Brian J.; David Woodroffe; David A. Hardy. *Elements.* Danbury, CT: Grolier Educational, 2000.

Mebane, Robert C. and Thomas R. Rybolt. *Air and Other Gases.* Illustrations by Anni Matsick. New York: Twenty-First Century Books, 1995.

"Noble Gases" *Xrefer.com* (Web site). <http://www.xrefer.com/entry/643259> (May 13, 2001).

Rare Gases. *Praxair* (Web site). <http://www.praxair.com/Praxair.nsf/X1/gase_rarega?openDocument> (May 13, 2001).

Stwertka, Albert. *Superconductors: The Irresistible Future.* New York: F. Watts, 1991.

Taylor, Ron. *Facts on Radon and Asbestos.* Illustrated by Ian Moores. New York: F. Watts, 1990.

CARBON

CONCEPT

The phrase "carbon-based life forms," often used in science-fiction books and movies by aliens to describe the creatures of Earth, is something of a cliché. It is also a redundancy when applied to creatures on Earth, the only planet known to support life: all living things contain carbon. Carbon is also in plenty of things that were once living, which makes it useful for dating the remains of past settlements on Earth. Of even greater usefulness is petroleum, a substance containing carbon-based forms that died long ago, became fossilized, and ultimately changed chemically into fuels. Then again, not all materials containing carbon were once living creatures; yet because carbon is a common denominator to all living things on Earth, the branch of study known as organic chemistry is devoted to the study of compounds containing carbon. Among the most important organic compounds are the many carboxylic acids that are vital to life, but carbon is also present in numerous important inorganic compounds—most notably carbon dioxide and carbon monoxide.

HOW IT WORKS

THE BASICS OF CARBON

Carbon's name comes from the Latin word *carbo,* or charcoal—which, indeed, is almost pure carbon. Its chemical symbol is C, and it has an atomic number of 6, meaning that there are six protons in its nucleus. Its two stable isotopes are ^{12}C, which constitutes 98.9% of all carbon found in nature, and ^{13}C, which accounts for the other 1.1%.

The mass of the ^{12}C atom is the basis for the atomic mass unit (amu), by which mass figures for all other elements are measured: the amu is defined as exactly 1/12 the mass of a single ^{12}C atom. The difference in mass between ^{12}C and ^{13}C, which is heavier because of its extra neutron, account for the fact that the atomic mass of carbon is 12.01 amu: were it not for the small quantities of ^{13}C present in a sample of carbon, the mass would be exactly 12.00 amu.

WHERE CARBON IS FOUND. Carbon makes up only a small portion of the known elemental mass in Earth's crust, oceans, and atmosphere—just 0.08%, or 1/1250 of the whole—yet it is the fourteenth most abundant element on the planet. In the human body, carbon is second only to oxygen in abundance, and accounts for 18% of the body's mass. Thus if a person weighs 100 lb (45.3 kg), she is carrying around 18 lb (8.2 kg) of carbon—the same material from which diamonds are made!

Present in the inorganic rocks of the ground and in the living creatures above it, carbon is everywhere. Combined with other elements, it forms carbonates, most notably calcium carbonate ($CaCO_3$), which appears in the form of limestone, marble, and chalk. In combination with hydrogen, it creates hydrocarbons, present in deposits of fossil fuels: natural gas, petroleum, and coal. In the environment, carbon—in the form of carbon dioxide (CO_2)—is taken in by plants, which undergo the process of photosynthesis and release oxygen to animals. Animals breathe in oxygen and release carbon dioxide to the atmosphere.

CARBON AND BONDING

Located in Group 4 of the periodic table of elements (Group 14 in the IUPAC system), carbon has a valence electron configuration of $2s^22p^2$;

likewise, all the members of Group 4—sometimes known as the "carbon family"—have configurations of ns^2np^2, where n is the number of the period or row that the element occupies on the table.

There are two elements noted for their ability to form long strings of atoms and seemingly endless varieties of molecules: one is carbon, and the other is silicon, directly below it on the periodic table. Silicon, found in virtually all types of rocks except the calcium carbonates (mentioned above), is to the inorganic world what carbon is to the organic. Yet silicon atoms are about one and a half times as large as those of carbon; thus not even silicon can compete with carbon's ability to form a seemingly limitless array of molecules in various shapes and sizes, and having various chemical properties.

BASICS OF CHEMICAL BONDING. Carbon is further distinguished by its high value of electronegativity, the relative ability of an atom to attract valence electrons. Electronegativity increases with an increase in group number, and decreases with an increase in period number. In other words, the elements with the highest electronegativity values lie in the upper right-hand corner of the periodic table.

Actually, the previous statement requires one significant qualification: the extreme right-hand side of the periodic table is occupied by elements with negligible electronegativity values. These are the noble gases, which have eight valence electrons each. Eight, as it turns out, is the "magic number" for chemical bonding: most elements follow what is known as the octet rule, meaning that when one element bonds to another, the two atoms have eight valence electrons.

If the two atoms have an electric charge and thus are ions, they form strong ionic bonds. Ionic bonding occurs when a metal bonds with a nonmetal. The other principal type of bond is a covalent bond, in which two uncharged atoms share eight valence electrons. If the electronegativity values of the two elements involved are equal, they share the electrons equally; but if one element has a higher electronegativity value, the electrons will be more drawn to that element.

ELECTRONEGATIVITY OF CARBON. To return to electronegativity and the periodic table, let us ignore the noble gases, which are the chemical equivalent of snobs.

(Hence the term "noble," meaning that they are set apart.) To the left of the noble gases are the halogens, a wildly gregarious bunch—none more so than the element that occupies the top of the column, fluorine. With an electronegativity value of 4.0, fluorine is the most reactive of all elements, and the only one capable of bonding even to a few of the noble gases.

So why is fluorine—capable of forming multitudinous bonds—not as chemically significant as carbon? There are a number of answers, but a simple one is this: because fluorine is too strong, and tends to "overwhelm" other elements, precluding the possibility of forming long chains, it is less chemically significant than carbon. Carbon, on the other hand, has an electronegativity value of 2.5, which places it well behind fluorine. Yet it is still at sixth place (in a tie with iodine and sulfur) on the periodic table, behind only fluorine; oxygen (3.5); nitrogen and chlorine (3.0); and bromine (2.8). In addition, with four valence electrons, carbon is ideally suited to find other elements (or other carbon atoms) for forming covalent bonds according to the octet rule.

MULTIPLE BONDS. Normally, an element does not necessarily have the ability to bond with as many other elements as it has valence electrons, but carbon—with its four valence electrons—happens to be tetravalent, or capable of bonding to four other atoms at once. Additionally, carbon is capable of forming not only a single bond, but also a double bond, or even a triple bond, with other elements.

Suppose a carbon atom bonds to two oxygen atoms to form carbon dioxide. Let us imagine these three atoms side by side, with the oxygen in the middle. (This, in fact, is how these bonds are depicted in the Couper and Lewis systems of chemical symbolism, discussed in the Chemical Bonding essay.) We know that the carbon has four valence electrons, that the oxygens have six, and that the goal is for each atom to have eight valence electrons—some of which it will share covalently.

Two of the valence electrons from the carbon bond with two valence electrons each from the oxygen atoms on either side. This means that the carbon is doubly bonded to each of the oxygen atoms. Therefore, the two oxygens each have four other unbonded valence electrons, which might bond to another atom. It is theoretically possible, also, for the carbon to form a triple bond with one of the oxygens by sharing three of its valence electrons. It would then have one electron free to share with the other oxygen.

REAL-LIFE APPLICATIONS

ORGANIC CHEMISTRY

We have stated that carbon forms tetravalent bonds, and makes multiple bonds with a single atom. In addition, we have mentioned the fact that carbon forms long chains of atoms and varieties of shapes. But how does it do these things, and why? These are good questions, but not ones we will attempt to answer here. In fact, an entire branch of chemistry is devoted to answering these theoretical questions, as well as to determining solutions to a host of other, more practical problems.

Organic chemistry is the study of carbon, its compounds, and their properties. (There are carbon-containing compounds that are not considered organic, however. Among these are oxides such as carbon dioxide and monoxide; as well as carbonates, most notably calcium carbonate.) At one time, chemists thought that "organic" was synonymous with "living," and even as recently as the early nineteenth century, they believed that organic substances contained a supernatural "life force." Then, in 1828, German chemist Friedrich Wöhler (1800-1882) cracked the code that distinguished the living from the nonliving, and the organic from the inorganic.

Wöhler took a sample of ammonium cyanate (NH_4OCN), and by heating it, converted it into urea ($H_2N\text{-}CO\text{-}NH_2$), a waste product in the urine of mammals. In other words, he had turned an inorganic material into a organic one, and he did so, as he observed, "without benefit of a kidney, a bladder, or a dog." It was almost as though he had created life. In fact, what Wöhler had glimpsed—and what other scientists who followed came to understand, was this: what separates the organic from the inorganic is the manner in which the carbon chains are arranged.

Ammonium cyanate and urea have exactly the same numbers and proportions of atoms, yet they are different compounds. They are thus isomers: substances which have the same formula, but are different chemically. In urea, the carbon forms an organic chain, and in ammonium

cyanate, it does not. Thus, to reduce the specifics of organic chemistry even further, it can be said that this area of the field constitutes the study of carbon chains, and ways to rearrange them in order to create new substances.

Rubber, vitamins, cloth, and paper are all organically based compounds we encounter in our daily lives. In each case, the material comes from something that once was living, but what truly makes these substance organic in nature is the common denominator of carbon, as well as the specific arrangements of the atoms. We have organic chemistry to thank for any number of things: aspirins and all manner of other drugs; preservatives that keep food from spoiling; perfumes and toiletries; dyes and flavorings, and so on.

ALLOTROPES OF CARBON

GRAPHITE. Carbon has several allotropes—different versions of the same element, distinguished by molecular structure. The first of these is graphite, a soft material with an unusual crystalline structure. Graphite is essentially a series of one-atom-thick sheets of carbon, bonded together in a hexagonal pattern, but with only very weak attractions between adjacent sheets. A piece of graphite is thus like a big, thick stack of carbon paper: on the one hand, the stack is heavy, but the sheets are likely to slide against one another.

Actually, people born after about 1980 may have little experience with carbon paper, which was gradually phased out as photocopiers became cheaper and more readily available. Today, carbon paper is most often encountered when signing a credit-card receipt: the signature goes through the graphite-based backing of the receipt, onto a customer copy.

In such a situation, one might notice that the copied image of the signature looks as though it were signed in pencil. This is not surprising, considering that pencil "lead" is, in fact, a mixture of graphite, clay, and wax. In ancient times, people did indeed use lead—the heaviest member of Group 4, the "carbon family"—for writing, because it left gray marks on a surface. Lead, of course, is poisonous, and is not used today in pencils or in most applications that would involve prolonged exposure of humans to the element. Nonetheless, people still use the word "lead" in reference to pencils, much as they still refer to a galvanized steel roof with a zinc coating as a "tin roof."

In graphite the atoms of each "sheet" are tightly bonded in a hexagonal, or six-sided, pattern, but the attractions between the sheets are not very strong. This makes it highly useful as a lubricant for locks, where oil would tend to be messy. A good conductor of electricity, graphite is also utilized for making high-temperature electrolysis cells. In addition, the fact that graphite resists temperatures of up to about 6,332°F (3,500°C) makes it useful in electric motors and generators.

DIAMOND. The second allotrope of carbon is also crystalline in structure. This is diamond, most familiar in the form of jewelry, but in fact widely applied for a number of other purposes. According to the Moh scale, which measures the hardness of minerals, diamond is a 10—in other words, the hardest type of material. It is used for making drills that bore through solid rock; likewise, small diamonds are used in dentists' drills for boring through the ultra-hard enamel on teeth.

Neither diamonds nor graphite are, in the strictest sense of the term, formed of molecules. Their arrangement is definite, as with a molecule, but their size is not: they simply form repeating patterns that seem to stretch on forever. Whereas graphite is in the form of sheets, a diamond is basically a huge "molecule" composed of carbon atoms strung together by covalent bonds. The size of this "molecule" corresponds to the size of the diamond: a diamond of 1 carat, for instance, contains about 10^{22} (10,000,000,000,000,000,000,000 or 10 billion billion) carbon atoms.

The diamonds used in industry look quite different from the ones that appear in jewelry. Industrial diamonds are small, dark, and cloudy in appearance, and though they have the same chemical properties as gem-quality diamonds, they are cut with functionality (rather than beauty) in mind. A diamond is hard, but brittle: in other words, it can be broken, but it is very difficult to scratch or cut a diamond—except with another diamond.

The cutting of fine diamonds for jewelry is an art, exemplified in the alluring qualities of such famous gems as the jewels in the British Crown or the infamous Hope Diamond in Washington, D.C.'s Smithsonian Institution. Such diamonds—as well as the diamonds on an engagement ring—are cut to refract or bend light rays, and to disperse the colors of visible light.

BUCKMINSTERFULLERENE. Until 1985, carbon was believed to exist in only two crystalline forms, graphite and diamond. In that year, however, chemists at Rice University in Houston, Texas, and at the University of Sussex in England, discovered a third variety of carbon—and later jointly received a Nobel Prize for their work. This "new" carbon molecule composed of 60 bonded atoms in the shape of what is called a "hollow truncated icosahedron." In plain language, this is rather like a soccer ball, with interlocking pentagons and hexagons. However, because the surface of each geometric shape is flat, the "ball" itself is not a perfect sphere. Rather, it describes the shape of a geodesic dome, a design created by American engineer and philosopher R. Buckminster Fuller (1895-1983).

There are other varieties of buckminsterfullerene molecules, known as fullerenes. However, the 60-atom shape, designated as ^{60}C, is the most common of all fullerenes, the result of condensing carbon slowly at high temperatures. Fullerenes potentially have a number of applications, particularly because they exhibit a whole range of electrical properties: some are insulators, while some are conductors, semiconductors, and even superconductors. Due to the high cost of producing fullerenes artificially, however, the ways in which they are applied remain rather limited.

AMORPHOUS CARBON. There is a fourth way in which carbon appears, distinguished from the other three in that it is amorphous, as opposed to crystalline, in structure. An example of amorphous carbon is carbon black, obtained from smoky flames and used in ink, or for blacking rubber tires.

Though it retains some of the microscopic structures of the plant cells in the wood from which it is made, charcoal—wood or other plant material that has been heated without enough air present to make it burn—is mostly amorphous carbon. One form of charcoal is activated charcoal, in which steam is used to remove the sticky products of wood decomposition. What remains are porous grains of pure carbon with enormous microscopic surface areas. These are used in water purifiers and gas masks.

Coal and coke are particularly significant varieties of amorphous carbon. Formed by the decay of fossils, coal was one of the first "fossil fuels" (for example, petroleum) used to provide heat and power for industrial societies. Indeed, when the words "industrial revolution" are mentioned, many people picture tall black smokestacks belching smoke from coal fires. Fortunately— from an environmental standpoint—coal is not nearly so widely used today, and when it is (as for instance in electric power plants), the methods for burning it are much more efficient than those applied in the nineteenth century.

Actually, much of what those smokestacks of yesteryear burned was coke, a refined version of coal that contains almost pure carbon. Produced by heating soft coal in the absence of air, coke has a much greater heat value than coal, and is still widely used as a reducing agent in the production of steel and other alloys.

CARBON DIOXIDE

Carbon forms many millions of compounds, some families of which will be discussed below. Two others, formed by the bonding of carbon atoms with oxygen atoms, are of particular significance. In carbon dioxide, a single carbon joins with two oxygens to produce a gas essential to plant life. In carbon monoxide (CO), a single oxygen joins the carbon, creating a toxic—but nonetheless important—compound.

The first gas to be distinguished from ordinary air, carbon dioxide is an essential component in the natural balance between plant and animal life. Animals, including humans, produce carbon dioxide by breathing, and humans further produce it by burning wood and other fuels. Plants use carbon dioxide when they store energy in the form of food, and they release oxygen to be used by animals.

DISCOVERY. Flemish chemist and physicist Johannes van Helmont (1579-1644) discovered in 1630 that air was not, as had been thought up to that time, a single element: it contained a second substance, produced in the burning of wood, which he called "gas sylvestre." Thus he is recognized as the first scientist to note the existence of carbon dioxide.

More than a century later, in 1756, Scottish chemist Joseph Black (1728-1799) showed that carbon dioxide—which he called "fixed air"—combines with other chemicals to form compounds. This and other determinations Black made concerning carbon dioxide led to enor-

SOFT DRINKS ARE MADE POSSIBLE BY THE USE OF CARBONATED WATER. *(Sergio Dorantes/Corbis. Reproduced by permission.)*

mous progress in the discovery of gases by various chemists of the late eighteenth century.

By that time, chemists had begun to arrive at a greater degree of understanding with regard to the relationship between plant life and carbon dioxide. Up until that time, it had been believed that plants purify the air by day, and poison it at night. Carbon dioxide and its role in the connection between animal and plant life provided a much more sophisticated explanation as to the ways plants "breathe."

USES. Around the same time that Black made his observations on carbon dioxide, English chemist Joseph Priestley (1733-1804) became the first scientist to put the chemical to use. Dissolving it in water, he created carbonated water, which today is used in making soft drinks. Not only does the gas add bubbles to drinks, it also acts as a preservative.

Though the natural uses of carbon dioxide are by far the most important, the compound has numerous industrial and commercial applications. Used in fire extinguishers, carbon dioxide is ideal for controlling electrical and oil fires, which cannot be put out with water. Heavier than air, carbon dioxide blankets the flames and smothers them.

In the solid form of dry ice, carbon dioxide is used for chilling perishable food during transport. It is also one of the only compounds that experiences sublimation, or the instantaneous transformation of a solid to a gas without passing through an intermediate liquid state, at conditions of ordinary pressure and temperature. Dry ice has often been used in movies to generate "mists" or "smoke" in a particular scene.

CARBON MONOXIDE

During the late eighteenth century, Priestley discovered a carbon-oxygen compound different from carbon dioxide: carbon monoxide. Scientists had actually known of this toxic gas, released in the incomplete combustion of wood, from the Middle Ages onward, but Priestley was the first to identify it scientifically.

Industry uses carbon monoxide in a number of ways. By blowing air across very hot coke, the result is producer gas, which, along with water gas (made by passing hot steam over coal) is an important fuel. Producer gas constitutes a 6:1:18 mixture of carbon monoxide, carbon dioxide, and nitrogen, while water gas is 40% carbon monoxide, 50% hydrogen, and 10% carbon dioxide and other gases.

KEY TERMS

ALLOTROPES: Different versions of the same element, distinguished by molecular structure.

AMORPHOUS: Having no definite structure.

CARBOHYDRATES: Naturally occurring compounds of carbon, hydrogen, and oxygen. These are primarily produced by green plants through the process of photosynthesis.

CELLULAR RESPIRATION: The process whereby nutrients from plants are broken down in an animal's body to create carbon dioxide.

COVALENT BONDING: A type of chemical bonding in which two atoms share valence electrons.

CRYSTALLINE: A term describing a type of solid in which the constituent parts have a simple and definite geometric arrangement that is repeated in all directions.

DOUBLE BOND: A form of bonding in which two atoms share two pairs of valence electrons. Carbon is also capable of single bonds and triple bonds.

ELECTRONEGATIVITY: The relative ability of an atom to attract valence electrons.

ION: An atom or group of atoms that has lost or gained one or more electrons, and thus has a net electric charge.

IONIC BONDING: A form of chemical bonding that results from attractions between ions with opposite electric charges.

ISOMERS: Substances which have the same chemical formula, but which are different chemically due to differences in the arrangement of atoms.

ISOTOPES: Atoms that have an equal number of protons, and hence are of the same element, but differ in their number of neutrons. This results in a difference of mass. Isotopes may be either stable or unstable. The latter type, known as radioisotopes, are radioactive.

OCTET RULE: A term describing the distribution of valence electrons that takes place in chemical bonding for most elements, which usually end up with eight valence electrons.

ORGANIC CHEMISTRY: The study of carbon, its compounds, and their properties. (Many carbon-containing oxides and carbonates are not considered organic, however.)

PHOTOSYNTHESIS: The biological conversion of light energy (that is, electromagnetic energy) to chemical energy in plants.

RADIOACTIVITY: A term describing a phenomenon whereby certain isotopes known as radioisotopes are subject to a form of decay brought about by the emission of high-energy particles. "Decay" does not mean that the isotope "rots"; rather, it decays to form another isotope until eventually (though this may take a long time) it becomes stable.

SINGLE BOND: A form of bonding in which two atoms share one pair of valence electrons. Carbon is also capable of double bonds and triple bonds.

TETRAVALENT: Capable of bonding to four other elements.

KEY TERMS CONTINUED

TRIPLE BOND: A form of bonding in which two atoms share three pairs of valence electrons. Carbon is also capable of single bonds and double bonds.

VALENCE ELECTRONS: Electrons that occupy the highest principal energy level in an atom. These are the electrons involved in chemical bonding.

Not only are producer and water gas used for fuel, they are also applied as reducing agents. Thus, when carbon monoxide is passed over hot iron oxides, the oxides are reduced to metallic iron, while the carbon monoxide is oxidized to form carbon dioxide. Carbon monoxide is also used in reactions with metals such as nickel, iron, and cobalt to form some types of carbonyls.

Carbon monoxide—produced by burning petroleum in automobiles, as well as by the combustion of wood, coal, and other carbon-containing fuels—is extremely hazardous to human health. It bonds with iron in hemoglobin, the substance in red blood cells that transports oxygen throughout the body, and in effect fools the body into thinking that it is receiving oxygenated hemoglobin, or oxyhemoglobin. Upon reaching the cells, carbon monoxide has much less tendency than oxygen to break down, and therefore it continues to circulate throughout the body. Low concentrations can cause nausea, vomiting, and other effects, while prolonged exposure to high concentrations can result in death.

Carbon and the Environment

Carbon is released into the atmosphere by one of three means: cellular respiration; the burning of fossil fuels; and the eruption of volcanoes. When plants take in carbon dioxide from the atmosphere, they combine this with water and manufacture organic compounds using energy they have trapped from sunlight by means of photosynthesis—the conversion of light to chemical energy through biological means. As a by-product of photosynthesis, plants release oxygen into the atmosphere.

In the process of undergoing photosynthesis, plants produce carbohydrates, which are various compounds of carbon, hydrogen, and oxygen essential to life. The other two fundamental components of a diet are fats and proteins, both

carbon-based as well. Animals eat the plants, or eat other animals that eat the plants, and thus incorporate the fats, proteins, and sugars (a form of carbohydrate) from the plants into their bodies. Cellular respiration is the process whereby these nutrients are broken down to create carbon dioxide.

Photosynthesis and cellular respiration are thus linked in what is known as the carbon cycle. Cellular respiration also releases carbon into the atmosphere through the action of decomposers—bacteria and fungi that feed on the remains of plants and animals. The decomposers extract the energy in the chemical bonds of the decomposing matter, thus releasing more carbon dioxide into the atmosphere.

When creatures die and are buried in such a way that they cannot be reached by decomposers—for instance, at the bottom of the ocean, or beneath layers of rock—the carbon in their bodies is eventually converted to fossil fuels, including petroleum, natural gas, and coal. The burning of fossil fuels releases carbon (both monoxide and dioxide) into the atmosphere.

Because the rate of such burning has increased dramatically since the late nineteenth century, this has raised fears that carbon dioxide in the atmosphere may create a greenhouse effect, leading to global warming. On the other hand, volcanoes release tons of carbon into the atmosphere regardless of whether humans burn fossil fuels or not.

Radiocarbon Dating

Radiocarbon dating is used to date the age of charcoal, wood, and other biological materials. When an organism is alive, it incorporates a certain ratio of carbon-12 in proportion to the amount of the radioisotope (that is, radioactive isotope) carbon-14 that it receives from the atmosphere. As soon as the organism dies, however, it stops incorporating new carbon, and the ratio between carbon-12 and carbon-14 will begin to change as the carbon-14 decays to form nitrogen-14.

Carbon-14 has a half-life of 5,730 years, meaning that it takes that long for half the isotopes in a sample to decay to nitrogen-14. Therefore a scientist can use the ratios of carbon-12, carbon-14, and nitrogen-14 to guess the age of an organic sample. The problem with radiocarbon dating, however, is that there is a good likelihood the sample can become contaminated by additional carbon from the soil. Furthermore, it cannot be said with certainty that the ratio of carbon-12 to carbon-14 in the atmosphere has been constant throughout time.

WHERE TO LEARN MORE

Blashfield, Jean F. *Carbon.* Austin, TX: Raintree Steck-Vaughn, 1999.

"Carbon." *Xrefer* (Web site). <http://www.xrefer.com/entry/639742> (May 30, 2001).

"Diamonds." *American Museum of Natural History* (Web site). <http://www.amnh.org/exhibitions/diamonds/structure.html> (May 30, 2001).

Knapp, Brian J. *Carbon Chemistry.* Illustrated by David Woodroffe. Danbury, CT: Grolier Educational, 1998.

Loudon, G. Marc. *Organic Chemistry.* Menlo Park, CA: Benjamin/Cummings, 1988.

"Organic Chemistry" (Web site). <http://edie.cprost.sfu.ca/~rhlogan/organic.html> (May 30, 2001).

"Organic Chemistry." *Frostburg State University Chemistry Helper* (Web site). <http://www.chemhelper.com/> (May 30, 2001).

Sparrow, Giles. *Carbon.* New York: Benchmark Books, 1999.

Stille, Darlene. *The Respiratory System.* New York: Children's Press, 1997.

H Y D R O G E N

CONCEPT

First element on the periodic table, hydrogen is truly in a class by itself. It does not belong to any family of elements, and though it is a nonmetal, it appears on the left side of the periodic table with the metals. The other elements with it in Group 1 form the alkali metal family, but obviously, hydrogen does not belong with them. Indeed, if there is any element similar to hydrogen in simplicity and abundance, it is the only other one on the first row, or period, of the periodic table: helium. Together, these two elements make up 99.9% of all known matter in the entire universe, because hydrogen atoms in stars fuse to create helium. Yet whereas helium is a noble gas, and therefore chemically unreactive, hydrogen bonds with all sorts of other elements. In one such variety of bond, with carbon, hydrogen forms the backbone for a vast collection of organic molecules, known as hydrocarbons and their derivatives. Bonded with oxygen, hydrogen forms the single most important compound on Earth, and the most important complex substance other than air: water. Yet when it bonds with sulfur, it creates toxic hydrogen sulfide; and on its own, hydrogen is extremely flammable. The only element whose isotopes have names, hydrogen has long been considered as a potential source of power and transportation: once upon a time for airships, later as a component in nuclear reactions—and, perhaps in the future, as a source of abundant clean energy.

HOW IT WORKS

THE ESSENTIALS

The atomic number of hydrogen is 1, meaning that it has a single proton in its nucleus. With its single electron, hydrogen is the simplest element of all. Because it is such a basic elemental building block, figures for the mass of other elements were once based on hydrogen, but the standard today is set by ^{12}C or carbon-12, the most common isotope of carbon.

Hydrogen has two stable isotopes—forms of the element that differ in mass. The first of these, protium, is simply hydrogen in its most common form, with no neutrons in its nucleus. Protium (the name is only used to distinguish it from the other isotopes) accounts for 99.985% of all the hydrogen that appears in nature. The second stable isotope, deuterium, has one neutron, and makes up 0.015% of all hydrogen atoms. Tritium, hydrogen's one radioactive isotope, will be discussed below.

The fact that hydrogen's isotopes have separate names, whereas all other isotopes are designated merely by element name and mass number (for example, "carbon-12") says something about the prominence of hydrogen as an element. Not only is its atomic number 1, but in many ways, it is like the number 1 itself—the essential piece from which all others are ultimately constructed. Indeed, nuclear fusion of hydrogen in the stars is the ultimate source for the 90-odd elements that occur in nature.

The mass of this number-one element is not, however, 1: it is 1.008 amu, reflecting the small quantities of deuterium, or "heavy hydrogen," present in a typical sample. A gas at ordinary temperatures, hydrogen turns to a liquid at -423.2°F (-252.9°C), and to a solid at –434.°F (–259.3°C). These figures are its boiling point and melting point respectively; only the figures for helium are lower. As noted earlier, these two elements make up all but 0.01% of the known

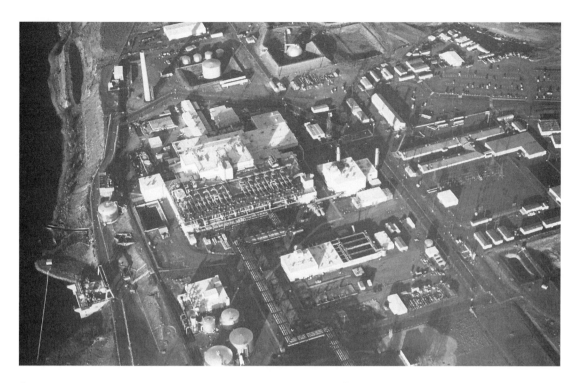

AMONG THE USES OF TRITIUM IS IN FUEL FOR HYDROGEN BOMBS. SHOWN HERE IS THE HANFORD PLANT IN THE STATE OF WASHINGTON, WHERE SUCH FUEL IS PRODUCED. *(Roger Ressmeyer/Corbis. Reproduced by permission.)*

elemental mass of the universe, and are the principal materials from which stars are formed.

Normally hydrogen is diatomic, meaning that its molecules are formed by two atoms. At the interior of a star, however, where the temperature is many millions of degrees, H_2 molecules are separated into atoms, and these atoms become ionized. In other words, the electron separates from the proton, resulting in an ion with a positive charge, along with a free electron. The positive ions experience fusion—that is, their nuclei bond, releasing enormous amounts of energy as they form new elements.

Because the principal isotopic form of helium has two protons in the nucleus, it is natural that helium is the element usually formed; yet it is nonetheless true—amazing as it may seem—that all the elements found on Earth were once formed in stars. On Earth, however, hydrogen ranks ninth in its percentage of the planet's known elemental mass: just 0.87%. In the human body, on the other hand, it is third, after oxygen and carbon, making up 10% of human elemental body mass.

HYDROGEN AND BONDING

Having just one electron, hydrogen can bond to other atoms in one of two ways. The first option is to combine its electron with one from the atom of a nonmetallic element to make a covalent bond, in which the two electrons are shared. Hydrogen is unusual in this regard, because most atoms conform to the octet rule, ending up with eight valence electrons. The bonding behavior of hydrogen follows the duet rule, resulting in just two electrons for bonding.

Examples of this first type of bond include water (H_2O), hydrogen sulfide (H_2S), and ammonia (NH_3), as well as the many organic compounds formed on a hydrogen-carbon backbone. But hydrogen can form a second type of bond, in which it gains an extra electron to become the negative ion H-, or hydride. It is then able to combine with a metallic positive ion to form an ionic bond. Ionic hydrides are convenient sources of hydrogen gas: for instance, calcium hydride, or CaH_2, is sold commercially, and provides a very convenient means of hydrogen generation. The hydrogen gas produced by the reaction of calcium hydride with water can be used to inflate life rafts.

The presence of hydrogen in certain types of molecules can also be a factor in intermolecular bonding. Intermolecular bonding is the attraction between molecules, as opposed to the bond-

ing within molecules, which is usually what chemists mean when they talk about "bonding."

HYDROGEN'S EARLY HISTORY

Because it bonds so readily with other elements, hydrogen almost never appears in pure elemental form on Earth. Yet by the late fifteenth century, chemists recognized that by adding a metal to an acid, hydrogen was produced. Only in 1766, however, did English chemist and physicist Henry Cavendish (1731-1810) recognize hydrogen as a substance distinct from all other "airs," as gases then were called.

Seventeen years later, in 1783, French chemist Antoine Lavoisier (1743-1794) named the substance after two Greek words: *hydro* (water) and *genes* (born or formed). It was another two decades before English chemist John Dalton (1766-1844) formed his atomic theory of matter, and despite the great strides he made for science, Dalton remained convinced that hydrogen and oxygen in water formed "water atoms." Around the same time, however, Italian physicist Amedeo Avogadro (1776-1856) clarified the distinction between atoms and molecules, though this theory would not be generally accepted until the 1850s.

Contemporary to Dalton and Avogadro was Swedish chemist Jons Berzelius (1779-1848), who developed a system of comparing the mass of various atoms in relation to hydrogen. This method remained in use for more than a century, until the discovery of neutrons, protons, and isotopes pointed the way toward a means of making more accurate determinations of atomic mass. In 1931, American chemist and physicist Harold Urey (1893-1981) made the first separation of an isotope: deuterium, from ordinary water.

REAL-LIFE APPLICATIONS

DEUTERIUM AND TRITIUM

Designated as ^{2}H, deuterium is a stable isotope, whereas tritium—^{3}H—is unstable, or radioactive. Not only do these two have names; they even have chemical symbols (D and T, respectively), as though they were elements on the periodic table. Just as hydrogen represents the most basic proton-electron combination against which other atoms are compared, these two are respectively the most basic isotope containing a single neutron, and the most basic radioisotope, or radioactive isotope.

Deuterium is sometimes called "heavy hydrogen," and its nucleus is called a deuteron. In separating deuterium—an achievement for which he won the 1934 Nobel Prize—Urey collected a relatively large sample of liquid hydrogen: 4.2 qt (4 l). He then allowed the liquid to evaporate very slowly, predicting that the more abundant protium would evaporate more quickly than the heavier isotope. When all but 0.034 oz (1 ml) of the sample had evaporated, he submitted the remainder to a form of analysis called spectroscopy, adding a burst of energy to the atoms and then analyzing the light spectrum they emitted for evidence of differing varieties of atoms.

With an atomic mass of 2.014102 amu, deuterium is almost exactly twice as heavy as protium, which has an atomic mass of 1.007825. Its melting points and boiling points, respectively –426°F (–254°C) and –417°F (–249°C), are higher than for protium. Often, deuterium is applied as a tracer, an atom or group of atoms whose participation in a chemical, physical, or biological reaction can be easily observed.

DEUTERIUM IN WAR AND PEACE. In nuclear power plants, deuterium is combined with oxygen to form "heavy water" (D_2O), which likewise has higher boiling and melting points than ordinary water. Heavy water is often used in nuclear fission reactors to slow down the fission process, or the splitting of atoms. Deuterium is also present in nuclear fusion, both on the Sun and in laboratories.

During the period shortly after World War II, physicists developed a means of duplicating the thermonuclear fusion process. The result was the hydrogen bomb—more properly called a fusion bomb—whose detonating device was a compound of lithium and deuterium called lithium deuteride. Vastly more powerful than the "atomic" (that is, fission) bombs dropped by the United States over Japan (Nagaski and Hiroshima) in 1945, the hydrogen bomb greatly increased the threat of worldwide nuclear annihilation in the postwar years.

Yet the power that could destroy the world also has the potential to provide safe, abundant fusion energy from power plants—a dream as yet unrealized. Physicists studying nuclear fusion are attempting several approaches, including a process involving the fusion of two deuterons. This fusion would result in a triton, the nucleus of tritium, along with a single proton. Theoretically, the triton and deuteron would then be fused to create a helium nucleus, resulting in the production of vast amounts of energy.

TRITIUM. Whereas deuterium has a single neutron, tritium—as its mass number of 3 indicates—has two. And just as deuterium has approximately twice the mass of protium, tritium has about three times the mass, or 3.016 amu. Its melting and boiling points are higher still than those of deuterium: thus tritium heavy water (T_2O) melts at 40°F (4.5°C), as compared with 32°F (0°C) for H_2O.

Tritium has a half-life (the length of time it takes for half the radioisotopes in a sample to become stable) of 12.26 years. As it decays, its nucleus emits a low-energy beta particle, which is either an electron or a subatomic particle called a positron, resulting in the creation of the helium-3 isotope. Due to the low energy levels involved, the radioactive decay of tritium poses little danger to humans. In fact, there is always a small quantity of tritium in the atmosphere, and this quantity is constantly being replenished by cosmic rays.

Like deuterium, tritium is applied in nuclear fusion, but due to its scarcity, it is usually combined with deuterium. Sometimes it is released in small quantities into the groundwater as a means of monitoring subterranean water flow. It is also used as a tracer in biochemical processes, and as an ingredient in luminous paints.

HYDROGEN AND OXYGEN

WATER. Water, of course, is the most well-known compound involving hydrogen. Nonetheless, it is worthwhile to consider the interaction between hydrogen and oxygen, the two ingredients in water, which provides an interesting illustration of chemistry in action.

Chemically bonded as water, hydrogen and oxygen can put out any type of fire except an oil or electrical fire; as separate substances, however, hydrogen and oxygen are highly flammable. In an oxyhydrogen torch, the potentially explosive reaction between the two gases is controlled by a gradual feeding process, which produces combustion instead of the more violent explosion that sometimes occurs when hydrogen and oxygen come into contact.

HYDROGEN PEROXIDE. Aside from water, another commonly used hydrogen-oxygen compound is hydrogen peroxide, or H_2O_2. A colorless liquid, hydrogen peroxide is chemically unstable (not "unstable" in the way that a radioisotope is), and decomposes slowly to form water and oxygen gas. In high concentrations, it can be used as rocket fuel.

By contrast, the hydrogen peroxide used in homes as a disinfectant and bleaching agent is only a 3% solution. The formation of oxygen gas molecules causes hydrogen peroxide to bubble, and this bubbling is quite rapid when the peroxide is placed on cuts, because the enzymes in blood act as a catalyst to speed up the reaction.

HYDROGEN CHLORIDE

Another significant compound involving hydrogen is hydrogen chloride, or HCl—in other words, one hydrogen atom bonded to chlorine, a member of the halogens family. Dissolved in water, it produces hydrochloric acid, used in laboratories for analyses involving other acids. Normally, hydrogen chloride is produced by the reaction of salt with sulfuric acid, though it can also be created by direct bonding of hydrogen and chlorine at temperatures above 428°F (250°C).

Hydrogen chloride and hydrochloric acid have numerous applications in metallurgy, as well as in the manufacture of pharmaceuticals, dyes, and synthetic rubber. They are used, for instance, in making pharmaceutical hydrochlorides, water-soluble drugs that dissolve when ingested. Other applications include the production of fertilizers, synthetic silk, paint pigments, soap, and numerous other products.

Not all hydrochloric acid is produced by industry, or by chemists in laboratories. Active volcanoes, as well as waters from volcanic mountain sources, contain traces of the acid. So, too, does the human body, which generates it during digestion. However, too much hydrochloric acid in the digestive system can cause the formation of gastric ulcers.

HYDROGEN SULFIDE

It may not be a pleasant subject, but hydrogen—in the form of hydrogen sulfide—is also present in intestinal gas. The fact that hydrogen sulfide is an extremely malodorous substance once again illustrates the strange things that happen when elements bond: neither hydrogen nor sulfur has any smell on its own, yet together they form an extremely noxious—and toxic—substance.

Pockets of hydrogen sulfide occur in nature. If a person were to breathe the vapors for very long, it could be fatal, but usually, the foul odor keeps people away. The May 2001 *National Geographic* included two stories relating to such natural hydrogen-sulfide deposits, on opposite sides of the Earth, and in both cases the presence of these toxic fumes created interesting results.

In southern Mexico is a system of caves known as Villa Luz, through which run some 20 underground springs, many of them carrying large quantities of hydrogen sulfide. The National Geographic Society's team had to enter the caves wearing gas masks, yet the area teems with strange varieties of life. Among these are fish that are red from high concentrations of hemoglobin, or red blood cells. The creatures need this extra dose of hemoglobin, necessary to move oxygen through the body, in order to survive on the scant oxygen supplies. The waters of the cave are further populated by microorganisms that oxidize the hydrogen sulfide and turn it into sulfuric acid, which dissolves the rock walls and continually enlarges the cave.

Thousands of miles away, in the Black Sea, explorers supported by a grant from the National Geographic Society examined evidence suggesting that there indeed had been a great ancient flood in the area, much like the one depicted in the Bible. In their efforts, they had an unlikely ally: hydrogen sulfide, which had formed at the bottom of the sea, and was covered by dense layers of salt water. Because the Black Sea lacks the temperature differences that cause water to circulate from the bottom upward, the hydrogen sulfide stayed at the bottom.

Under normal circumstances, the wreck of a 1,500-year-old wooden ship would not have been preserved; but because oxygen could not reach the bottom of the Black Sea—and thus wood-boring worms could not live in the toxic environment—the ship was left undisturbed. Thanks to the presence of hydrogen sulfide, explorers were able to study the ship, the first fully intact ancient shipwreck to be discovered.

HYDROCARBONS

Together with carbon, hydrogen forms a huge array of organic materials known as hydrocarbons—chemical compounds whose molecules are made up of nothing but carbon and hydrogen atoms. Theoretically, there is no limit to the number of possible hydrocarbons. Not only does carbon form itself into seemingly limitless molecular shapes, but hydrogen is a particularly good partner. Because it has the smallest atom of any element on the periodic table, it can bond to one of carbon's valence electrons without getting in the way of the others.

Hydrocarbons may either be saturated or unsaturated. A saturated hydrocarbon is one in which the carbon atom is already bonded to four other atoms, and thus cannot bond to any others. In an unsaturated hydrocarbon, however, not all the valence electrons of the carbon atom are bonded to other atoms.

Hydrogenation is a term describing any chemical reaction in which hydrogen atoms are added to carbon multiple bonds. There are many applications of hydrogenation, but one that is particularly relevant to daily life involves its use in turning unsaturated hydrocarbons into saturated ones. When treated with hydrogen gas, unsaturated fats (fats are complex substances that involve hydrocarbons bonded to other molecules) become saturated fats, which are softer and more stable, and stand up better to the heat of frying. Many foods contain hydrogenated vegetable oil; however, saturated fats have been linked with a rise in blood cholesterol levels—and with an increased risk of heart disease.

PETROCHEMICALS AND FUNCTIONAL GROUPS. One important variety of hydrocarbons is described under the collective heading of petrochemicals—that is, derivatives of petroleum. These include natural gas; petroleum ether, a solvent; naphtha, a solvent (for example, paint thinner); gasoline; kerosene; fuel for heating and diesel fuel; lubricating oils; petroleum jelly; paraffin wax; and pitch, or tar. A host of other organic chemicals, including various drugs, plastics, paints, adhesives, fibers, detergents, synthetic rubber, and agricultural chemicals, owe their existence to petrochemicals.

Then there are the many hydrocarbon derivatives formed by the bonding of hydrocarbons to various functional groups—broad arrays of molecule types involving other elements. Among these are alcohols—both ethanol (the alcohol in beer and other drinks) and methanol, used in adhesives, fibers, and plastics, and as a fuel. Other functional groups include aldehydes, ketones, carboxylic acids, and esters. Products of these functional groups range from aspirin to butyric acid, which is in part responsible for the smell both of rancid butter and human sweat. Hydrocarbons also form the basis for polymer plastics such as Nylon and Teflon.

HYDROGEN FOR TRANSPORTATION AND POWER

We have already seen that hydrogen is a component of petroleum, and that hydrogen is used in creating nuclear power—both deadly and peaceful varieties. But hydrogen has been applied in many other ways in the transportation and power industries.

There are only three gases practical for lifting a balloon: hydrogen, helium, and hot air. Each is much less dense than ordinary air, and this gives them their buoyancy. Because hydrogen is the lightest known gas and is relatively cheap to produce, it initially seemed the ideal choice, particularly for airships, which made their debut near the end of the nineteenth century.

For a few decades in the early twentieth century, airships were widely used, first in warfare and later as the equivalent of luxury liners in the skies. One of the greatest such craft was Germany's *Hindenburg*, which used hydrogen to provide buoyancy. Then, on May 6, 1937, the *Hindenburg* caught fire while mooring at Lakehurst, New Jersey, and 36 people were killed—a tragic and dramatic event that effectively ended the use of hydrogen in airships.

Adding to the pathos of the *Hindenburg* crash was the voice of radio announcer Herb Morrison, whose audio report has become a classic of radio history. Morrison had come to Lakehurst to report on the landing of the famous airship, but ended up with the biggest—and most horrifying—story of his career. As the ship burst into flames, Morrison's voice broke, and he uttered words that have become famous: "Oh, the humanity!"

Half a century later, a hydrogen-related disaster destroyed a craft much more sophisticated than the *Hindenburg*, and this time, the medium of television provided an entire nation with a view of the ensuing horror. The event was the explosion of the space shuttle *Challenger* on January 28, 1986, and the cause was the failure of a rubber seal in the shuttle's fuel tanks. As a result,

KEY TERMS

COVALENT BONDING: A type of chemical bonding in which two atoms share valence electrons.

DIATOMIC: A term describing an element that exists as molecules composed of two atoms.

DUET RULE: A term describing the distribution of valence electrons when hydrogen atoms—which end up with only two valence electrons—experience chemical bonding with other atoms. Most other elements follow the octet rule.

ELECTROLYSIS: The use of an electric current to cause a chemical reaction.

FISSION: A nuclear reaction involving the splitting of atoms.

FUSION: A nuclear reaction that involves the joining of atomic nuclei.

HYDROCARBON: Any chemical compound whose molecules are made up of nothing but carbon and hydrogen atoms.

HYDROGENATION: A chemical reaction in which hydrogen atoms are added to carbon multiple bonds, as in a hydrocarbon.

ION: An atom or group of atoms that has lost or gained one or more electrons, and thus has a net electrical charge.

IONIC BONDING: A form of chemical bonding that results from attractions between ions with opposite electric charges. The bonding of a metal to a nonmetal such as hydrogen is ionic.

ISOTOPES: Atoms that have an equal number of protons, and hence are of the same element, but differ in their number of neutrons. This results in a difference of mass. An isotope may either be stable or radioactive.

NUCLEUS: The center of an atom, a region where protons and neutrons are located, and around which electrons spin. The plural of "nucleus" is nuclei.

OCTET RULE: A term describing the distribution of valence electrons that takes place in chemical bonding for most elements, which end up with eight valence electrons. Hydrogen is an exception, and follows the duet rule.

ORGANIC: At one time, chemists used the term "organic" only in reference to living things. Now the word is applied to most compounds containing carbon, with the exception of calcium carbonate (limestone) and oxides such as carbon dioxide.

RADIOISOTOPE: An isotope subject to the decay associated with radioactivity. A radioisotope is thus an unstable isotope.

SATURATED: A term describing a hydrocarbon in which each carbon is already bound to four other atoms.

TRACER: An atom or group of atoms whose participation in a chemical, physical, or biological reaction can be easily observed. Radioisotopes are often used as tracers.

UNSATURATED: A term describing a hydrocarbon, in which the carbons involved in a multiple bond are free to bond with other atoms.

VALENCE ELECTRONS: Electrons that occupy the highest principal energy level in an atom. These are the electrons involved in chemical bonding.

hydrogen gas flooded out of the craft and straight into the jet of flame behind the rocket. All seven astronauts aboard were killed.

THE FUTURE OF HYDROGEN POWER. Despite the misfortunes that have occurred as a result of hydrogen's high flammability, the element nonetheless holds out the promise of cheap, safe power. Just as it made possible the fusion, or hydrogen, bomb—which fortunately has never been dropped in wartime, but is estimated to be many hundreds of times more lethal than the fission bombs dropped on Japan—hydrogen may be the key to the harnessing of nuclear fusion, which could make possible almost unlimited power.

A number of individuals and agencies advocate another form of hydrogen power, created by the controlled burning of hydrogen in air. Not only is hydrogen an incredibly clean fuel, producing no by-products other than water vapor, it is available in vast quantities from water. In order to separate it from the oxygen atoms, electrolysis would have to be applied—and this is one of the challenges that must be addressed before hydrogen fuel can become a reality.

Electrolysis requires enormous amounts of electricity, which would have to be produced before the benefits of hydrogen fuel could be realized. Furthermore, though the burning of hydrogen could be controlled, there are the dangers associated with transporting it across country in pipelines. Nonetheless, a number of advocacy groups—some of whose Web sites are listed below—continue to promote efforts toward realizing the dream of nonpolluting, virtually limitless, fuel.

WHERE TO LEARN MORE

American Hydrogen Association (Web site). <http://www.clean-air.org> (June 1, 2001).

Blashfield, Jean F. *Hydrogen.* Austin, TX: Raintree Steck-Vaughn, 1999.

Farndon, John. *Hydrogen.* New York: Benchmark Books, 2001.

"Hydrogen" (Web site). <http://pearl1.lanl.gov/periodic/elements/1.html> (June 1, 2001).

Hydrogen Energy Center (Web site). <http://www.h2eco.org/> (June 1, 2001).

Hydrogen Information Network (Web site). <http://www.eren.doe.gov/hydrogen/> (June 1, 2001).

Knapp, Brian J. *Carbon Chemistry.* Illustrated by David Woodroffe. Danbury, CT: Grolier Educational, 1998.

Knapp, Brian J. *Elements.* Illustrated by David Woodroffe and David Hardy. Danbury, CT: Grolier Educational, 1996.

National Hydrogen Association (Web site). <http://www.ttcorp.com/nha/> (June 1, 2001).

Uehling, Mark. *The Story of Hydrogen.* New York: Franklin Watts, 1995.

BONDING AND REACTIONS

CHEMICAL BONDING

CONCEPT

Almost everything a person sees or touches in daily life—the air we breathe, the food we eat, the clothes we wear, and so on—is the result of a chemical bond, or, more accurately, many chemical bonds. Though a knowledge of atoms and elements is essential to comprehend the subjects chemistry addresses, the world is generally not composed of isolated atoms; rather, atoms bond to one another to form molecules and hence chemical compounds. Not all chemical bonds are created equal: some are weak, and some very strong, a difference that depends primarily on the interactions of electrons between atoms.

HOW IT WORKS

EARLY IDEAS OF BONDING

The theory that all of matter is composed of atoms did not originate in modern times: the atomic model actually dates back to the fifth century B.C. in Greece. The leading exponent of atomic theory in ancient times was Democritus (c. 460-370 B.C.), who proposed that matter could not be infinitely subdivided. At its deepest substructure, Democritus maintained, the material world was made up of tiny fragments he called *atomos*, a Greek term meaning "no cut" or "indivisible"

Forward-thinking though it was, Democritus's idea was not what modern scientists today would describe as a proper scientific hypothesis. His "atoms" were not purely physical units, but rather idealized philosophical constructs, and thus, he was not really approaching the subject from the perspective of a scientist. In any case,

there was no way for Democritus to test his hypothesis even if he had wanted to: by their very nature, the atoms he described were far too small to observe. Even today, what scientists know about atomic behavior comes not from direct observation, but indirect means.

Hence, Democritus and the few other ancients who subscribed to atomic theory went more on instinct than by scientific methods. Yet, some of them were remarkably prescient in their description of the bonding of atoms, in view of the primitive scientific methods they had at their disposal. No other scientist came close to the accuracy of their theory for about 2,000 years.

ASCLEPIADES AND LUCRETIUS DISCUSS BONDS. The physician Asclepiades of Prusa (c.130-40 B.C.) drew on the ideas of the Greek philosopher Epicurus (341-270 B.C., another proponent of atomism. Asclepiades speculated on the ways in which atoms interact, and discussed "clusters of atoms," though, of course, he had no idea what force attracted the atoms to one another.

A few years after Asclepiades, the Roman philosopher and poet Lucretius (c.95-c.55 B.C. espoused views that combined atomism with the idea of the "four elements"—earth, air, fire, and water. In his great work *De rerum natura* ("On the Nature of Things"), Lucretius described atoms as tiny spheres attached to each other by fishhook-like appendages that became entangled with one another.

LACK OF PROGRESS UNTIL 1800. Unfortunately, the competing idea of the four elements, handed down by the great philosopher Aristotle (384-322 B.C.), prevailed over the atomic model. As the Roman Empire

ELECTRON-DOT DIAGRAMS OF SOME ELEMENTS IN THE PERIODIC TABLE. *(Robert L. Wolke. Reproduced by permission.)*

began to decline after A.D. 200, the pace of scientific inquiry slowed and—in Western Europe at least—eventually came to a virtual halt. Hence, the four elements theory, which had its own fanciful explanations as to why certain "elements" bonded with one another, held sway in Europe until the beginning of the modern era.

During the seventeenth century, a mounting array of facts from the realms of astronomy and physics collectively disproved the Aristotelian model. In the area of chemistry, English physicist and chemist Robert Boyle (1627-1691) showed that the four elements were not elements at all, because they could be broken down into simpler substances. Yet, no one really understood what constituted an element until the very beginning of the nineteenth century, and until that question was addressed, it was difficult to move on to the mystery of why certain atoms bonded with one another.

EARLY MODERN ADVANCES IN BONDING THEORY

DALTON'S ATOMIC THEORY. The birth of atomic theory in modern times occurred in 1803, when English chemist John Dalton (1766-1844) formulated the idea that all elements are composed of tiny, indestructible particles. These he called by the name Democritus had given them nearly 23 centuries earlier: atoms. All known substances, he said, are composed of some combination of atoms, which differ from one another only in mass.

Though Dalton's theory paved the way for enormous advances in the years that followed, there were a number of flaws in it. Mass alone, for instance, is not really what differentiates one atom from another: differences in mass reflect the presence of subatomic particles—protons and neutrons—of whose existence scientists were unaware at the time.

Furthermore, the properties of atoms that cause them to bond relate to a third subatomic particle, the electron, which, though it contributes little to the mass of the atom, is all-important to the energy it possesses. As for how atoms bond to one another, Dalton had little to say: in his conception of the atomic model, atoms simply sit adjacent to one another without forming true bonds, as such.

AVOGADRO AND THE MOLECULE. Though Dalton recognized that the structure of atoms in a particular element or compound is uniform, he maintained that compounds are made up of compound atoms: thus,

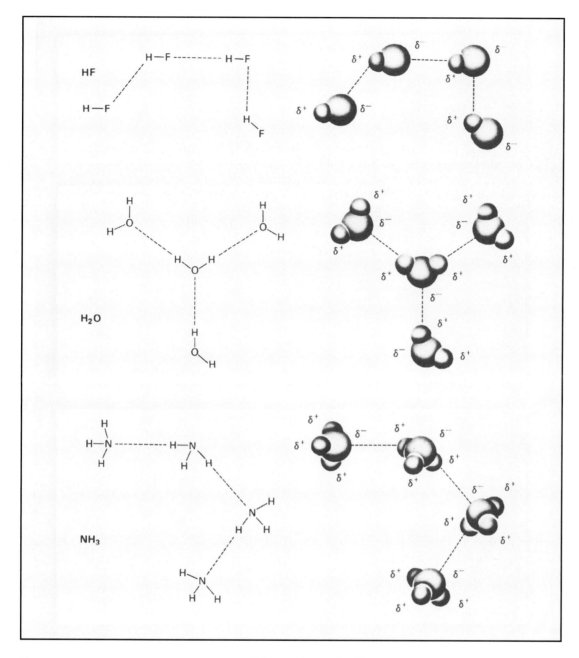

HYDROGEN BONDING IN HF, H₂O, AND NH₃. *(Robert L. Wolke. Reproduced by permission.)*

water is a compound of "water atoms." However, water is not an element, and therefore, there had to be some structure—still very small, but larger than the atom—in which atoms coalesced to form the basic materials of a compound.

That structure was the molecule, first described by Italian physicist Amedeo Avogadro (1776-1856). For several decades, Avogadro, who originated the idea of the mole as a means of comparing large groups of atoms or molecules, remained a more or less unsung hero. Only in 1860, four years after his death, was his idea of

the molecule resurrected by Italian chemist Stanislao Cannizzaro (1826-1910). Cannizzaro's work was occasioned by disagreement among scientists regarding the determination of atomic mass; however, the establishment of the molecular model had far-reaching implications for theories of bonding.

SYMBOLIZING ATOMIC BONDS. In 1858, German chemist Friedrich August Kekulé (1829-1896) made the first attempt to define the concept of valency, or the property an atom of one element possesses that determines

"PURE" WATER FROM A MOUNTAIN STREAM IS ACTUALLY FILLED WITH TRACES OF THE ROCKS OVER WHICH IT HAS FLOWED. IN FACT, WATER IS ALMOST IMPOSSIBLE TO FIND IN PURE FORM EXCEPT BY PURIFYING IT IN A LABO-RATORY. *(David Muench/Corbis. Reproduced by permission.)*

its ability to bond with atoms of other elements. A pioneer in organic chemistry, which deals with chemical structures containing carbon, Kekulé described the carbon atom as tetravalent, mean-ing that it can bond to four other atoms. (The Latin prefix *tetra-* means "four.") He also specu-lated that carbon atoms are capable of bonding with one another in long chains.

This was one of the first attempts to examine the subject of bonding using modern scientific terminology, complete with hypotheses that could be tested by experimentation. Kekulé also recognized that in order to discuss bonds understandably, there needed to be some means of representing those bonds with symbols. He even went so far as to develop a system for showing the arrangement of bonds in space; however, his system was so elaborate that it was replaced in favor of a simpler one developed by Scottish chemist Archibald Scott Couper (1831-1892).

Couper, who also studied valency and the tetravalent carbon bond—he is usually given equal credit with Kekulé for these ideas—created an extremely straightforward schematic representation still in use by chemists today. In Couper's system, short dashed lines serve to designate chemical bonds. Hence, the bond between two hydrogen atoms and an oxygen atom in a water molecule would be represented thus: H-O-H. As the understanding of bonds progressed in modern times, this system was modified to take into account multiple bonds, discussed below.

REAL-LIFE APPLICATIONS

ATOMS, ELECTRONS, AND IONS

Today, chemical bonding is understood as the joining of atoms through electromagnetic force. Before that understanding could be achieved, however, scientists had to unlock the secret of the electromagnetic interactions that take place within an atom.

The key to bonding is the electron, discovered in 1897 by English physicist J. J. Thomson (1856-1940). Atomic structure in general, and the properties of the electron in particular, are discussed at length elsewhere in this volume. However, because these specifics are critical to bonding, they will be presented here in the shortest possible form.

At the center of an atom is a nucleus, consisting of protons, with a positive electrical charge; and neutrons, which have no charge. These form the bulk of the atom's mass, but they have little to do with bonding. In fact, the neutron has nothing to do with it, while the proton plays only a passive role, rather like a flower being pollinated by a bee. The "bee" is the electron,

and, like a bee, it buzzes to and fro, carrying a powerful "sting"—its negative electric charge, which attracts it to the positively charged proton.

ELECTRONS AND IONS. Though the electron weighs much, much less than a proton, it possesses enough electric charge to counterbalance the positive charge of the proton. All atoms have the same number of protons as electrons, and hence the net electric charge is zero. However, as befits their highly active role, electrons are capable of moving from one atom to another under the proper circumstances. An atom that loses or acquires electrons has an electric charge, and is called an ion.

The atom that has lost an electron or electrons becomes a positively charged ion, or cation. On the other hand, an atom that gains an electron or electrons becomes a negatively charged ion, or anion. As we shall see, ionic bonds, such as those that join sodium and chlorine atoms to form NaCl, or salt, are extremely powerful.

ELECTRON CONFIGURATION. Even in covalent bonding, which does not involve ions, the configurations of electrons in two atoms are highly important. The basics of electron configuration are explained in the Electrons essay, though even there, this information is presented with the statement that the student should consult a chemistry textbook for a more exhaustive explanation.

In the simplest possible terms, electron configuration refers to the distribution of electrons at various positions in an atom. However, because the behavior of electrons cannot be fully predicted, this distribution can only be expressed in terms of probability. An electron moving around the nucleus of an atom can be compared to a fly buzzing around some form of attractant (e.g., food or a female fly, if the moving fly is male) at the center of a sealed room. We can state positively that the fly is in the room, and we can predict that he will be most attracted to the center, but we can never predict his location at any given moment.

As one moves along the periodic table of elements, electron configurations become ever more complex. The reason is that with an increase in atomic number, there is an increase in the energy levels of atoms. This indicates a greater range of energies that electrons can occupy, as well as a greater range of motion. Electrons occupying the highest energy level in an atom are called valence electrons, and these are the only ones involved in chemical bonding. By contrast, the core electrons, or the ones closest to the nucleus, play no role in the bonding of atoms.

IONIC AND COVALENT BONDS

THE GOAL OF EIGHT VA-LENCE ELECTRONS. The above discussion of the atom, and the electron's place in it, refers to much that was unknown at the time Thomson discovered the electron. Protons were not discovered for several more years, and neutrons several decades after that. Nonetheless, the electron proved the key to solving the riddle of how substances bond, and not long after Thomson's discovery, German chemist Richard Abegg (1869-1910) suggested as much.

While studying noble gases, noted for their tendency not to bond, Abegg discovered that these gases always have eight valence electrons. His observation led to one of the most important principles of chemical bonding: atoms bond in such a way that they achieve the electron configuration of a noble gas. This has been shown to be the case in most stable chemical compounds.

TWO DIFFERENT TYPES OF BONDS. Perhaps, Abegg hypothesized, atoms combine with one another because they exchange electrons in such a way that both end up with eight valence electrons. This was an early model of ionic bonding, which results from attractions between ions with opposite electric charges: when they bond, these ions "complete" one another.

Ionic bonds, which occur when a metal bonds with a nonmetal are extremely strong. As noted earlier, salt is an example of an ionic bond: the metal sodium loses an electron, forming a cation; meanwhile, the nonmetal chlorine gains the electron to become an anion. Their ionic bond results from the attraction of opposite charges.

Ionic bonding, however, could not explain all types of chemical bonds for the simple reason that not all compounds are ionic. A few years after Abegg's death, American chemist Gilbert Newton Lewis (1875-1946) discovered a very different type of bond, in which nonionic compounds share electrons. The result, once again, is eight valence electrons for each atom, but in this case, the nuclei of the two atoms share electrons.

In ionic bonding, two ions start out with different charges and end up forming a bond in which both have eight valence electrons. In the type of bond Lewis described, a covalent bond, two atoms start out as atoms do, with a net charge of zero. Each ends up possessing eight valence electrons, but neither atom "owns" them; rather, they share electrons.

LEWIS STRUCTURES. In addition to discovering the concept of covalent bonding, Lewis developed the Lewis structure, a means of showing schematically how valence electrons are arranged among the atoms in a molecule. Also known as the electron-dot system, Lewis structures represent the valence electrons as dots surrounding the chemical symbols of the atoms involved. These dots, which look rather like a colon, may be above or below, or on either side of, the chemical symbol. (The dots above or below the chemical symbol are side-by-side, like a colon turned at a 90°-angle.)

To obtain the Lewis structure representing a chemical bond, it is first necessary to know the number of valence electrons involved. One pair of electrons is always placed between elements, indicating the bond between them. Sometimes this pair of valence electrons is symbolized by a dashed line, as in the system developed by Couper. The remaining electrons are distributed according to the rules by which specific elements bond.

MULTIPLE BONDS. Hydrogen bonds according to what is known as the duet rule, meaning that a hydrogen atom has only two valence electrons. In most other elements—there are exceptions, but these will not be discussed here—atoms end up with eight valence electrons, and thus are said to follow the octet rule. If the bond is covalent, the total number of valence electrons will not be a multiple of eight, however, because the atoms share some electrons.

When carbon bonds to two oxygen atoms to form carbon dioxide (CO_2), it is represented in the Couper system as O-C-O. The Lewis structure also uses dashed lines, which stand for two valence electrons shared between atoms. In this case, then, the dashed line to the left of the carbon atom indicates a bond of two electrons with the oxygen atom to the left, and the dashed line to the right of it indicates a bond of two electrons with the oxygen atom on that side.

The non-bonding valence electrons in the oxygen atoms can be represented by sets of two dots above, below, and on the outside of each atom, for a total of six each. Combined with the two dots for the electrons that bond them to carbon, this gives each oxygen atom a total of eight valence electrons. So much for the oxygen atoms, but something is wrong with the representation of the carbon atom, which, up to this point, is shown only with four electrons surrounding it, not eight.

In fact carbon in this particular configuration forms not a single bond, but a double bond, which is represented by two dashed lines—a symbol that looks like an equals sign. By showing the double bonds joining the carbon atom to the two oxygen atoms on either side, the carbon atom has the required number of eight valence electrons. The carbon atom may also form a triple bond (represented by three dashed lines, one above the other) with an oxygen atom, in which case the oxygen atom would have only two other valence electrons.

ELECTRONEGATIVITY AND POLAR COVALENT BONDS

Today, chemists understand that most bonds are neither purely ionic nor purely covalent; rather, there is a wide range of hybrids between the two extremes. Credit for this discovery belongs to American chemist Linus Pauling (1901-1994), who, in the 1930s, developed the concept of electronegativity—the relative ability of an atom to attract valence electrons.

Elements capable of bonding are assigned an electronegativity value ranging from a minimum of 0.7 for cesium to a maximum of 4.0 for fluorine. Fluorine is capable of bonding with some noble gases, which do not bond with any other elements or each other. The greater the electronegativity value, the greater the tendency of an element to draw valence electrons to itself.

If fluorine and cesium bond, then, the bond would be purely ionic, because the fluorine exerts so much more attraction for the valence electrons. But if two elements have equal electronegativity values—for instance, cobalt and silicon, both of which are rated at 1.9—the bond is purely covalent. Most bonds, as stated earlier, fall somewhere in between these two extremes.

POLAR COVALENT BONDING. When substances of differing electronegativity values form a covalent bond, this is described as polar covalent bonding. Sometimes these are simply called "polar bonds," but that is not as accurate: all ionic bonds, after all, are polar, due to the extreme differences in electronegativity. The term "polar covalent bond" is much more specific, describing a bond, for instance, between hydrogen (2.1) and sulfur (2.6). Because sulfur has a slightly greater electronegativity value, the valence electrons will be slightly more attracted to the sulfur atom than to the hydrogen atom.

Another example of a polar covalent bond is the one that forms between hydrogen and oxygen (3.5) to form H_2O or water, which has a number of interesting properties. For instance, the polar quality of a water molecule gives it a great attraction for ions, and thus ionic substances such as salt dissolve easily in water. "Pure" water from a mountain stream is actually filled with traces of the rocks over which it has flowed. In fact, water—sometimes called the "universal solvent"—is almost impossible to find in pure form, except when it is purified in a laboratory.

By contrast, molecules of petroleum (CH_2) tend to be nonpolar, because carbon and hydrogen have almost identical electronegativity values—2.5 and 2.1 respectively. Thus, an oil molecule offers no electric charge to bond it with a water molecule, and for this reason, oil and water do not mix. It is a good thing that water molecules attract each other so strongly, because this means that a great amount of energy is required to change water from a liquid to a gas. If this were not so, the oceans and rivers would vaporize, and life on Earth could not exist as it does.

BOND ENERGY. The last two paragraphs allude to attractions between molecules, which is not the same as (nor is it as strong as) the attraction between atoms within a molecule. In fact, the bond energy—the energy required to pull apart the atoms in a chemical bond—is low for water. This is due to the presence of hydrogen atoms, with their two (rather than eight) valence electrons. It is thus relatively easy to separate water into its constituent parts of hydrogen and oxygen, through a process known as electrolysis.

Covalent bonds that involve hydrogen are among the weakest bonds between atoms. (Again, this is different from bonds between molecules.) Stronger than hydrogen bonds are regular, octet-rule covalent bonds: as one might expect, double covalent bonds are stronger than

KEY TERMS

ANION: The negative ion that results when an atom gains one or more electrons. An anion (pronounced "AN-ie-un") of an element is never called, for instance, the chlorine anion. Rather, for an anion involving a single element, it is named by adding the suffix -ide to the name of the original element—hence, "chloride." Other rules apply for more complex anions.

ATOM: The smallest particle of an element. An atom can exist either alone or in combination with other atoms in a molecule.

ATOMIC NUMBER: The number of protons in the nucleus of an atom. Since this number is different for each element, elements are listed on the periodic table of elements in order of atomic number.

BOND ENERGY: The energy required to pull apart the atoms in a chemical bond.

CATION: The positive ion that results when an atom loses one or more electrons. A cation (pronounced "KAT-ie-un") is named after the element of which it is an ion and thus is called, for instance, the aluminum ion or the aluminum cation.

CHEMICAL BONDING: The joining, through electromagnetic force, of atoms representing different elements. The principal types of bonds are covalent bonding and ionic bonding, though few bonds are purely one or the other. Rather, there is a wide range of "hybrid" bonds, in accordance with the electronegativity values of the elements involved.

CHEMICAL SYMBOL: A one-or two-letter abbreviation for the name of an element.

COMPOUND: A substance made up of atoms of more than one element. These atoms are usually joined in molecules.

COVALENT BONDING: A type of chemical bonding in which two atoms share valence electrons. Atoms may bond by single, double, or triple covalent bonds, which, in representations of Lewis structures, are shown by single, double, or triple dashed lines. (The double dashed line looks like an equals sign.) When atoms have differing values of electronegativity, they form polar covalent bonds.

DUET RULE: A term describing the distribution of valence electrons when hydrogen atoms—which end up with only two valence electrons—experience chemical bonding with other atoms. Most other elements follow the octet rule.

ELECTRON: Negatively charged particles in an atom. Electrons, which spin around the protons and neutrons that make up the atom's nucleus, are essential to chemical bonding.

ELECTRONEGATIVITY: The relative ability of an atom to attract valence electrons.

ELEMENT: A substance made up of only one kind of atom. Unlike compounds, elements cannot be broken down chemically into other substances.

ION: An atom that has lost or gained one or more electrons, and thus has a net electrical charge. Ions may either be anions or cations.

IONIC BONDING: A form of chemical bonding that results from attractions between ions with opposite electrical charges.

LEWIS STRUCTURE: A means of showing schematically how valence electrons are distributed among the atoms in a molecule. Also known as the electron-dot system, Lewis structure represents pairs of electrons with a symbol rather like a colon, which—depending on the situation—can be placed above, below, or on either side of the chemical symbol. In the Lewis structure, the pairs of electrons involved in chemical bonds are usually represented by a dashed line.

MOLECULE: A group of atoms, usually, but not always, representing more than one element, joined in a structure. Compounds are typically made of up molecules.

NEUTRON: A subatomic particle that has no electrical charge. Neutrons are found at the nucleus of an atom, alongside protons.

NUCLEUS: The center of an atom, a region where protons and neutrons are located, and around which electrons spin.

OCTET RULE: A term describing the distribution of valence electrons that takes place in chemical bonding for most elements, which end up with eight valence electrons. Hydrogen is an exception, and follows the duet rule. A few elements follow other rules, and some (most notably the noble gases) do not typically bond with other elements.

PERIODIC TABLE OF ELEMENTS: A chart showing the elements arranged in order of atomic number. Vertical columns within the periodic table indicate groups or "families" of elements with similar chemical characteristics.

POLAR COVALENT BONDING: The type of chemical bonding between atoms that have differing values of electronegativity. If the difference is extreme, of course, the bond is not a covalent bond at all, but an ionic bond. Thus, although these are sometimes called polar bonds, they are more properly identified as polar covalent bonds.

PROTON: A positively charged particle in an atom.

VALENCE ELECTRONS: Electrons that occupy the highest energy levels in an atom. These are the only electrons involved in chemical bonding. By contrast, the core electrons, or the ones at lower energy levels, play no role in the bonding of atoms.

VALENCY: The property of the atom of one element that determines its ability to bond with atoms of other elements.

single ones, and triple covalent bonds are stronger still. Strongest of all are ionic bonds, involved in the bonding of a metal to a metal, or a metal to a nonmetal, as in salt. The strength of the bond energy in salt is reflected by its boiling point of 1,472°F (800°C), much higher than that of water, at 212°F (100°C).

WHERE TO LEARN MORE

"Chemical Bond" (Web site). <http://www.science.uwaterloo.ca/~cchieh/cact/c120/chembond.html> (May 18, 2001).

"Chemical Bonding" Oklahoma State University (Web site). <http://www.okstate.edu/jgelder/bondtable.html> (May 19, 2001).

"Chemical Bonding" (Web site). <http://pc65.frontier.osrhe.edu/hs/science/pbond.htm> (May 19, 2001).

"Chemical Bonding" (Web site). <http://users.senet.com.au/~rowanb/chem/chembond.htm> (May 19, 2001).

"The Colored Periodic Table and Chemical Bonding" (Web site). <http://students.washington.edu/manteca/> (May 19, 2001).

Ebbing, Darrell D.; R. A. D. Wentworth; and James P. Birk. Introductory Chemistry. Boston: Houghton Mifflin, 1995.

Linus Pauling and the Twentieth Century: An Exhibition (Web site). <http://www.paulingexhibit.org> (May 19, 2001).

"Valence Shell Electron Pair Repulsion (VSEPR)." <http://www.shef.ac.uk/~chem/vsepr/chime/vsepr.html> (May 18, 2001).

White, Florence Meiman. Linus Pauling, Scientist and Crusader. New York: Walker, 1980.

Zumdahl, Steven S. Introductory Chemistry: A Foundation, 4th ed. Boston: Houghton Mifflin, 2000.

COMPOUNDS

CONCEPT

A compound is a chemical substance in which atoms combine in such a way that the compound always has the same composition, unless it is chemically altered in some way. Elements make up compounds, and although there are only about 90 elements that occur in nature, there are literally many millions of compounds. A compound is not the same as a mixture, which has a variable composition, but until chemists understood the atomic and molecular substructure of compounds, the distinction was not always clear. When atoms of one element bond to atoms of another, they form substances quite different from either element. Sugar, for instance, is made up of carbon, the material in coal and graphite, along with two gases, hydrogen and oxygen. None of these is sweet, yet when they are brought together in just the right way, they make the compound that sweetens everything from breakfast cereals to colas. A compound cannot be understood purely in terms of its constituent elements; an awareness of the structure is also needed. Likewise, it is important to know just how to name a compound, using a uniform terminology, since there are far too many substances in the world to give each an individual name.

HOW IT WORKS

ELEMENTS AND COMPOUNDS

A compound is a substance made up of atoms representing more than one element, and these atoms are typically joined in molecules. The composition of a compound is always the same: for instance, water always contains molecules composed of two hydrogen atoms bonded to a single oxygen atom. It can be frozen or boiled, but the molecules themselves are unchanged, because freezing and boiling are merely physical processes. To transform one compound into another, on the other hand, requires a chemical change or reaction.

Likewise, a chemical change is required to break down a compound into its constituent elements. Water can be broken down by passing a powerful electric current through it. This process, known as electrolysis, separates water into hydrogen and oxygen, both of which are highly flammable gases. The fact that they can be joined to form water, a substance used for putting out most kinds of fires, illustrates the types of changes elements undergo when they join to form compounds.

The atoms in a compound do not change into other atoms; or to put it another way, the elements do not change into other elements. Though two hydrogen atoms bond with an oxygen atom to form water, they are still hydrogens, and the oxygen is still an oxygen. The atoms' elemental identity thus remains intact, and if these elements are separated, they can join with other elements to form entirely different compounds.

LETTERS AND ELEMENTS; WORDS AND COMPOUNDS

The nature of a compound is almost paradoxical: the constituent elements remain the same, yet the compound typically bears little resemblance to the elements that form it. How can this be? Perhaps an analogy to language, and the symbols used to express it, will serve to show that this

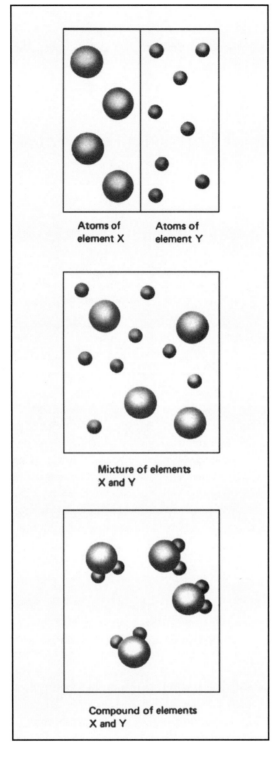

Atoms of
element X

Atoms of
element Y

Mixture of elements
X and Y

Compound of elements
X and Y

A MIXTURE VERSUS A COMPOUND. *(Robert L. Wolke. Reproduced by permission.)*

apparent contradiction is not a contradiction at all—it is merely evidence of the complexities that occur when a simple particle contributes to a larger combination.

There are between 88 and 92 elements that appear in nature; figures vary, because a handful of the elements with atomic numbers less than 92 have not actually been found in nature, but were produced in laboratories. (All elements with an atomic number higher than that of uranium, which is 92, are artificial.) In any case, there are far more elements than there are letters in the English alphabet; yet even with 26 characters, it is possible to form an almost limitless vocabulary.

Let us focus on a single letter, *g*. Phonetically, it can make a hard sound, as in great, or a soft one, as in geranium. It may be silent, as in light, or it may combine with another letter to make either a typical "g" sound (edge) or a totally unexpected one (rough). It can slide into a vowel sound smoothly, as in singer, or with an almost imperceptible pause, as in finger. The complexities multiply when we consider the range of possibilities for the letter *g*, and all the resultant meanings involved: from God to dog, or from glory to degeneration.

WHAT THIS ANALOGY SHOWS ABOUT CHEMISTRY. One could go on endlessly in this vein, and all with just one letter; however, a few points need to be made regarding the analogy between letters of the alphabet and elements. First of all, in all of the examples given above, or any number of others, the *g* still remained a *g*. In the same way, an atom of hydrogen (another g-word!) is always a hydrogen, whether it combines with oxygen to form water, nitrogen to make ammonia, or carbon to produce petroleum.

Secondly, note that this illustration would have been more difficult if the letter chosen had been *q* or *z*, since these are used more rarely. Likewise, there are elements such as technetium or praseodymium that seldom come up in discussions of compounds. On the other hand, one cannot go far in the study of chemistry without running across compounds involving hydrogen, carbon, oxygen, nitrogen, various metals, or halogens such as fluorine.

A third point to consider is the fact that there is nothing about *g* itself that provides any information as to the meaning of the word it helps to form. Here the analogy to elements is imperfect, because the nature of the element's subatomic structure—particularly the configuration of electrons on the outer shell of the atom—

provides a great deal of information as to how it will combine with other atoms.

Still, it is evident, as we have seen, that the properties of a compound cannot easily be predicted by studying the properties of its constituent elements—just as one cannot begin to define a word simply because it contains a *g*. On the other hand, knowledge of the sounds that *g* makes may help us to pronounce a word containing that letter. In the same way, the fact that a compound includes carbon provides a clue that the compound might be organic.

Even when we have all the letters to form a word, we still need to know how they are ordered. Loge and ogle contain the same letters, but one is a noun referring to a theater box, whereas the other is a verb meaning the act of looking at someone in an improper fashion. In chemistry, these are called isomers: two substances having identical chemical formulas, but differing chemical structures.

It so happens that *g* is also part of the suffix -ing, used in forming a gerund. Aside from being yet another *g*-word, a gerund is a substantive, or noun form of a verb: for instance, going. Once again, there is a similarity between words and compounds. Just as certain classes of words are formed by the addition of regular suffixes, so the naming of whole classes of compounds can be achieved by means of a uniform system of nomenclature.

REAL-LIFE APPLICATIONS

DISTINGUISHING BETWEEN COMPOUNDS AND MIXTURES

To continue the analogy used above just one step further, a word is not just a collection of letters; it has to have a meaning. Likewise, the fact that various substances are mixed together does not necessarily make them a compound. Actually, the difference between a word and a mere collection of letters is somewhat greater than the difference between a compound and a mixture—a substance in which elements are not chemically bonded, and in which the composition is variable. A nonsensical string of characters serves no linguistic purpose; on the other hand, mixtures are an integral part of life.

Tea, whether iced or hot, is a mixture. So is coffee, or even wine. In each case, substances are added together and subjected to a process, but the result is not a compound. We know this because the composition varies. Depending on the coffee beans used, for instance, coffee can have a wide variety of flavors. If, in the brewing process, too much coffee is used in proportion to the water, the resulting mixture will be strong or bitter; on the hand, an insufficient coffee-to-water ratio will produce coffee that is too weak.

Note that a number of terms have been used here that, from a scientific standpoint at least, are vague. How weak is "too weak"? That all depends on the tastes of the person making the coffee. But as long as coffee beans and hot water are used, no matter what the proportion, the mixture is still coffee. On the other hand, when two oxygen atoms, rather than one, are chemically combined with two hydrogen atoms, the result is not "strong water." Nor is it "oxygenated water": it is hydrogen peroxide, a substance no one should drink.

Three principal characteristics serve to differentiate a compound from a mixture. First, as we have seen, a compound has a definite and constant composition, whereas a mixture can exist with virtually any proportion between its constituent parts. Second, elements lose their characteristic elemental properties once they form a compound, but the parts of a mixture do not. (For example, when mixed with water, sugar is still sweet.) Third, the formation of a compound involves a chemical reaction, whereas a mixture can be created simply by the physical act of stirring items together.

THE ATOMIC AND MOLECULAR KEYS

The means by which compounds are formed are discussed numerous times, and in various ways, throughout this book. A few of those particulars will be mentioned briefly below, in relation to the naming of compounds, but for the most part, there will be no attempt to explain the details of the processes involved in chemical bonding. The reader is therefore encouraged to consult the essays on Chemical Bonding and Electrons.

It is important, nonetheless, to recognize that chemists' knowledge is based on their understanding of the atom and the ways that electrons, negatively charged particles in the atom, bring about bonds between elements. Awareness of these specifics emerged only at the beginning of the twentieth century, with the discovery of subatomic particles. Another important threshold had been crossed a century before, with the development of atomic theory by English chemist John Dalton (1766-1844), and of molecular theory by Italian physicist Amedeo Avogadro (1776-1856).

Around the same time, French chemists Antoine Lavoisier (1743-1794) and Joseph-Louis Proust (1754-1826), respectively, clarified the definitions of "element" and "compound." Until then, the idea of a compound had little precise meaning for chemists, who often used the term to describe a mixture. Thus, French chemist Claude Louis Berthollet (1748-1822) asserted that compounds have variable composition, and for evidence he pointed to the fact that when some metals are heated, they form oxides, in which the percentage of oxygen increases with temperature.

Proust, on the other hand, maintained that compounds must have a constant composition, an argument supported by Dalton's atomic theory. Proust worked to counter Berthollet's positions on a number of particulars, but was still unable to explain why metals form variable alloys, or combinations of metals; no chemist at the time understood that an alloy is a mixture, not a compound.

Nonetheless, Proust was right in his theory of constant composition, and Berthollet was incorrect on this score. With the discovery of subatomic structures, it became possible to develop highly sophisticated theories of chemical bonding, which in turn facilitated understanding of compounds.

TYPES OF COMPOUNDS

ORGANIC COMPOUNDS. Though there are millions of compounds, these can be grouped into just a few categories. Organic compounds, of which there are many millions, are compounds containing carbon. The only major groupings of carbon-containing compounds that are not considered organic are carbonates, such as limestone, and oxides, such as carbon dioxide. Organic compounds can be further subdivided into a number of functional groups, such as alcohols. Within the realm of organic compounds, whether natural or artificial, are petroleum and its many products, as well as plastics and other synthetic materials.

The term "organic," as applied in chemistry, does not necessarily refer to living things, since the definition is based on the presence of carbon. Nonetheless, all living organisms are organic, and the biochemical compounds in living things form an important subset of organic compounds. Biochemical compounds are, in turn, divided into four families: carbohydrates, proteins, nucleic acids, and lipids.

INORGANIC COMPOUNDS. Inorganic compounds can be classified according to five major groupings: acids, bases, salts, oxides, and others. An acid is a compound which, when dissolved in water, produces positive ions (atoms with an electric charge) of hydrogen. Bases are substances that produce negatively charged hydroxide (OH^-) ions when dissolved in water. An oxide is a compound in which the only negatively charged ion is an oxygen, and a salt is formed by the reaction of an acid with a base. Generally speaking, a salt is any combination of a metal and a nonmetal, and it can contain ions of any element but hydrogen.

The remaining inorganic compounds, classified as "others," are those that do not fit into any of the groupings described above. An important subset of this broad category are the coordination compounds, formed when one or more ions or molecules contributes both electrons necessary for a bonding pair, in order to bond with a metallic ion or atom.

NAMING COMPOUNDS

In the early days of chemistry as a science, common names were applied to compounds. Water is an example of a common name; so is sugar, as well as salt. These names work well enough in everyday life, and in fact, chemists still refer to water simply as "water." (The only other common name still used in chemistry is ammonia.)

But as the number of compounds discovered and developed by chemists began to proliferate, the need for a systematic means of naming them became apparent. With millions of compounds, it would be nearly impossible to come up with names for every one. Furthermore, common names tell chemists nothing about the chemical properties of a particular substance.

Today, chemists use a system of nomenclature that is rather detailed but fairly easy to understand, once the rules are understood. We will examine this system briefly, primarily as it relates to binary compounds—compounds containing just two elements. Binary compounds are divided into three groups. The first two are ionic compounds, involving metals that form positively charged ions, or cations (pronounced KAT-ieunz). The third consists of compounds that contain only nonmetals. These groups are:

- Type I: Ionic compounds involving a metal that always forms a cation of a certain electric charge.
- Type II: Ionic compounds involving a metal (typically a transition metal) that forms cations with differing charges.
- Type III: Compounds containing only nonmetals.

CATIONS AND ANIONS. Cations are represented symbolically thus: H^+, or Mg^{2+}. The first, a hydrogen cation, has a positive charge of 1, but note that the 1 is not shown—just as the first power of a number is never designated in mathematics. In the second example, a magnesium cation, the superscript number, combined with the plus sign (which can either follow the number, as is shown here, or proceed it) indicates that the atom has a positive charge of two. Thus, even if one were not told that this is a cation, it would be easy enough to discern from the notation.

Anions (AN-ie-unz), or negatively charged ions, are represented in a similar way: H^- for hydride, an anion of hydrogen; or O^{2-} for oxide. Note, however, that the naming of anions and cations is different. Cations are always called, for example, a hydrogen cation, or a magnesium cation. On the other hand, names of simple anions (involving a single atom) are formed by taking the root of the element name and adding an -ide: for example, fluoride (F^-).

TYPE I BINARY COMPOUNDS. In a binary ionic compound, a metal combines with a nonmetal. The metal loses one or more electrons to become a cation, while the nonmetal gains one or more electrons to become an anion. Thus, table salt is formed by the joining of a cation of the metal sodium (Na^+) and an anion of the nonmetal chlorine (Cl^-). Instead of the common name "salt," which can apply to a range of substances, its chemical name is sodium chloride.

In naming Type I binary compounds, of which sodium chloride is an example, the cation is always represented first by the name of the element. The anion follows, with the root name of the element attached to the -ide suffix, as described above. Another example is calcium sulfide, formed by a cation of the metal calcium (Ca^{2+}) and an anion of the nonmetal sulfur (S^{2-}).

TYPE II BINARY COMPOUNDS. The chemical nomenclature for type II binary compounds is somewhat more complicated, because they involve metals that can have multiple positive charges. This is particularly true of the transition metals, a family of 40 elements in the middle of the periodic table distinguished from other elements by a number of characteristics. The name "transition" thus implies a break in the even pattern of the periodic table.

Iron (Fe), for instance, is a transition metal, and it can form cations with charges of 2+ or 3+, while copper (Cu) can form cations with charges of 1+ or 2+. When encountering positively charged cations, it is not enough to say, for instance, "iron oxide," or "copper sulfide," because it is not clear which iron cation is involved. To solve the problem, chemists use a system of Roman numerals.

According to this system, the cations referred to above are expressed in the name of a Type II binary compound thus: iron (II), iron (III), copper (I), and copper (II). This is followed with the name of the anion, as before, using the -ide suffix. Note that the Roman numeral is usually the same as the number of positive charges in the cation.

It should be noted, also, that there is an older system for naming Type II binary compounds with terms that incorporate the element name—often the Latin original, reflected in the chemical symbol—with a suffix. For instance, this system uses the word "ferrous" for iron (II) and "ferric" for iron (III). However, this method of nomenclature is increasingly being replaced by the one we have described here.

TYPE III BINARY COMPOUNDS. In a Type III binary compound involving only nonmetals, the first element in the formula is referred to simply by its element name, as though it were a cation, while the second element is given an -ide suffix, as though it were an anion. If there is more than one atom present, prefixes are used to indicate the number of atoms. These prefixes are listed below. It should be noted that mono- is never used for naming the first element in a type III binary compound.

- mono-: 1
- di-: 2
- tri-: 3
- tetra-: 4
- penta-: 5
- hexa-: 6
- hepta-: 7
- octa-: 8

Thus CO_2 is called carbon dioxide, indicating one carbon atom and two oxygens. Again, mono- is not used for the first element, but it is used for the second: hence, the name of the compound with the formula CO is carbon monoxide. It is also possible to know the formula for a compound simply from the name: if confronted with a name such as "dinitrogen pentoxide," for instance, it is fairly easy to apply the rules governing these prefixes to discern that the substance is N_2O_5. Note that vowels at the end of a prefix are dropped when the name of the element that follows it also begins with a vowel: we say monoxide, not "monooxide"; and pentoxide, not "pentaoxide." This makes pronunciation much easier.

POLYATOMIC IONS AND ACIDS. More complicated rules apply for polyatomic ions, which are charged groupings of atoms such as NH_4NO_3, or ammonium nitrate. The only way to learn how to name polyatomic ions is by memorizing the names of the constituent parts. In the above example, for instance, the first part of the formula, which has a positive charge, is always called ammonium, while the second part, which has a negative charge, is always called nitrate. A good chemistry textbook should provide a table listing the names of common polyatomic ions.

A number of polyatomic ions are called oxyanions, meaning that they include varying numbers of oxygen atoms combined with atoms of other elements. There are rules for designating the names of these polyatomic ions, some of which are listed in the essay on Ions and Ionization.

Still other rules govern the naming of acids; here again, the operative factor is the presence of oxygen. Thus, if the anion does not contain oxy-gen, the name of the acid is created by adding the prefix hydro- and the suffix -ic, as in hydrochloric acid. If the anion does contain oxygen, the root name of the principal element is joined to a suffix: depending on the relative numbers of oxygen atoms, this may be -ic or -ous.

ISOMERS

As observed much earlier, isomers are like two words with the same letters, but arranged in different ways. Specifically, isomers are chemical compounds having the same formula, but in which the atoms are arranged differently, thereby forming different compounds. There are two principal types of isomer: structural isomers, which differ according to the attachment of atoms on the molecule, and stereoisomers, which differ according to the locations of the atoms in space.

An example of structural isomerism is the difference between propyl alcohol and isopropyl (rubbing) alcohol: these two have differing properties, because their alcohol functional groups are not attached to the same carbon atom in the carbon chain to which the functional group is attached.

In a stereoisomer, on the other hand, atoms are attached in the same order, but have different spatial relationships. If functional groups are aligned on the same side of a double bond between two carbon atoms, this is called a cis isomer, from a Latin word meaning "on this side." If they are on opposite sides, it is called a trans ("across") isomer. Hence the term "trans fats," which are saturated fats that improve certain properties of foods—including taste—but which may contribute to heart disease.

WHERE TO LEARN MORE

"Chemical Compounds" (Web site). <http://www.netaccess.on.ca/~dbc/cic_hamilton/comp.html> (June 2, 2001).

"Chemtutor Compounds." Chemtutor (Web site). <http://www.chemtutor.com/compoun.htm> (June 2, 2001).

Fullick, Ann. Matter. Des Plaines, IL: Heinemann Library, 1999.

"Glossary of Products with Hazards A to Z" Environmental Protection Agency (Web site). <http://www.epa.gov/grtlakes/seahome/housewaste/house/products.htm> (June 2, 2001).

Knapp, Brian J. Elements, Compounds, and Mixtures, Volume 2. Edited by Mary Sanders. Danbury, CT: Grolier

ACID: An inorganic compound that, when dissolved in water, produces positive ions—that is, cations—of hydrogen, designated symbolically as H^+ ions. (This is just one definition; see Acids and Bases; Acid-Base Reactions for more.)

ANION: The negatively charged ion that results when an atom gains one or more electrons. The word is pronounced "AN-ie-un."

BASE: An inorganic compound that produces negative hydroxide ions when it is dissolved in water. These anions are designated by the symbol OH^-. (This is just one definition; see Acids and Bases; Acid-Base Reactions for more.)

BINARY COMPOUND: A compound that contains just two elements. For the purposes of establishing compound names, binary compounds are divided into Type I, Type II, and Type III.

CATION: The positively charged ion that results when an atom loses one or more electrons. The word "cation" is pronounced "KAT-ie-un."

CHEMICAL BONDING: The joining, through electromagnetic force, of atoms representing different elements.

COMPOUND: A substance made up of atoms of more than one element, which are chemically bonded and usually joined in molecules. The composition of a compound is always the same, unless it is changed chemically.

COORDINATION COMPOUNDS: Inorganic compounds formed when one or more ions or molecules contribute both electrons necessary for a bonding pair in order to bond with a metallic ion or atom.

ELECTRON: A negatively charged particle in an atom.

INORGANIC COMPOUND: For the most part, inorganic compounds are any compounds that do not contain carbon. However, carbonates and carbon oxides are also inorganic compounds. Compare with organic compounds.

ION: An atom or group of atoms that has lost or gained one or more electrons, and thus has a net electric charge.

IONIC BONDING: A form of chemical bonding that results from attractions between ions with opposite electric charges.

IONIC COMPOUND: A compound in which ions are present. Ionic compounds contain at least one metal joined to another element by an ionic bond.

ISOMERS: Substances which have the same chemical formula, but which have different chemical properties due to differences in the arrangement of atoms.

MIXTURE: A substance in which elements are not chemically bonded, and in which the composition is variable. A mixture is distinguished from a compound.

ORGANIC COMPOUND: Generally speaking, a compound containing carbon. The only exceptions are the carbonates (for example, calcium carbonate or limestone) and oxides, such as carbon dioxide.

OXIDE: An inorganic compound in which the only negatively charged ion is an oxygen.

KEY TERMS CONTINUED

SALT: An inorganic compound formed by the reaction of an acid with a base. Generally speaking, a salt is a combination of a metal and a nonmetal, and it can contain ions of any element but hydrogen.

TYPE I BINARY COMPOUNDS: Ionic compounds involving a metal that always forms a cation of a certain electric charge.

TYPE II BINARY COMPOUNDS: Ionic compounds involving a metal (typically a transition metal) that forms cations with differing charges.

TYPE III BINARY COMPOUNDS: Compounds containing only nonmetals.

VALENCE ELECTRONS: Electrons that occupy the highest energy levels in an atom. These are the only electrons involved in chemical bonding.

Educational, 1998.

"*List of Compounds*" (Web site). <http://www.speclab.com/compound/chemabc.htm> (June 2, 2001).

Maton, Anthea. *Exploring Physical Science.* Upper Saddle River, N.J.: Prentice Hall, 1997.

"*Molecules and Compounds.*" General Chemistry Online

(Web site). <http://antoine.fsu.umd.edu/chem/senese/101/compounds/index.shtml> (June 2, 2001).

Oxlade, Chris. *Elements and Compounds.* Chicago, IL: Heinemann Library, 2001.

Zumdahl, Steven S. *Introductory Chemistry: A Foundation,* 4th ed. Boston: Houghton Mifflin, 2000.

CHEMICAL REACTIONS

CONCEPT

If chemistry were compared to a sport, then the study of atomic and molecular properties, along with learning about the elements and how they relate on the periodic table, would be like going to practice. Learning about chemical reactions, which includes observing them and sometimes producing them in a laboratory situation, is like stepping out onto the field for the game itself. Just as every sport has its "vocabulary"—the concepts of offense and defense, as well as various rules and strategies—the study of chemical reactions involves a large set of terms. Some aspects of reactions may seem rather abstract, but the effects are not. Every day, we witness evidence of chemical reactions—for instance, when a fire burns, or metal rusts. To an even greater extent, we are surrounded by the products of chemical reactions: the colors in the clothes we wear, or artificial materials such as polymers, used in everything from nylon running jackets to plastic milk containers.

HOW IT WORKS

WHAT IS A CHEMICAL REACTION?

If liquid water is boiled, it is still water; likewise frozen water, or ice, is still water. Melting, boiling, or freezing simply by the application of a change in temperature are examples of physical changes, because they do not affect the internal composition of the item or items involved. A chemical change, on the other hand, occurs when the actual composition changes—that is, when one substance is transformed into another. Water can be chemically changed, for instance, when an elec-

tric current is run through a sample, separating it into oxygen and hydrogen gas.

Chemical change requires a chemical reaction, a process whereby the chemical properties of a substance are altered by a rearrangement of the atoms in the substance. Of course we cannot see atoms with the naked eye, but fortunately, there are a number of clues that tell us when a chemical reaction has occurred. In many chemical reactions, for instance, the substance may experience a change of state or phase—as for instance when liquid water turns into gaseous oxygen and hydrogen as a result of electrolysis.

HOW DO WE KNOW WHEN A CHEMICAL REACTION HAS OCCURRED? Changes of state may of course be merely physical—as for example when liquid water is boiled to form a vapor. (These and other examples of physical changes resulting from temperature changes are discussed in the essays on Properties of Matter; Temperature and Heat.) The vapor produced by boiling water, as noted above, is still water; on the other hand, when liquid water is turned into the elemental gases hydrogen and oxygen, a more profound change has occurred.

Likewise the addition of liquid potassium chromate (K_2CrO_4) to a solution of barium nitrate ($Ba[NO_3]_2$ forms solid barium chromate ($BaCrO_4$). In the reaction described, a solution is also formed, but the fact remains that the mixture of two solids has resulted in the formation of a solid in a different solution. Again, this is a far more complex phenomenon than the mere freezing of water to form ice: here the fundamental properties of the materials involved have changed.

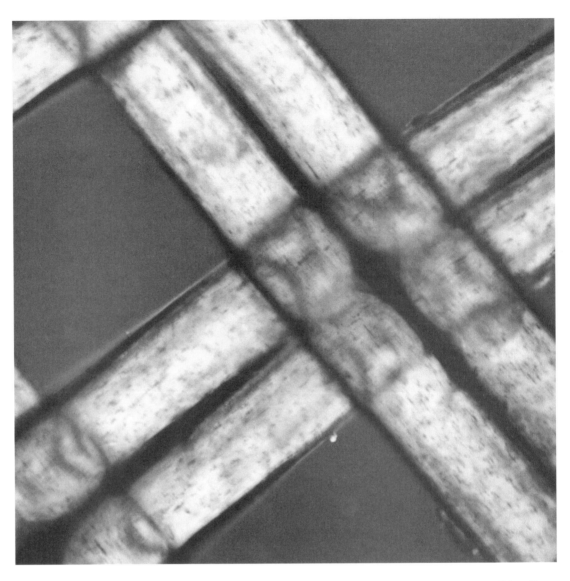

A MICROSCOPIC CLOSE-UP OF NYLON FIBERS. NYLON IS FORMED THROUGH A SYNTHESIS REACTION. *(Science Pictures Limited/Corbis. Reproduced by permission.)*

The physical change of water to ice or steam, of course, involves changes in temperature; likewise, chemical changes are often accompanied by changes in temperature, the crucial difference being that these changes are the result of alterations in the chemical properties of the substances involved. Such is the case, for instance, when wood burns in the presence of oxygen: once wood is turned to ash, it has become an entirely different mixture than it was before. Obviously, the ashes cannot be simply frozen to turn them back into wood again. This is an example of an irreversible chemical reaction.

Chemical reactions may also involve changes in color. In specific proportions and under the right conditions, carbon—which is black—can be combined with colorless hydrogen and oxygen to produce white sugar. This suggests another kind of change: a change in taste. (Of course, not every product of a chemical reaction should be tasted—some of the compounds produced may be toxic, or at the very least, extremely unpleasant to the taste buds.) Smell, too, can change. Sulfur is odorless in its elemental form, but when combined with hydrogen to form hydrogen sulfide (H_2S), it becomes an evil-smelling, highly toxic gas.

The bubbling of a substance is yet another clue that a chemical reaction has occurred. Though water bubbles when it boils, this is merely because heat has been added to the water, increasing the kinetic energy of its molecules. But when hydrogen peroxide bubbles when exposed to oxygen, no heat has been added. As with many of the characteristics of a chemical reaction described above, bubbling does not always occur when two chemicals react; however, when one of these clues is present, it tells us that a chemical reaction may have taken place.

REAL-LIFE APPLICATIONS

CHEMICAL EQUATIONS

In every chemical reaction, there are participants known as reactants, which, by chemically reacting to one another, result in the creation of a product or products. As stated earlier, a chemical reaction involves changes in the arrangement of atoms. The atoms in the reactants (or, if the reactant is a compound, the atoms in its molecules) are rearranged. The atomic or molecular structure of the product is different from that of either reactant.

Note, however, that the number of atoms does not change. Atoms themselves are neither created nor destroyed, and in a chemical reaction, they merely change partners, or lose partners altogether as they return to their elemental form. This is a critical principle in chemistry, one that proves that medieval alchemists' dream of turning lead into gold was based on a fallacy. Lead and gold are both elements, meaning that each has different atoms. To imagine a chemical reaction in which one becomes the other is like saying "one plus one equals one."

SYMBOLS IN A CHEMICAL EQUATION. In a mathematical equation, the sums of the numbers on one side of the equals sign must be the same as the sum of the numbers on the other side. The same is true of a chemical equation, a representation of a chemical reaction in which the chemical symbols on the left stand for the reactants, and those on the right are the product or products. Instead of an equals sign separating them, an arrow, pointing to the right to indicate the direction of the reaction, is used.

Chemical equations usually include notation indicating the state or phase of matter for the

A CLOSE-UP OF RUST ON A PIECE OF METAL. RUST IS AN EXAMPLE OF A COMBINATION REACTION, NOT A DECOMPOSITION REACTION. *(Marko Modic/Corbis. Reproduced by permission.)*

reactants and products. These symbols are as follows:

- *(s)*: solid
- *(l)*: liquid
- *(g)*: gas
- *(aq)*: dissolved in water (an aqueous solution)

The fourth symbol, of course, does not indicate a phase of matter per se (though obviously it appears to be a liquid); but as we shall see, aqueous solutions play a role in so many chemical reactions that these have their own symbol. At any rate, using this notation, we begin to symbolize the reaction of hydrogen and oxygen to form water thus: $H(g) + O(g) \rightarrow H_2O(l)$.

This equation as written, however, needs to be modified in several ways. First of all, neither hydrogen nor oxygen is monatomic. In other words, in their elemental form, neither appears as a single atom; rather, these form diatomic (two-atom) molecules. Therefore, the equation must be rewritten as $H_2(g) + O_2(g) \rightarrow H_2O(l)$. But this is still not correct, as a little rudimentary analysis will show.

When checking a chemical equation, one should always break it down into its constituent elements, to determine whether all the atoms on the left side reappear on the right side; otherwise, the result may be an incorrect equation, along the lines of "1 + 1 = 1." That is exactly what has happened here. On the left side, we have two hydrogen atoms and two oxygen atoms; on the right side, however, there is only one oxygen atom to go with the two hydrogens.

Obviously, this equation needs to be corrected to account for the second oxygen atom, and the best way to do that is to show a second water molecule on the right side. This will be represented by a 2 before the H_2O, indicating that two water molecules now have been created. The 2, or any other number used for showing more than one of a particular chemical species in a chemical equation, is called a coefficient. Now we have $H_2(g) + O_2(g) \rightarrow 2H_2O(l)$.

Is this right? Once again, it is time to analyze the equation, to see if the number of atoms on the left equals the number on the right. Such analysis can be done in a number of ways: for instance, by symbolizing each chemical species as a circle with chemical symbols for each element in it. Thus a single water molecule would be shown as a circle containing two H's and one O.

Whatever the method used, analysis will reveal that the problem of the oxygen imbalance has been solved: now there are two oxygens on the left, and two on the right. But solving that problem has created another, because now there are four hydrogen atoms on the right, as compared with two on the left. Obviously, another coefficient of 2 is needed, this time in front of the hydrogen molecule on the left. The changed equation is thus written as: $2H_2(g) + O_2(g) \rightarrow 2H_2O(l)$. Now, finally, the equation is correct.

THE PROCESS OF BALANCING CHEMICAL EQUATIONS. What we have done is to balance an unbalanced equation. An unbalanced equation is one in which the numbers of atoms on the left are not the same as the number of atoms on the right. Though an unbalanced equation is incorrect, it is sometimes a necessary step in the process of finding the balanced equation—one in which the number of atoms in the reactants and those in the product are equal.

In writing and balancing a chemical equation, the first step is to ascertain the identities, by formula, of the chemical species involved, as well as their states of matter. After identifying the reactants and product, the next step is to write an unbalanced equation. After that, the unbalanced equation should be subjected to analysis, as demonstrated above.

The example used, of course, involves a fairly simple substance, but often, much more complex molecules will be part of the equation. In performing analysis to balance the equation, it is best to start with the most complex molecule, and determine whether the same numbers and proportions of elements appear in the product or products. After the most complicated molecule has been dealt with, the second-most complex can then be addressed, and so on.

Assuming the numbers of atoms in the reactant and product do not match, it will be necessary to place coefficients before one or more chemical species. After this has been done, the equation should again be checked, because as we have seen, the use of a coefficient to straighten out one discrepancy may create another. Note that only coefficients can be changed; the formulas of the species themselves (assuming they were correct to begin with) should not be changed.

After the equation has been fully balanced, one final step is necessary. The coefficients must be checked to ensure that the smallest integers possible have been used. Suppose, in the above exercise, we had ended up with an equation that looked like this: $12H_2(g) + 6O_2(g) \rightarrow 12H_2O(l)$. This is correct, but not very "clean." Just as a fraction such as 12/24 needs to be reduced to its simplest form, 1/2, the same is true of a chemical equation. The coefficients should thus always be the smallest number that can be used to yield a correct result.

TYPES OF CHEMICAL REACTIONS

Note that in chemical equations, one of the symbols used is (aq), which indicates a chemical species that has been dissolved in water—that is, an aqueous solution. The fact that this has its own special symbol indicates that aqueous solutions are an important part of chemistry. Examples of reactions in aqueous solutions are discussed, for instance, in the essays on Acid-Base Reactions; Chemical Equilibrium; Solutions.

Another extremely important type of reaction is an oxidation-reduction reaction. Sometimes called a redox reaction, an oxidation-reduction reaction occurs during the transfer of electrons. The rusting of iron is an example of an oxidation-reduction reaction; so too is combustion. Indeed, combustion reactions—in which oxygen produces energy so rapidly that a flame or even an explosion results—are an important subset of oxidation-reduction reactions.

REACTIONS THAT FORM WATER, SOLIDS, OR GASES.

Another type of reaction is an acid-base reaction, in which an acid is mixed with a base, resulting in the formation of water along with a salt.

Other reactions form gases, as for instance when water is separated into hydrogen and oxygen. Similarly, heating calcium carbonate (limestone) to make calcium oxide or lime for cement also yields gaseous carbon dioxide: $CaCO_3(s) +$ heat $\rightarrow CaO(s) + CO_2(g)$.

There are also reactions that form a solid, such as the one mentioned much earlier, in which solid $BaCrO_4(s)$ is formed. Such reactions are called precipitation reactions. But this is also a reaction in an aqueous solution, and there is another product: $2KNO_3(aq)$, or potassium nitrate dissolved in water.

SINGLE AND DOUBLE DISPLACEMENT.

The reaction referred to in the preceding paragraph also happens to be an example of another type of reaction, because two anions (negatively charged ions) have been exchanged. Initially K^+ and CrO_4^{2-} were together, and these reacted with a compound in which Ba^{2+} and NO_3^- were combined. The anions changed places, an instance of a double-displacement reaction, which is symbolized thus: AB + CD $\rightarrow$ AD + CB.

It is also possible to have a single-displacement reaction, in which an element reacts with a compound, and one of the elements in the compound is released as a free element. This can be represented symbolically as A + BC $\rightarrow$ B + AC. Single-displacement reactions often occur with metals and with halogens. For instance, a metal (A) reacts with an acid (BC) to produce hydrogen (B) and a salt (AC).

COMBINATION AND DECOMPOSITION.

A synthesis, or combination, reaction is one in which a compound is formed from simpler materials—whether those materi-

als be elements or simple compounds. A basic example of this is the reaction described earlier in relation to chemical equations, when hydrogen and oxygen combine to form water. On the other hand, some extremely complex substances, such as the polymers in plastics and synthetic fabrics such as nylon, also involve synthesis reactions.

When iron rusts (in other words, it oxidizes in the presence of air), this is both an oxidation-reduction and a synthesis reaction. This also represents one of many instances in which the language of science is quite different from everyday language. If a piece of iron—say, a railing on a balcony—rusts due to the fact that the paint has peeled off, it would seem from an unscientific standpoint that the iron has "decomposed." However, rust (or rather, metal oxide) is a more complex substance than the iron, so this is actually a synthesis or combination reaction.

A true decomposition reaction occurs when a compound is broken down into simpler compounds, or even into elements. When water is subjected to electrolysis such that the hydrogen and oxygen are separated, this is a decomposition reaction. The fermentation of grapes to make wine is also a form of decomposition.

And then, of course, there are the processes that normally come to mind when we think of "decomposition": the decay or rotting of a formerly living thing. This could also include the decay of something, such as an item of food, made from a formerly living thing. In such instances, an organic substance is eventually broken down through a number of processes, most notably the activity of bacteria, until it ultimately becomes carbon, nitrogen, oxygen, and other elements that are returned to the environment.

SOME OTHER PARAMETERS.

Obviously, there are numerous ways to classify chemical reactions. Just to complicate things a little more, they can also be identified as to whether they produce heat (exothermic) or absorb heat (endothermic). Combustion is clearly an example of an exothermic reaction, while an endothermic reaction can be exemplified by the process that takes place in a cold pack. Used for instance to prevent swelling on an injured ankle, a cold pack contains an ampule that absorbs heat when broken.

Still another way to identify chemical reactions is in terms of the phases of matter involved.

We have already seen that some reactions form gases, some solids, and some yield water as one of the products. If reactants in one phase of matter produce a substance or substances in the same phase (liquid, solid, or gas), this is called a homogeneous reaction. On the other hand, if the reactants are in different phases of matter, or if they produce a substance or substances that are in a different phase, this is called a heterogeneous reaction.

An example of a homogeneous reaction occurs when gaseous nitrogen combines with oxygen, also a gas, to produce nitrous oxide, or "laughing gas." Similarly, nitrogen and hydrogen combine to form ammonia, also a gas. But when hydrogen and oxygen form water, this is a heterogeneous reaction. Likewise, when a metal undergoes an oxidation-reduction reaction, a gas and a solid react, resulting in a changed form of the metal, along with the production of new gases.

Finally, a chemical reaction can be either reversible or irreversible. Much earlier, we described how wood experiences combustion, resulting in the production of ash. This is clearly an example of an irreversible reaction. The atoms in the wood and the air that oxidized it have not been destroyed, but it would be impossible to put the ash back together to make a piece of wood. By contrast, the formation of water by hydrogen and oxygen is reversible by means of electrolysis.

KEEPING IT ALL STRAIGHT. The different classifications of reactions discussed above are clearly not mutually exclusive; they simply identify specific aspects of the same thing. This is rather like the many physical characteristics that describe a person: gender, height, weight, eye color, hair color, race, and so on. Just because someone is blonde, for instance, does not mean that the person cannot also be brown-eyed; these are two different parameters that are more or less independent.

On the other hand, there is some relation between these parameters in specific instances: for example, females over six feet tall are rare, simply because women tend to be shorter than men. But there are women who are six feet tall, or even considerably taller. In the same way, it is unlikely that a reaction in an aqueous solution will be a combustion reaction—yet it does happen, as for instance when potassium reacts with water.

STUDYING CHEMICAL REACTIONS

Several aspects or subdisciplines of chemistry are brought to bear in the study of chemical reactions. One is stoichiometry (stoy-kee-AH-muh-tree), which is concerned with the relationships among the amounts of reactants and products in a chemical reaction. The balancing of the chemical equation for water earlier in this essay is an example of basic stoichiometry.

Chemical thermodynamics is the area of chemistry that addresses the amounts of heat and other forms of energy associated with chemical reactions. Thermodynamics is also a branch of physics, but in that realm, it is concerned purely with physical processes involving heat and energy. Likewise physicists study kinetics, associated with the movement of objects. Chemical kinetics, on the other hand, involves the study of the collisions between molecules that produce a chemical reaction, and is specifically concerned with the rates and mechanisms of reaction.

SPEEDING UP A CHEMICAL REACTION. Essentially, a chemical reaction is the result of collisions between molecules. According to this collision model, if the collision is strong enough, it can break the chemical bonds in the reactants, resulting in a rearrangement of the atoms to form products. The more the molecules collide, the faster the reaction. Increase in the numbers of collisions can be produced in two ways: either the concentrations of the reactants are increased, or the temperature is increased. In either case, more molecules are colliding.

Increases of concentration and temperature can be applied together to produce an even faster reaction, but rates of reaction can also be increased by use of a catalyst, a substance that speeds up the reaction without participating in it either as a reactant or product. Catalysts are thus not consumed in the reaction. One very important example of a catalyst is an enzyme, which speeds up complex reactions in the human body. At ordinary body temperatures, these reactions are too slow, but the enzyme hastens them along. Thus human life can be said to depend on chemical reactions aided by a wondrous form of catalyst.

WHERE TO LEARN MORE

Bender, Hal. "*Chemical Reactions.*" *Clackamas Community College* (Web site). <http://dl.clackamas.cc.or.us/ch104-01/chemical.htm> (June 3, 2001).

KEY TERMS

ACID-BASE REACTION: A chemical reaction in which an acid is mixed with a base, resulting in the formation of water along with a salt.

AQUEOUS SOLUTIONS: A mixture of water and any substance that is solvent in it.

BALANCED EQUATION: A chemical equation in which the numbers of atoms in the reactants and those in the product are equal. In the course of balancing an equation, coefficients may need to be applied to one or more of the chemical species involved; however, the actual formulas of the species cannot be changed.

CATALYST: A substance that speeds up a chemical reaction without participating in it either as a reactant or product. Catalysts are thus not consumed in the reaction.

CHEMICAL EQUATION: A representation of a chemical reaction in which the chemical symbols on the left stand for the reactants, and those on the right for the product or products. On paper, a chemical equation looks much like a mathematical one; however, instead of an equals sign, a chemical equation uses an arrow to show the direction of the reaction.

CHEMICAL KINETICS: the study of the rate at which chemical reactions occur.

CHEMICAL REACTION: A process whereby the chemical properties of a substance are changed by a rearrangement of the atoms in the substance.

CHEMICAL SPECIES: A generic term used for any substance studied in chem-

istry—whether it be an element, compound, mixture, atom, molecule, ion, and so forth.

CHEMICAL THERMODYNAMICS: The study of the amounts of heat and other forms of energy associated with chemical reactions.

COEFFICIENT: A number used to indicate the presence of more than one unit—typically, more than one molecule—of a chemical species in a chemical equation. For instance, $2H_2O$ indicates two water molecules. (Note that 1 is never used as a coefficient.)

COLLISION MODEL: The theory that chemical reactions are the result of collisions between molecules that are strong enough to break bonds in the reactants, resulting in a rearrangement of atoms to form a product or products.

DECOMPOSITION REACTION: A chemical reaction in which a compound is broken down into simpler compounds, or even into elements. This is the opposite of a synthesis or combination reaction.

DOUBLE-DISPLACEMENT REACTION: A chemical reaction in which the partners in two compounds change places. This can be symbolized as AB + CD →AD + CB. Compare single-displacement reaction.

ENDOTHERMIC: A term describing a chemical reaction in which heat is absorbed or consumed.

EXOTHERMIC: A term describing a chemical reaction in which heat is produced.

HETEROGENEOUS: A term describing a chemical reaction in which the reactants are in different phases of matter (liquid, solid, or gas), or one in which the product is in a different phase from that of the reactants.

HOMOGENEOUS: A term describing a chemical reaction in which the reactants and the product are all in the same phase of matter (liquid, solid, or gas).

OXIDATION-REDUCTION REACTION: A chemical reaction involving the transfer of electrons.

PRECIPITATION REACTION: A chemical reaction in which a solid is formed.

PRODUCT: The substance or substances that result from a chemical reaction.

REACTANT: A substance that interacts with another substance in a chemical reaction, resulting in the formation of a product.

SINGLE-DISPLACEMENT REACTION: A chemical reaction in which an element reacts with a compound, and one of the elements in the compound is released as a free element. This can be represented symbolically as $A + BC \rightarrow B + AC$. Compare double-displacement reaction.

STOICHIOMETRY: The study of the relationships among the amounts of reactants and products in a chemical reaction. Producing a balanced equation requires application of stoichiometry (pronounced "stoy-kee-AH-muh-tree").

SYNTHESIS OR COMBINATION REACTION: A chemical reaction in which a compound is formed from simpler materials—either elements or simple compounds. It is the opposite of a decomposition reaction.

UNBALANCED EQUATION: A chemical equation in which the sum of atoms in the product or products does not equal the sum of atoms in the reactants. Initial observations of a chemical reaction usually produce an unbalanced equation, which needs to be analyzed and corrected (by the use of coefficients) to yield a balanced equation.

"Catalysis, Separations, and Reactions." Accelrys (Web site). <http://www.accelrys.com/chemicals/catalysis/> (June 3, 2001).

Goo, Edward. "Chemical Reactions" (Web site). <http://www-classes.usc.edu/engr/ms/125/MDA125/reactions/> (June 3, 2001).

Knapp, Brian J. Oxidation and Reduction. Illustrated by David Woodroffe. Danbury, CT: Grolier Educational, 1998.

Knapp, Brian J. Energy and Chemical Change. Illustrated by David Woodroffe. Danbury, CT: Grolier Educational, 1998.

Newmark, Ann. Chemistry. New York: Dorling Kindersley, 1993.

"Periodic Table: Chemical Reaction Data." WebElements (Web site). <http://www.webelements.com/webelements/elements/text/periodic-table/chem.html> (June 3, 2001).

Richards, Jon. Chemicals and Reactions. Brookfield, CT: Copper Beech Books, 2000.

"Types of Chemical Reactions" (Web site). <http://www.usoe.k12.ut.us/curr/science/sciber00/8th/matter/sciber/chemtype.htm> (June 3, 2001).

Zumdahl, Steven S. Introductory Chemistry: A Foundation, 4th ed. Boston: Houghton Mifflin, 2000.

OXIDATION-REDUCTION REACTIONS

CONCEPT

Most people have heard the term "oxidation" at some point or another, and, from the sound of the word, may have developed the impression that it has something to do with oxygen. Indeed it does, because oxygen has a tendency to draw electrons to itself. This tendency, rather than the presence of oxygen itself, is actually what identifies oxidation, defined as a process in which a substance loses electrons. The oxidation of one substance is always accompanied by reduction, or the gaining of electrons, on the part of another substance—hence the term "oxidation-reduction reaction," sometimes called a redox reaction. The world is full of examples of this highly significant form of chemical reaction. One such example is combustion, or an even more rapid form of combustion, explosion. Likewise the metabolism of food, as well as other biological processes, involves oxidation and reduction reactions. So, too, do a number of processes that take place on the surfaces of metals: when iron rusts; when copper turns green; or when aluminum forms a coating of aluminum oxide that prevents it from rusting. Oxidation-reduction reactions also play a major role in electrochemistry, which has a highly useful application to daily life in the form of batteries.

HOW IT WORKS

CHEMICAL REACTIONS

A chemical reaction is a process whereby the chemical properties of a substance are changed by a rearrangement its atoms. The change produced by a chemical reaction is quite different from a purely physical change, which does not affect the fundamental properties of the substance itself. A piece of copper can be heated, melted, beaten into different shapes, and so forth, yet throughout all those changes, it remains pure copper, an element of the transition metals family.

But suppose a copper roof is exposed to the elements for many years. Copper is famous for its highly noncorrosive quality, and this, combined with its beauty, has made it a favored material for use in the roofs of imposing buildings. (Because it is relatively expensive, few middle-class people today can afford a roof entirely made of copper, but sometimes it is used as a decorative touch—for instance, over the entryway of a house.) Eventually, however, copper does begin to corrode when exposed to air for long periods of time.

Over the years, exposed copper develops a thin layer of black copper oxide, and as time passes, traces of carbon dioxide in the air contribute to the formation of greenish copper carbonate. This explains why the Statue of Liberty, covered in sheets of copper, is green, rather than having the reddish-golden hue of new, uncorroded copper.

EXTERNAL VS. INTERNAL CHANGE. The preceding paragraphs describe two very different phenomena. The first was a physical change in which the chemical properties of a substance—copper—remained unaltered. The second, on the other hand, involved a chemical change on the surface of the copper, as copper atoms bonded with carbon and oxygen atoms in the air to form something different from copper. The difference between these two types of changes can be likened to varieties of changes in a person's life—an external change on the one hand, and a deeply rooted change on the other.

A person may move to another house, job, school, or town, yet the person remains the same. Many sayings in the English language express this fact: for instance, "Wherever you go, there you are," or "You can take the boy out of the country, but you can't take the country out of the boy." Moving is simply a physical change. On the other hand, if a person changes belief systems, overcomes old feelings (or succumbs to new ones), changes lifestyles in a profound manner, or in any other way changes his or her mind about something important—this is analogous to a chemical change. In these instances, the person, like the surface of the copper described above, has changed not merely in external properties, but in inner composition.

"LEO THE LION SAYS 'GER'"

Chemical reactions are addressed in depth within the essay devoted to that subject, which discusses—among other subjects—many ways of classifying chemical reactions. These varieties of chemical reaction are not all mutually exclusive, as they relate to different aspects of the reaction. As noted in the review of various reaction types, one of the most significant is an oxidation-reduction reaction (sometimes called a redox reaction) involving the transfer of electrons.

As its name implies, an oxidation-reduction reaction is really two processes: oxidation, in which electrons are lost, and reduction, in which electrons are gained. Though these are defined separately here, they do not occur independently; hence the larger reaction of which each is a part is called an oxidation-reduction reaction. In order to keep the two straight, chemistry teachers long ago developed a useful, if nonsensical, mnemonic device: "LEO the lion says 'GER.'" LEO stands for "Loss of Electrons, Oxidation," and "GER" means "Gain of Electrons, Reduction."

Many, though not all, oxidation-reduction reactions involve oxygen. Oxygen combines readily with other elements, and in so doing, it tends to grab electrons from those other elements' atoms. As a result, the oxygen atom becomes an ion (an atom with an electric charge)—specifically, an anion, or negatively charged ion.

In interacting with another element, oxygen becomes reduced, while the other element is oxidized to become a cation, or a positively charged ion. This, too, is easy to remember: oxygen itself, obviously, cannot be oxidized, so it must be the one being reduced. But since not all oxidation-reduction reactions involve oxygen, perhaps the following is a better way to remember it. Electrons are negatively charged, and the element that takes them on in an oxidation-reduction reaction is reduced—just as a person who thinks negative thoughts are "reduced" if those negative thoughts overcome positive ones.

OXIDATION NUMBERS

An oxidation number (sometimes called an oxidation state) is a whole-number integer assigned to each atom in an oxidation-reduction reaction. This makes it easier to keep track of the electrons involved, and to observe the ways in which they change positions. Here are some rules for determining oxidation number.

- 1. The oxidation number for an atom of an element not combined with other elements in a compound is always zero.
- 2. For an ion of any element, the oxidation number is the same as its charge. Thus a sodium ion, which has a charge of +1 and is designated symbolically as Na^+, has an oxidation number of +1.
- 3. Certain elements or families form ions in predictable ways:
- a. Alkali metals, such as sodium, always form a +1 ion; oxidation number = +1.
- b. Alkaline earth metals, such as magnesium, always form a +2 ion; oxidation number = +2.
- c. Halogens, such as fluorine, form −1 ions; oxidation number = −1.
- d. Other elements have predictable ways to form ions; but some, such as nitrogen, can have numerous oxidation numbers.
- 4. The oxidation number for oxygen is −2 for most compounds involving covalent bonds.
- 5. When hydrogen is involved in covalent bonds with nonmetals, its oxidation number is +1.
- 6. In binary compounds (compounds with two elements), the element having greater electronegativity is assigned a negative oxidation number that is the same as its charge

BATTERIES USE OXIDATION-REDUCTION REACTIONS TO PRODUCE ELECTRICAL CURRENT. *(Lester V. Bergman/ Corbis. Reproduced by permission.)*

when it appears as an anion in ionic compounds.

- 7. When a compound is electrically neutral, the sum of its elements' oxidation states is zero.

- 8. In an ionic chemical species, the sum of the oxidation states for its constituent elements must equal the overall charge.

These rules will not be discussed here; rather, they are presented to show some of the complexities involved in analyzing an oxidation-reduction reaction from a structural standpoint—that is, in terms of the atomic or molecular reactions. For the most part, we will be observing oxidation-reductions phenomenologically, or in terms of their outward effects. A good chemistry textbook should provide a more detailed review of these rules, along with a table showing oxidation numbers of elements and binary compounds.

Oxidation-reduction reactions are easier to understand if they are studied as though they were two half-reactions. Half the reaction involves what happens to the substances and electrons in the oxidizing portion, while the other half-reaction indicates the activities of substances and electrons in the reduction portion.

REAL-LIFE APPLICATIONS

COMBUSTION AND EXPLOSIONS

As with any type of chemical reaction, combustion takes place when chemical bonds are broken and new bonds are formed. It so happens that combustion is a particularly dramatic type of oxidation-reduction reaction: whereas we cannot watch iron rust, combustion is a noticeable event. Even more dramatic is combustion that takes place at a rate so rapid that it results in an explosion.

Coal is almost pure carbon, and its combustion in air is a textbook example of oxidation-reduction. Although there is far more nitrogen than oxygen in air (which is a mixture rather than a compound), nitrogen is very unreactive at low temperatures. For this reason, it can be used to clean empty fuel tanks, a situation in which the presence of pure oxygen is extremely dangerous. In any case, when a substance burns, it is reacting with the oxygen in air.

As one might expect from what has already been said about oxidation-reduction, the oxygen is reduced while the carbon is oxidized. In terms of oxidation numbers, the oxidation number of

OXIDATION-REDUCTION REACTIONS FUEL THE
SPACE-SHUTTLE AT TAKE-OFF. (*Corbis. Reproduced by
permission.*)

carbon jumps from 0 to 4, while that of oxygen is
reduced to –2. As they burn, these two form carbon dioxide or CO_2, in which the two –2 charges
of the oxygen atoms cancel out the +4 charge of
the carbon atom to yield a compound that is
electrically neutral.

COMBUSTION IN HUMAN EX-
PERIENCE.

Combustion has been a signifi-
cant part of human life ever since our prehistoric
ancestors learned how to harness the power of
fire to cook food and light their caves. We tend to
think of premodern times—to use the memorable title of a book by American historian
William Manchester, about the Middle Ages—as
A World Lit Only By Fire. In fact, our modern age
is even more combustion-driven than that of our
forebears.

For centuries, burning animal fat—in torches, lamps, and eventually in candles—provided
light for humans. Wood fires supplied warmth, as
well as a means to cook meals. These were the
main uses of combustion, aside from the occasional use of fire in warfare or for other purposes (including that ghastly medieval form of exe-
cution, burning at the stake). One notable mili-

tary application, incidentally, was "Greek fire,"
created by the Byzantines in the seventh century
A.D. A mixture of petroleum, potassium nitrate,
and possibly quicklime, Greek fire could burn on
water, and was used in naval battles to destroy
enemy ships.

For the most part, however, the range of
activities to which combustion could be applied
was fairly narrow until the development of the
steam engine in the period from the late seventeenth century to the early nineteenth century.
The steam engine applied the combustion of coal
to the production of heat for boiling water, which
in turn provided the power to run machinery. By
the beginning of the twentieth century, combustion had found a new application in the internal
combustion engine, used to power automobiles.

EXPLOSIONS AND EXPLO-
SIVES.

An internal combustion engine does
not simply burn fuel; rather, by the combined
action of the fuel injectors (in a modern vehicle),
in concert with the pistons, cylinders, and spark
plugs, it actually produces small explosions in the
molecules of gasoline. These produce the output
of power necessary to turn the crankshaft, and
ultimately the wheels.

An explosion, in simple terms, is a sped-up
form of combustion. The first explosives were
invented by the Chinese during the Middle Ages,
and these included not only fireworks and
explosive rockets, but gunpowder. Ironically,
however, China rejected the use of gunpowder in
warfare for many centuries, while Europeans
took to it with enthusiasm. Needless to say, Europeans' possession of firearms aided their conquest of the Americas, as well as much of Africa,
Asia, and the Pacific, during the period from
about 1500 to 1900.

The late nineteenth and early twentieth centuries saw the development of new explosives,
such as TNT or trinitrotoluene, a hydrocarbon.
Then in the mid-twentieth century came the
most fearsome explosive of all: the nuclear bomb.
A nuclear explosion is not itself the result of an
oxidation-reduction reaction, but of something
much more complex—either the splitting of
atoms (fission) or the forcing together of atomic
nuclei (fusion).

Nuclear bombs release far more energy than
any ordinary explosive, but the resulting blast
also causes plenty of ordinary combustion. When
the United States dropped atomic bombs on the

IF NOT POLISHED REGULARLY, SILVER RESPONDS TO OXIDATION BY TARNISHING—FORMING A SURFACE OF
SILVER SULFIDE, OR Ag_2S. *(Mimmo Jodice/Corbis. Reproduced by permission.)*

Japanese cities of Hiroshima and Nagasaki in August 1945, those cities suffered not only the effects of the immediate blast, but also massive fires resulting from the explosion itself.

FUELING THE SPACE SHUT-TLE. Oxidation-reduction reactions also fuel the most advanced form of transportation known today, the space shuttle. The actual orbiter vehicle is relatively small compared to its external power apparatus, which consists of two solid rocket boosters on either side, along with an external fuel tank.

Inside the solid rocket boosters are ammonium perchlorate (NH_4ClO_4) and powdered aluminum, which undergo an oxidation-reduction reaction that gives the shuttle enormous amounts of extra thrust. As for the larger single external fuel tank, this contains the gases that power the rocket: hydrogen and oxygen.

Because these two are extremely explosive, they must be kept in separate compartments.

When they react, they form water, of course, but in doing so, they also release vast quantities of energy. The chemical equation for this is: $2H_2 + O_2 \rightarrow 2H_2O$ + energy.

On January 28, 1986, something went terribly wrong with this arrangement on the space shuttle *Challenger*. Cold weather had fatigued the O-rings that sealed the hydrogen and oxygen compartments, and the gases fed straight into the flames behind the shuttle itself. This produced a powerful and uncontrolled oxidation-reduction reaction, an explosion that took the lives of all seven astronauts aboard the shuttle.

THE ENVIRONMENT AND HUMAN HEALTH

Combustion, though it can do much good, can also do much harm. This goes beyond the obvious: by burning fossil fuels or hydrocarbons, excess carbon (in the form of carbon dioxide and carbon monoxide) is released to the atmosphere, with a damaging effect on the environment.

In fact, oxidation-reduction reactions are intimately connected with the functioning of the natural environment. For example, photosynthesis, the conversion of light to chemical energy by plants, is a form of oxidation-reduction reaction that produces two essentials of human life: oxygen and carbohydrates. Likewise cellular respiration, which along with photosynthesis is discussed in the Carbon essay, is an oxidation-reduction reaction in which living things break down molecules of food to produce energy, carbon dioxide, and water.

Enzymes in the human body regulate oxidation-reduction reactions. These complex proteins, of which several hundred are known, act as catalysts, speeding up chemical processes in the body. Oxidation-reduction reactions also take place in the metabolism of food for energy, with substances in the food broken down into components the body can use.

OXIDATION: SPOILING AND AGING.
At the same time, oxidation-reduction reactions are responsible for the spoiling of food, the culprit here being the oxidation portion of the reaction. To prevent spoilage, manufacturers of food items often add preservatives, which act as reducing agents.

Oxidation may also be linked with the effects of aging in humans, as well as with other conditions such as cancer, hardening of the arteries, and rheumatoid arthritis. It appears that oxygen molecules and other oxidizing agents, always hungry for electrons, extract these from the membranes in human cells. Over time, this can cause a gradual breakdown in the body's immune system.

To forestall the effects of oxidation, some doctors and scientists recommend antioxidants—natural reducing agents such as vitamin C and vitamin E. The vitamin C in lemon juice can be used to prevent oxidizing on the cut surface of an apple, to keep it from turning brown. Perhaps, some experts maintain, natural reducing agents can also slow the pace of oxidation in the human body.

FORMING A NEW SURFACE ON METAL

Clearly, oxidization can have a corrosive effect, and nowhere is this more obvious than in the corrosion of metals by exposure to oxidizing agents—primarily oxygen itself. Most metals react with O_2, and might corrode so quickly that they become useless, were it not for the formation of a protective coating—an oxide.

Iron forms an oxide, commonly known as rust, but this in fact does little to protect it from corrosion, because the oxide tends to flake off, exposing fresh surfaces to further oxidation. Every year, businesses and governments devote millions of dollars to protecting iron and steel from oxidation by means of painting and other measures, such as galvanizing with zinc. In fact, oxidation-reduction reactions virtually define the world of iron. Found naturally only in ores, the element is purified by heating the ore with coke (impure carbon) in the presence of oxygen, such that the coke reduces the iron.

COINAGE METALS.
Copper, as we have seen, responds to oxidation by corroding in a different way: not by rusting, but by changing color. A similar effect occurs in silver, which tarnishes, forming a surface of silver sulfide, or Ag_2S. Copper and silver are two of the "coinage metals," so named because they have often been used to mint coins. They have been used for this purpose not only because of their beauty, but also due to their relative resistance to corrosion. This resistance has, in fact, earned them the nickname "noble metals."

KEY TERMS

ANION: An ion with a negative charge; pronounced "AN-ie-un."

BINARY COMPOUND: A compound involving two elements.

CHEMICAL REACTION: A process whereby the chemical properties of a substance are changed by a rearrangement of the atoms in the substance.

CHEMICAL SPECIES: A generic term used for any substance studied in chemistry—whether it be an element, compound, mixture, atom, molecule, ion, and so forth.

COVALENT BONDING: A type of chemical bonding in which two atoms share valence electrons.

ELECTROCHEMISTRY: The study of the relationship between chemical and electrical energy.

ELECTROLYSIS: The use of an electric current to produce a chemical change.

ELECTRONEGATIVITY: The relative ability of an atom to attract valence electrons.

CATION: An ion with a positive charge; pronounced "KAT-ie-un."

ION: An atom or group of atoms that has acquired an electric charge due to the loss or gain of electrons.

OXIDATION: A chemical reaction in which a substance loses electrons. It is always accompanied by reduction; hence the term oxidation-reduction reaction.

OXIDATION NUMBER: A number assigned to each atom in an oxidation-reduction reaction as a means of keeping track of the electrons involved. Another term for oxidation number is "oxidation state."

OXIDATION-REDUCTION REACTION: A chemical reaction involving the transfer of electrons.

REDOX REACTION: Another name for an oxidation-reduction reaction.

REDUCTION: A chemical reaction in which a substance gains electrons. It is always accompanied by oxidation; hence the term oxidation-reduction reaction.

VALENCE ELECTRONS: Electrons that occupy the highest energy levels in an atom. These are the only electrons involved in chemical bonding.

The third member of this mini-family is gold, which is virtually noncorrosive. Wonderful as gold is in this respect, however, no one is likely to use it as a roofing material, or for any such large-scale application involving its resistance to oxidation. Aside from the obvious expense, gold is soft, and not very good for structural uses, even if it were much cheaper. Yet there is such a "wonder metal": one that experiences virtually no corrosion, is cheap, and strong enough in alloys to be used for structural purposes. Its name is aluminum.

ALUMINUM. There was a time, in fact, when aluminum was even more expensive than gold. When the French emperor Napoleon III wanted to impress a dinner guest, he arranged for the person to be served with aluminum utensils, while less distinguished personages had to settle for "ordinary" gold and silver.

In 1855, aluminum sold for $100,000 a pound, whereas in 1990, the going rate was about $0.74. Demand did not go down—in fact, it increased exponentially—but rather, supply increased, thanks to the development of an inexpensive aluminum-reduction process. Two men, one American and one French, discovered this process at the same time: interestingly, their years of birth and death were the same.

Aluminum was once a precious metal because it proved extremely difficult to separate from oxygen. The Hall-Heroult process overcame the problem by applying electrolysis—the use of an electric current to produce a chemical change—as a way of reducing Al^{3+} ions (which have a high affinity for oxygen) to neutral aluminum atoms. In the United States today, 4.5% of the total electricity output is used for the production of aluminum through electrolysis.

The foregoing statistic is staggering, considering just how much electricity Americans use, and it indicates the importance of this once-precious metal. Actually, aluminum oxidizes just like any other metal—and does so quite quickly, as a matter of fact, by forming a coating of aluminum oxide (Al_2O_3). But unlike rust, the aluminum oxide is invisible, and acts as a protective coating. Chromium, nickel, and tin react to oxygen in a similar way, but these are not as inexpensive as aluminum.

ELECTROCHEMISTRY AND BATTERIES

Electrochemistry is the study of the relationship between chemical and electrical energy. Among its applications is the creation of batteries, which use oxidation-reduction reactions to produce an electric current.

A basic battery can be pictured schematically as two beakers of solution connected by a wire. In one solution is the oxidizing agent; in the other, a reducing agent. The wire allows electrons to pass back and forth between the two solutions, but to ensure that the flow goes both ways, the two solutions are also connected by a "salt bridge." The salt bridge contains a gel or solution that permits ions to pass back and forth, but a porous membrane prevents the solutions from actually mixing.

In the lead storage battery of an automobile, lead itself is the reducing agent, while lead (IV)

oxide (PbO_2) acts as the oxidizing agent. A highly efficient type of battery, able to withstand wide extremes in temperature, the lead storage battery has been in use since 1915. Along the way, features have been altered, but the basic principles have remained—a testament to the soundness of its original design.

The batteries people use for powering all kinds of portable appliances, from flashlights to boom boxes, are called dry cell batteries. In contrast to the model described above, using solutions, a dry cell (as its name implies) involves no liquid components. Instead, it utilizes various elements in a range of combinations, including zinc, magnesium, mercury, silver, nickel, and cadmium. The last two are applied in the nickel-cadmium battery, which is particularly useful because it can be recharged over and over again by an external current. The current turns the products of the chemical reactions in the battery back into reactants.

WHERE TO LEARN MORE

"Batteries." *Oregon State University Department of Chemistry* (Web site). <http://www.chem.orst.edu/ch411/scbatt.htm> (June 4, 2001).

"Batteries and Fuel Cells" (Web site). <http://vectorsite.com/ttfuelc.html> (June 4, 2001).

Borton, Paula and Vicky Cave. *The Usborne Book of Batteries and Magnets.* Tulsa, OK: EDC Publishing, 1995.

Craats, Rennay. *The Science of Fire.* Milwaukee, WI: Gareth Stevens Publishing, 2000.

Knapp, Brian J. *Oxidation and Reduction.* Danbury, CT: Grolier Educational, 1998.

"Oxidation Reduction Links" (Web site). <http://users.erols.com/merosen/redox.htm> (June 4, 2001).

"Oxidation-Reduction Reactions." *General Chemistry* (Web site). <http://ull.chemistry.uakron.edu/genchem/11/> (June 4, 2001).

"Oxidation-Reduction Reactions: Redox." *UNC-Chapel Hill Chemistry Fundamentals* (Web site). <http://www.shodor.org/unchem/advanced/redox/> (June 4, 2001).

Yount, Lisa. *Antoine Lavoisier: Founder of Modern Chemistry.* Springfield, NJ: Enslow Publishers, 1997.

Zumdahl, Steven S. *Introductory Chemistry: A Foundation,* fourth edition. Boston: Houghton Mifflin, 2000.

CHEMICAL EQUILIBRIUM

CONCEPT

Reactions are the "verbs" of chemistry—the activity that chemists study. Many reactions move to their conclusion and then stop, meaning that the reactants have been completely transformed into products, with no means of returning to their original state. In some cases, the reaction truly is irreversible, as for instance when combustion changes both the physical and chemical properties of a substance. There are plenty of other circumstances, however, in which a reverse reaction is not only possible but an ongoing process, as the products of the first reaction become the reactants in a second one. This dynamic state, in which the concentrations of reactants and products remains constant, is referred to as equilibrium. It is possible to predict the behavior of substances in equilibrium through the use of certain laws, which are applied in industries seeking to lower the costs of producing specific chemicals. Equilibrium is also useful in understanding processes that preserve—or potentially threaten—human health.

HOW IT WORKS

CHEMICAL REACTIONS IN BRIEF

What follows is a highly condensed discussion of chemical reactions, and particularly the methods for writing equations to describe them. For a more detailed explanation of these principles, the reader is encouraged to consult the Chemical Reactions essay.

A chemical reaction is a process whereby the chemical properties of a substance are altered by a rearrangement of the atoms in the substance.

The changes produced by a chemical reaction are fundamentally different from physical changes, such as boiling or melting liquid water, changes that alter the physical properties of water without affecting its molecular structure.

INDICATIONS THAT A CHEMICAL REACTION HAS OCCURRED. Though chemical reactions are most effectively analyzed in terms of molecular properties and behaviors, there are numerous indicators that suggest to us when a chemical reaction has occurred. It is unlikely that all of these will result from any one reaction, and in fact chances are that a particular reaction will manifest only one or two of these effects. Nonetheless, these offer us hints that a reaction has taken place:

Signs that a substance has undergone a chemical reaction:

- Water is produced
- A solid forms
- Gases are produced
- Bubbles are formed
- There is a change in color
- The temperature changes
- The taste of a consumable substance changes
- The smell changes

CHEMICAL CHANGES CONTRASTED WITH PHYSICAL CHANGES OF TEMPERATURE. Many of these effects can be produced simply by changing the temperature of a substance, but again, the mere act of applying heat from outside (or removing heat from the substance itself) does not constitute a chemical change. Water can be "produced" by melting ice, but the water was already there—it only changed form. By contrast, when an acid

MANY MOUNTAIN CLIMBERS USE PRESSURIZED OXYGEN
AT VERY HIGH ALTITUDES TO MAINTAIN PROPER HEMO-
GLOBIN-OXYGEN EQUILIBRIUM. *(Galen Rowell/Corbis. Repro-
duced by permission.)*

and a base react to form water and a salt, that is a
true chemical reaction.

Similarly, the freezing of water forms a solid,
but no new substance has been formed; in a
chemical reaction, by contrast, two liquids can
react to form a solid. When water boils through
the application of heat, bubbles form, and a gas
or vapor is produced; yet in chemical changes,
these effects are not the direct result of applying
heat.

In this context, a change in temperature,
noted as another sign that a reaction has taken
place, is a change of temperature from within the
substance itself. Chemical reactions can be classified as heat-producing (exothermic) or heat-
absorbing (endothermic). In either case, the
transfer of heat is not accomplished simply by
creating a temperature differential, as would
occur if heat were transferred merely through
physical means.

WHY DO CHEMICAL REACTIONS OCCUR?

At one time, chemists could only study reactions
from "outside," as it were—purely in terms of
effects noticeable through the senses. Between

the early nineteenth and the early twentieth centuries, however, the entire character of chemistry
changed, as did the terms in which chemists discussed reactions. Today, those reactions are analyzed primarily in terms of subatomic, atomic,
and molecular properties and activities.

Despite all this progress, however, chemists
still do not know exactly what happens in a
chemical reaction—but they do have a good
approximation. This is the collision model,
which explains chemical reactions in terms of
collisions between molecules. If the collision is
strong enough, it can break the chemical bonds
in the reactants, resulting in a re-formation of
atoms within different molecules. The more the
molecules collide, the more bonds are being broken, and the faster the reaction.

Increase in the numbers of collisions can be
produced in two ways: either the concentrations
of the reactants are increased, or the temperature
is increased. By raising the temperature, the
speeds of the molecules themselves increase, and
the collisions possess more energy. A certain
energy threshold, the activation energy (symbolized E_a) must be crossed in order for a reaction to
occur. A temperature increase raises the likelihood that a given collision will cross the activation-energy threshold, producing the energy to
break the molecular bonds and promote the
chemical reaction.

Raising the temperature and the concentrations of reactants can increase the energy and
hasten the reactions, but in some cases it is not
possible to do either. Fortunately, the rate of
reaction can be increased in a third way, through
the introduction of a catalyst, a substance that
speeds up the reaction without participating in it
either as a reactant or product. Catalysts are thus
not consumed in the reaction. Many chemistry
textbooks discuss catalysts within the context of
equilibrium; however, because catalysts play such
an important role in human life, in this book
they are the subject of a separate essay.

CHEMICAL EQUATIONS INVOLVING EQUILIBRIUM

A chemical equation, like a mathematical equation, symbolizes an interaction between entities
that produces a particular result. In the case of a
chemical equation, the entities are not numbers
but reactants, and they interact with each other
not through addition or multiplication, but by

chemical reaction. Yet just as a product is the result of multiplication in mathematics, a product in a chemical equation is the substance or substances that result from the reaction.

Instead of an equals sign between the reactants and the product, an arrow is used. When the arrow points to the right, this indicates a forward reaction; conversely, an arrow pointing to the left symbolizes a reverse reaction. In a reverse reaction, the products of a forward reaction have become the reactants, and the reactants of the forward reaction are now the products. This is indicated by an arrow that points toward the left.

Chemical equilibrium, which occurs when the ratio between the reactants and products is constant, and in which the forward and reverse reactions take place at the same rate, is symbolized thus: $\rightleftharpoons$. Note that the arrows, the upper one pointing right and the lower one pointing left, are of the same length. There may be certain cases, discussed below, in which it is necessary to show these arrows as unequal in length as a means of indicating the dominance of either the forward or reverse reaction.

Chemical equations usually include notation indicating the state or phase of matter for the reactants and products: *(s)* for a solid; *(l)* for a liquid; *(g)* for a gas. A fourth symbol, *(aq)*, indicates a substance dissolved in water—that is, an aqueous solution. In the following paragraphs, we will apply a chemical equation to the demonstration of equilibrium, but will not discuss the balancing of equations. The reader is encouraged to consult the passage in the Chemical Reactions essay that addresses that process, vital to the recording of accurate data.

A SIMPLE EQUILIBRIUM EQUATION. Let us now consider a simple equation, involving the reaction between water and carbon monoxide (CO) at high temperatures in a closed container. The initial equation is written thus: $H_2O(g) + CO(g) \rightarrow H_2(g) + CO_2g)$. In plain English, water in the gas phase (steam) has reacted with carbon monoxide to produce hydrogen gas and carbon dioxide.

As the reaction proceeds, the amount of reactants decreases, and the concentration of products increases. At some point, however, there will be a balance between the numbers of products and reactants—a state of chemical equilibrium represented by changing the right-pointing arrow to an equilibrium symbol: $H_2O(g) +$

$CO(g) \rightleftharpoons H_2(g) + CO_2g)$. Assuming that the system is not disturbed (that is, that the container is kept closed and no outside substances are introduced), equilibrium will continue to be maintained, because the reverse reaction is occurring at the same rate as the forward one.

Note what has been said here: reactions are still occurring, but the forward and rearward reactions balance one another. Thus equilibrium is not a static condition, but a dynamic one, and indeed, chemical equilibrium is sometimes referred to as "dynamic equilibrium." On the other hand, some chemists refer to chemical equilibrium simply as equilibrium, but here the qualifier chemical has been used to distinguish this from the type of equilibrium studied in physics. Physical equilibrium, which involves factors such as center of gravity, does help us to understand chemical equilibrium, but it is a different phenomenon.

REAL-LIFE APPLICATIONS

HOMOGENEOUS AND HETEROGENEOUS EQUILIBRIA

It should be noted that the equation used above identifies a situation of homogeneous equilibrium, in which all the substances are in the same phase or state of matter—gas, in this case. It is also possible to achieve chemical equilibrium in a reaction involving substances in more than one phase of matter.

An example of such heterogeneous equilibrium is the decomposition of calcium carbonate for the production of lime, a process that involves the application of heat. Here the equation would be written thus: $CaCO_3(s) \rightleftharpoons CaO(s) + CO_2(g)$. Both the calcium carbonate ($CaCO_3$) and the lime (CaO) are solids, whereas the carbon dioxide produced in this reaction is a gas.

THE EQUILIBRIUM CONSTANT

In 1863, Norwegian chemists Cato Maximilian Guldberg (1836-1902) and Peter Waage (1833-1900)—who happened to be brothers-in-law—formulated what they called the law of mass action. Today, this is called the law of chemical equilibrium, which states that the direction taken by a reaction is dependant not merely on the mass of the various components of the reaction, but also upon the concentration—that is, the mass present in a given volume.

This can be expressed by the formula $aA + bB \rightleftharpoons cC + dD$, where the capital letters represent chemical species, and the italicized lowercase letters indicate their coefficients. The equation $[C]^c[D]^d/[A]^a[B]^b$ yields what is called an equilibrium constant, symbolized K.

The above formula expresses the equilibrium constant in terms of molarity, the amount of solute in a given volume of solution, but in the case of gaseous reactants and products, the equilibrium constant can also be expressed in terms of partial pressures. In the reaction of water and carbon monoxide to produce hydrogen molecules and carbon dioxide ($H_2O + CO \rightleftharpoons H_2 + CO_2$). In chemical reactions involving solids, however, the concentration of the solid—because it is considered to be invariant—does not appear in the equilibrium constant. In the reaction described earlier, in which calcium carbonate was in equilibrium with solid lime and gaseous carbon dioxide, K = pressure of CO_2.

We will not attempt here to explore the equilibrium constant in any depth, but it is important to recognize its usefulness. For a particular reaction at a specific temperature, the ratio of concentrations between reactants and products will always have the same value—the equilibrium constant, or K. Because it is not dependant on the amounts of reactants and products mixed together initially, K remains the same: the concentrations themselves may vary, but the ratios between the concentrations in a given situation do not.

LE CHÂTELIER'S PRINCIPLE

Not all situations of equilibrium are alike: depending on certain factors, the position of equilibrium may favor one side of the equation or the other. If a company is producing chemicals for sale, for example, its production managers will attempt to influence reactions in such a way as to favor the forward reaction. In such a situation, it is said that the equilibrium position has been shifted to the right. In terms of physical equilibrium, mentioned above, this would be analogous to what would happen if you were holding your arms out on either side of your body, with a heavy lead weight in your left hand and a much smaller weight in the right hand.

Your center of gravity, or equilibrium position, would shift to the left to account for the greater force exerted by the heavier weight.

A value of K significantly above 1 causes a shift to the right, meaning that at equilibrium, there will be more products than reactants. This is a situation favorable to a chemical company's managers, who desire to create more of the product from less of the reactants. However, nature abhors an imbalance, as expressed in Le Châtelier's principle. Named after French chemist Henri Le Châtelier (1850-1936), this principle maintains that whenever a stress or change is imposed on a chemical system in equilibrium, the system will adjust the amounts of the various substances to reduce the impact of that stress.

Suppose we add more of a particular substance to increase the rate of the forward reaction. In an equation for this reaction, the equilibrium symbol is altered, with a longer arrow pointing to the right to indicate that the forward reaction is favored. Again, the equilibrium position has shifted to the right—just as one makes physical adjustments to account for an imbalanced weight. The system responds by working to consume more of the reactant, thus adjusting to the stress that was placed on it by the addition of more of that substance. By the same token, if we were to remove a particular reactant or product, the system would shift in the direction of the detached component.

Note that Le Châtelier's principle is mathematically related to the equilibrium constant. Suppose we have a basic equilibrium equation of $A + B \rightleftharpoons C$, with A and B each having molarities of 1, and C a molarity of 4. This tells us that K is equal to the molarity of C divided by that of A multiplied by B = $4/(1 \cdot 1)$. Suppose, now, that enough of C were added to bring its concentration up to 6. This would mean that the system was no longer at equilibrium, because $C/(A \cdot B)$ no longer equals 4. In order to return the ratio to 4, the numerator (C) must be decreased, while the denominator (A · B) is increased. The reaction thus shifts from right to left.

CHANGES IN VOLUME AND TEMPERATURE. If the volume of gases in a closed container is decreased, the pressure increases. An equilibrium system will therefore shift in the direction that reduces the pressure; but if the volume is increased, thus reducing the pressure, the system will respond by shifting to increase pressure. Note, however, that not all increases in pressure lead to a shift in the equilibrium. If the pressure were increased by the addition of a noble gas, the gas itself—since these elements are noted for their lack of reactivity—would not be part of the reaction. Thus the species added would not be part of the equilibrium constant expression, and there would be no change in the equilibrium.

In any case, no change in volume alters the equilibrium constant K; but where changes in temperature are involved, K is indeed altered. In an exothermic, or heat-producing reaction, the heat is treated as a product. Thus, when nitrogen and hydrogen react, they produce not only ammonia, but a certain quantity of heat. If this system is at equilibrium, Le Châtelier's principle shows that the addition of heat will induce a shift in equilibrium to the left—in the direction that consumes heat or energy.

The reverse is true in an endothermic, or heat-absorbing reaction. As in the process described earlier, the thermal decomposition of calcium carbonate produces lime and carbon dioxide. Because heat is used to cause this reaction, the amount of heat applied is treated as a reactant, and an increase in temperature will cause the equilibrium position to shift to the right.

EQUILIBRIUM AND HEALTH

Discussions of chemical equilibrium tend to be rather abstract, as the foregoing sections on the equilibrium constant and Le Châtelier's principle illustrate. (The reader is encouraged to consult additional sources on these topics, which involve a number of particulars that have been touched upon only briefly here.) Despite the challenges involved in addressing the subject of equilibrium, the results of chemical equilibrium can be seen in processes involving human health.

The cooling of food with refrigerators, along with means of food preservation that do not involve changes in temperature, maintains chemical equilibrium in the foods and thereby prevents or at least retards spoilage. Even more important is the maintenance of equilibrium in reactions between hemoglobin and oxygen in human blood.

HEMOGLOBIN AND OXYGEN. Hemoglobin, a protein containing iron, is the material in red blood cells responsible for transporting oxygen to the cells. Each hemoglobin molecule attaches to four oxygen atoms, and the equilibrium conditions of the hemoglobin-oxygen interaction can be expressed thus: $Hb(aq) + 4O_2(g) \leftrightarrows Hb(O_2)_4(aq)$, where "Hb" stands for hemoglobin. As long as there is sufficient oxygen in the air, a healthy equilibrium is maintained; but at high altitudes, considerable changes occur.

At significant elevations above sea level, the air pressure is lowered, and thus it is more difficult to obtain the oxygen one needs. The result, in accordance with Le Châtelier's principle, is a shift in equilibrium to the left, away from the oxygenated hemoglobin. Without adequate oxygen fed to the body's cells and tissues, a person tends to feel light-headed.

When someone not physically prepared for the change is exposed to high altitudes, it may be necessary to introduce pressurized oxygen from an oxygen tank. This shifts the equilibrium to the right. For people born and raised at high altitudes, however, the body's chemistry performs the equilibrium shift—by producing more hemoglobin, which also shifts equilibrium to the right.

HEMOGLOBIN AND CARBON MONOXIDE. When someone is exposed to carbon monoxide gas, a frightening variation on the normal hemoglobin-oxygen interaction occurs. Carbon monoxide "fools" hemoglobin into mistaking it for oxygen because it also bonds to hemoglobin in groups of four, and the equilibrium expression thus becomes: $Hb(aq) + 4CO(g) \leftrightarrows Hb(CO)_4(aq)$. Instead of hemoglobin, what has been produced is called carboxyhemoglobin, which is even redder than hemoglobin. Therefore, one sign of carbon monoxide poisoning is a flushed face.

The bonds between carbon monoxide and hemoglobin are about 300 times as strong as those between hemoglobin and oxygen, and this means a shift in equilibrium toward the right side of the equation—the carboxyhemoglobin side. It also means that K for the hemoglobin-carbon monoxide reaction is much higher than for the hemoglobin-oxygen reaction. Due to the affinity of hemoglobin for carbon monoxide, the hemoglobin puts a priority on carbon monoxide bonds, and hemoglobin that has bonded with carbon monoxide is no longer available to carry oxygen.

KEY TERMS

ACTIVATION ENERGY: The minimal energy required to convert reactants into products, symbolized E_a.

AQUEOUS SOLUTION: A mixture of water and a substance that is dissolved in it.

CATALYST: A substance that speeds up a chemical reaction without participating in it, either as a reactant or product. Catalysts are thus not consumed in the reaction.

CHEMICAL EQUATION: A representation of a chemical reaction in which the chemical symbols on the left stand for the reactants, and those on the right are the product or products.

CHEMICAL EQUILIBRIUM: A situation in which the ratio between the reactants and products in a chemical reaction is constant, and in which the forward reactions and reverse reactions take place at the same rate.

CHEMICAL REACTION: A process whereby the chemical properties of a substance are changed by a rearrangement of the atoms in the substance.

CHEMICAL SPECIES: A generic term used for any substance studied in chemistry—whether it be an element, compound, mixture, atom, molecule, ion, and so forth.

COEFFICIENT: a number used to indicate the presence of more than one unit—typically, more than one molecule—of a chemical species in a chemical equation.

COLLISION MODEL: The theory that chemical reactions are the result of collisions between molecules that are strong enough to break bonds in the reactants, resulting in a reformation of atoms.

ENDOTHERMIC: A term describing a chemical reaction in which heat is absorbed or consumed.

Carbon monoxide in small quantities can cause headaches and dizziness, but larger concentrations can be fatal. To reverse the effects of the carbon monoxide, pure oxygen must be introduced to the body. It will react with the carboxyhemoglobin to produce properly oxygenated hemoglobin, along with carbon monoxide: $Hb(CO)_4(aq) + 4O_2(g) \leftrightharpoons Hb(O_2)_4(aq) + 4CO(g)$. The gaseous carbon monoxide thus produced is dissipated when the person exhales.

WHERE TO LEARN MORE

"Catalysts" (Web site). <http://edie.cprost.sfu.ca/~rhlogan/catalyst.html> (June 9, 2001).

Challoner, Jack. *The Visual Dictionary of Chemistry.* New York: DK Publishing, 1996.

"Chemical Equilibrium." *Davidson College Department of Chemistry* (Web site). <http://www.chm.davidson.edu/ronutt/che115/EquKin.htm> (June 9, 2001).

"Chemical Equilibrium in the Gas Phase." *Virginia Tech Chemistry Department* (Web site). <http://www.chem.vt.edu/RVGS/ACT/notes/chem-eqm.html> (June 9, 2001).

"Chemical Sciences: Mechanism of Catalysis." *University of Alberta Department of Chemistry* (Web site). <http://www.chem.ualberta.ca/~plambeck/che/p102/p02174.htm> (June 9, 2001).

Ebbing, Darrell D.; R. A. D. Wentworth; and James P. Birk. *Introductory Chemistry.* Boston: Houghton Mifflin, 1995.

Hauser, Jill Frankel. *Super Science Concoctions: 50 Mysterious Mixtures for Fabulous Fun.* Charlotte, VT: Williamson Publishing, 1996.

"Mark Rosen's Chemical Equilibrium Links" (Web site). <http://users.erols.com/merosen/equilib.htm> (June 9, 2001).

EXOTHERMIC: A term describing a chemical reaction in which heat is produced.

FORWARD REACTION: A chemical reaction symbolized by a chemical equation in which the reactants and product are separated by an arrow that points to the right, toward the products.

HETEROGENEOUS EQUILIBRIUM: Chemical equilibrium in which the substances involved are in different phases of matter (solid, liquid, gas.)

HOMOGENEOUS EQUILIBRIUM: Chemical equilibrium in which the substances involved are in the same phase of matter.

LE CHÂTELIER'S PRINCIPLE: A statement, formulated by French chemist Henri Le Châtelier (1850-1936), which holds that whenever a stress or change is imposed on a system in chemical equilibrium, the system will adjust the amounts of the various substances in such a way as to reduce the impact of that stress.

PRODUCT: The substance or substances that result from a chemical reaction.

REACTANT: A substance that interacts with another substance in a chemical reaction, resulting in a product.

REVERSE REACTION: A chemical reaction symbolized by a chemical equation in which the products of a forward reaction have become the reactants, and the reactants of the forward reaction are now the products. This is indicated by an arrow that points toward the left.

SYSTEM: In chemistry and other sciences, the term "system" usually refers to any set of interactions isolated from the rest of the universe. Anything outside of the system, including all factors and forces irrelevant to a discussion of that system, is known as the environment.

Oxlade, Chris. *Chemistry.* Illustrated by Chris Fairclough. Austin, TX: Raintree Steck-Vaughn, 1999.

Zumdahl, Steven S. *Introductory Chemistry: A Foundation,* 4th ed. Boston: Houghton Mifflin, 2000.

CATALYSTS

CONCEPT

In most of the processes studied within the physical sciences, the lesson again and again is that nature provides no "free lunch"; in other words, it is not possible to get something for nothing. A chemical reaction, for instance, involves the creation of substances different from those that reacted in the first place, but the number of atoms involved does not change. In view of nature's inherently conservative tendencies, then, the idea of a catalyst—a substance that speeds up a reaction without being consumed—seems almost like a magic trick. But catalysts are very real, and their presence in the human body helps to sustain life. Similarly, catalysts enable the synthesis of foods, and catalytic converters in automobiles protect the environment from dangerous exhaust fumes. Yet the presence of one particular catalyst in the upper atmosphere poses such a threat to Earth's ozone layer that production of certain chemicals containing that substance has been banned.

HOW IT WORKS

REACTIONS AND COLLISIONS

In a chemical reaction, substances known as reactants interact with one another to create new substances, called products. In the present context, our concern is not with the reactants and products themselves, but with an additional entity, an agent that enables the reaction to move forward at faster rates and lower temperatures.

According to the collision model generally accepted by chemists, chemical reactions are the result of collisions between molecules. Collisions that are sufficiently energetic break the chemical bonds that hold molecules together; as a result, the atoms in those molecules are free to recombine with other atoms to form new molecules. Hastening of a chemical reaction can be produced in one of three ways. If the concentrations of the reactants are increased, this means that more molecules are colliding, and potentially more bonds are being broken. Likewise if the temperature is increased, the speeds of the molecules themselves increase, and their collisions possess more energy.

Energy is an important component in the chemical reaction because a certain threshold, called the activation energy (E_a), must be crossed before a reaction can occur. A temperature increase raises the energy of the collisions, increasing the likelihood that the activation-energy threshold will be crossed, resulting in the breaking of molecular bonds.

CATALYSTS AND CATALYSIS

It is not always feasible or desirable, however, to increase the concentration of reactants, or the temperature of the system in which the reaction is to occur. Many of the processes that take place in the human body, for instance, "should" require high temperatures—temperatures too high to sustain human life. But fortunately, our bodies contain proteins called enzymes, discussed later in this essay, that facilitate the necessary reactions without raising temperatures or increasing the concentrations of substances.

An enzyme is an example of a catalyst, a substance that speeds up a reaction without participating in it either as a reactant or product. Catalysts are thus not consumed in the reaction. The

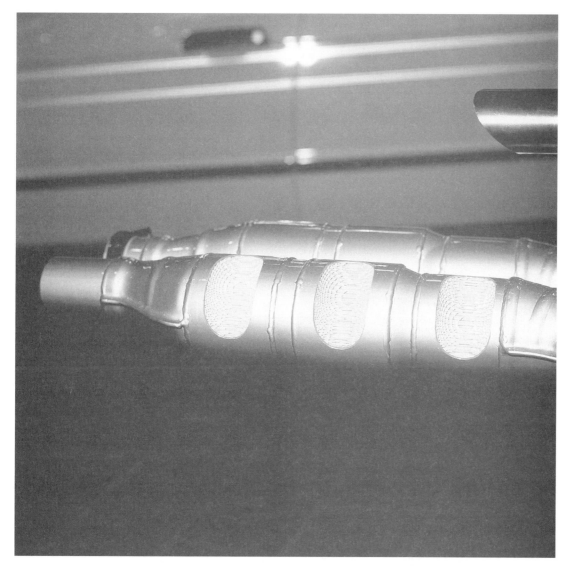

CATALYTIC CONVERTERS EMPLOY A CATALYST TO FACILITATE THE TRANSFORMATION OF POLLUTION-CAUSING EXHAUSTS TO LESS HARMFUL SUBSTANCES. *(Ian Harwood; Ecoscene/Corbis. Reproduced by permission.)*

catalyst does its work—catalysis—by creating a different path for the reaction, and though the means whereby it does this are too complex to discuss in detail here, the process of catalyst can at least be presented in general terms.

Imagine a graph whose x-axis is labeled "reaction progress," while the y-axis bears the legend "energy." There is some value of y equal to the normal activation energy, and in the course of experiencing the molecular collisions that lead to a reaction, the reactants reach this level. In a catalyzed reaction, however, the level of activation energy necessary for the reaction is represented by a lower y-value on the graph. The catalyzed substances do not need to have as much energy as they do without a catalyst, and there-

fore the reaction can proceed more quickly—without changing the temperature or concentrations of reactants.

REAL-LIFE APPLICATIONS

A BRIEF HISTORY OF CATALYSIS

Long before chemists recognized the existence of catalysts, ordinary people had been using the process of catalysis for a number of purposes: making soap, for instance, or fermenting wine to create vinegar, or leavening bread. Early in the nineteenth century, chemists began to take note of this phenomenon.

FRITZ HABER. *(Austrian Archives/Corbis. Reproduced by permission.)*

In 1812, Russian chemist Gottlieb Kirchhof was studying the conversion of starches to sugar in the presence of strong acids when he noticed something interesting. When a suspension of starch in water was boiled, Kirchhof observed, no change occurred in the starch. However, when he added a few drops of concentrated acid before boiling the suspension (that is, particles of starch suspended in water), he obtained a very different result. This time, the starch broke down to form glucose, a simple sugar, while the acid—which clearly had facilitated the reaction—underwent no change.

Around the same time, English chemist Sir Humphry Davy (1778-1829) noticed that in certain organic reactions, platinum acted to speed along the reaction without undergoing any change. Later on, Davy's star pupil, the great British physicist and chemist Michael Faraday (1791-1867), demonstrated the ability of platinum to recombine hydrogen and oxygen that had been separated by the electrolysis of water. The catalytic properties of platinum later found application in catalytic converters, as we shall see.

AN IMPROVED DEFINITION. In 1835, Swedish chemist Jons Berzelius (1779-1848) provided a name to the process Kirchhof and Davy had observed from very different perspectives: catalysis, derived from the Greek words

kata ("down") and *lyein* ("loosen.") As Berzelius defined it, catalysis involved an activity quite different from that of an ordinary chemical reaction. Catalysis induced decomposition in substances, resulting in the formation of new compounds—but without the catalyst itself actually entering the compound.

Berzelius's definition assumed that a catalyst manages to do what it does without changing at all. This was perfectly adequate for describing heterogeneous catalysis, in which the catalyst and the reactants are in different phases of matter. In the platinum-catalyzed reactions that Davy and Faraday observed, for instance, the platinum is a solid, while the reaction itself takes place in a gaseous or liquid state. However, homogeneous catalysis, in which catalyst and reactants are in the same state, required a different explanation, which English chemist Alexander William Williamson (1824-1904) provided in an 1852 study.

In discussing the reaction observed by Kirchhof, of liquid sulfuric acid with starch in an aqueous solution, Williamson was able to show that the catalyst does break down in the course of the reaction. As the reaction takes place, it forms an intermediate compound, but this too is broken down before the reaction ends. The catalyst thus emerges in the same form it had at the beginning of the reaction.

ENZYMES: HELPFUL CATALYSTS IN THE BODY

In 1833, French physiologist Anselme Payen (1795-1871) isolated a material from malt that accelerated the conversion of starch to sugar, as for instance in the brewing of beer. Payen gave the name "diastase" to this substance, and in 1857, the renowned French chemist Louis Pasteur (1822-1895) suggested that lactic acid fermentation is caused by a living organism.

In fact, the catalysts studied by Pasteur are not themselves separate organisms, as German biochemist Eduard Buchner (1860-1917) showed in 1897. Buchner isolated the catalysts that bring about the fermentation of alcohol from living yeast cells—what Payen had called "diastase," and Pasteur "ferments." Buchner demonstrated that these are actually chemical substances, not organisms. By that time, German physiologist Willy Kahne had suggested the name "enzyme" for these catalysts in living systems.

Enzymes are made up of amino acids, which in turn are constructed from organic compounds called proteins. About 20 amino acids make up the building blocks of the many thousands of known enzymes. The beauty of an enzyme is that it speeds up complex, life-sustaining reactions in the human body—reactions that would be too slow at ordinary body temperatures. Rather than force the body to undergo harmful increases in temperature, the enzyme facilitates the reaction by opening up a different reaction pathway that allows a lower activation energy.

One example of an enzyme is cytochrome, which aids the respiratory system by catalyzing the combination of oxygen with hydrogen within the cells. Other enzymes facilitate the conversion of food to energy, and make possible a variety of other necessary biological functions.

Because numerous interactions are required in their work of catalysis, enzymes are very large, and may have atomic mass figures as high as 1 million amu. However, it should be noted that reactions are catalyzed at very specific locations—called active sites—on an enzyme. The reactant molecule fits neatly into the active site on the enzyme, much like a key fitting in a lock; hence the name of this theory, the "lock-and-model."

CATALYSIS AND THE ENVIRONMENT

The exhaust from an automobile contains many substances that are harmful to the environment. As a result of increased concerns regarding the potential damage to the atmosphere, the federal government in the 1970s mandated the adoption of catalytic converters, devices that employ a catalyst to transform pollutants in the exhaust to less harmful substances.

Platinum and palladium are favored materials for catalytic converters, though some nonmetallic materials, such as ceramics, have been used as well. In any case, the function of a catalytic converter is to convert exhausts through oxidation-reduction reactions. Nitric oxide is reduced to molecular oxygen and nitrogen; at the same time, the hydrocarbons in petroleum, along with carbon monoxide, are oxidized to form carbon dioxide and water. Sometimes a reducing agent, such as ammonia, is used to make the reduction process more effective.

A DANGEROUS CATALYST IN THE ATMOSPHERE. Around the same time that automakers began rolling out models equipped with catalytic converters, scientists and the general public alike became increasingly concerned about another threat to the environment. In the upper atmosphere of Earth are traces of ozone, a triatomic (three-atom) molecular form of oxygen which protects the planet from the Sun's ultraviolet rays. During the latter part of the twentieth century, it became apparent that a hole had developed in the ozone layer over Antarctica, and many chemists suspected a culprit in chlorofluorocarbons, or CFCs.

CFCs had long been used in refrigerants and air conditioners, and as propellants in aerosol sprays. Because they were nontoxic and noncorrosive, they worked quite well for such purposes, but the fact that they are chemically unreactive had an extremely negative side-effect. Instead of reacting with other substances to form new compounds, they linger in Earth's atmosphere, eventually drifting to high altitudes, where ultraviolet light decomposes them. The real trouble begins when atoms of chlorine, isolated from the CFC, encounter ozone.

Chlorine acts as a catalyst to transform the ozone to elemental oxygen, which is not nearly as effective as ozone for shielding Earth from ultraviolet light. It does so by interacting also with monatomic, or single-atom oxygen, with which it produces ClO, or the hypochlorite ion. The end result of reactions between chlorine, monatomic oxygen, hypochlorite, and ozone is the production of chlorine, hypochlorite, and diatomic oxygen—in other words, no more ozone. It is estimated that a single chlorine atom can destroy up to 1 million ozone molecules per second.

Due to concerns about the danger to the ozone layer, an international agreement called the Montreal Protocol, signed in 1996, banned the production of CFCs and the coolant Freon that contains them. But people still need coolants for their homes and cars, and this has led to the creation of substitutes—most notably hydrochlorofluorocarbons (HCFCs), organic compounds that do not catalyze ozone.

OTHER EXAMPLES OF CATALYSTS

Catalysts appear in a number of reactions, both natural and artificial. For instance, catalysts are used in the industrial production of ammonia,

KEY TERMS

ACTIVATION ENERGY: The minimal energy required to convert reactants into products, symbolized E_a

AQUEOUS SOLUTIONS: A mixture of water and a substance that is dissolved in it.

CATALYST: A substance that speeds up a chemical reaction without participating in it, either as a reactant or product. Catalysts are thus not consumed in the reaction.

CHEMICAL REACTION: A process whereby the chemical properties of a substance are changed by a rearrangement of the atoms in the substance.

COLLISION MODEL: The theory that chemical reactions are the result of collisions between molecules strong enough to break bonds in the reactants, resulting in a reformation of atoms.

HETEROGENEOUS CATALYSIS: A reaction in which the catalyst and the reactants are in different phases of matter.

HOMOGENEOUS CATALYSIS: A reaction in which catalyst and reactants are in the same phase of matter.

PRODUCT: The substance or substances that result from a chemical reaction.

REACTANT: A substance that interacts with another substance in a chemical reaction, resulting in a product.

SYSTEM: In chemistry and other sciences, the term "system" usually refers to any set of interactions isolated from the rest of the universe. Anything outside of the system, including all factors and forces irrelevant to a discussion of that system, is known as the environment.

nitric acid (produced from ammonia), sulfuric acid, and other substances. The ammonia process, developed in 1908 by German chemist Fritz Haber (1868-1934), is particularly noteworthy. Using iron as a catalyst, Haber was able to combine nitrogen and hydrogen under pressure to form ammonia—one of the world's most widely used chemicals.

Eighteen ninety-seven was a good year for catalysts. In that year, it was accidentally discovered that mercury catalyzes the reaction by which indigo dye is produced; also in 1897, French chemist Paul Sabatier (1854-1941) found that nickel catalyzes the production of edible fats. Thanks to Sabatier's discovery, nickel is used to transform inedible plant oils to margarine and shortening.

Another good year for catalysts—particularly those involved in the production of polymers—was 1953. That was the year when German chemist Karl Ziegler (1898-1973) discovered a resin catalyst for the production of polyethylene, which produced a newer, tougher product with a much higher melting point than polyethylene as it was produced up to that time. Also in 1953, Italian chemist Giulio Natta (1903-1979) adapted Ziegler's idea, and developed a new type of plastic he called "isotactic" polymers. These could be produced easily, and in abundance, through the use of catalysts.

One of the lessons of chemistry, or indeed of any science, is that there are few things chemists can do that nature cannot achieve on a far more wondrous scale. No artificial catalyst can compete with enzymes, and no use of a catalyst in a laboratory can compare with the grandeur of that which takes place on the Sun. As German-American physicist Hans Bethe (1906-) showed in 1938, the reactions of hydrogen that form helium on the surface of the Sun are catalyzed by carbon—the same element, incidentally, found in all living things on Earth.

WHERE TO LEARN MORE

"Bugs in the News: What the Heck Is an Enzyme?" University of Kansas (Web site). <http://falcon.cc.ukans.edu/~jbrown/enzyme.html> (June 9, 2001).

"Catalysis." University of Idaho Department of Chemistry (Web site). <http://www.chem.uidaho.edu/~honors/rate4.html> (June 9, 2001).

"Catalysts" (Web site). <http://edie.cprost.sfu.ca/ ~rhlogan/catalyst.html> (June 9, 2001).

Ebbing, Darrell D.; R. A. D. Wentworth; and James P. Birk. *Introductory Chemistry.* Boston: Houghton Mifflin, 1995.

"Enzymes." *Strategis* (Web site). <http://strategis.ic.gc.ca/ SSG/tc00048e.html> (June 9, 2001).

"Enzymes: Classification, Structure, Mechanism." *The Hebrew University* (Web site). <http://www.md.huji. ac.il/MedChem/Mechanism-Chymotrypsin/> (June 9, 2001).

Oxlade, Chris. *Chemistry.* Illustrated by Chris Fairclough. Austin, TX: Raintree Steck-Vaughn, 1999.

"Ozone Depletion" (Web site). <http://www.energy. rochester.edu/iea/1992/p1/2-3.htm> (June 9, 2001).

"University Chemistry: Chemical Kinetics: Catalysis." *University of Alberta Department of Chemistry* (Web site). <http://www.chem.ualberta.ca/courses/plambeck/p10 2/p0217x.htm> (June 9, 2001).

Zumdahl, Steven S. *Introductory Chemistry: A Foundation,* 4th ed. Boston: Houghton Mifflin, 2000.

ACIDS AND BASES

CONCEPT

The name "acid" calls to mind vivid sensory images—of tartness, for instance, if the acid in question is meant for human consumption, as with the citric acid in lemons. On the other hand, the thought of laboratory- and industrial-strength substances with scary-sounding names, such as sulfuric acid or hydrofluoric acid, carries with it other ideas—of acids that are capable of destroying materials, including human flesh. The name "base," by contrast, is not widely known in its chemical sense, and even when the older term of "alkali" is used, the sense-impressions produced by the word tend not to be as vivid as those generated by the thought of "acid." In their industrial applications, bases too can be highly powerful. As with acids, they have many household uses, in substances such as baking soda or oven cleaners. From a taste standpoint, (as anyone who has ever brushed his or her teeth with baking soda knows), bases are bitter rather than sour. How do we know when something is an acid or a base? Acid-base indicators, such as litmus paper and other materials for testing pH, offer a means of judging these qualities in various substances. However, there are larger structural definitions of the two concepts, which evolved in three stages during the late nineteenth and early twentieth centuries, that provide a more solid theoretical underpinning to the understanding of acids and bases.

HOW IT WORKS

INTRODUCTION TO ACIDS AND BASES

Prior to the development of atomic and molecular theory in the nineteenth century, followed by the discovery of subatomic structures in the late nineteenth and early twentieth centuries, chemists could not do much more than make measurements and observations. Their definitions of substances were purely phenomenological—that is, the result of experimentation and the collection of data. From these observations, they could form general rules, but they lacked any means of "seeing" into the atomic and molecular structures of the chemical world.

The phenomenological distinctions between acids and bases, gathered by scientists from ancient times onward, worked well enough for many centuries. The word "acid" comes from the Latin term *acidus,* or "sour," and from an early period, scientists understood that substances such as vinegar and lemon juice shared a common acidic quality. Eventually, the phenomenological definition of acids became relatively sophisticated, encompassing such details as the fact that acids produce characteristic colors in certain vegetable dyes, such as those used in making litmus paper. In addition, chemists realized that acids dissolve some metals, releasing hydrogen in the process.

WHY "BASE" AND NOT "ALKA-LI"? The word "alkali" comes from the Arabic *al-qili,* which refers to the ashes of the seawort plant. The latter, which typically grows in marshy areas, was often burned to produce soda ash, used in making soap. In contrast to acids, bases—caffeine, for example—have a bitter taste, and many of them feel slippery to the touch. They also produce characteristic colors in the vegetable dyes of litmus paper, and can be used to promote certain chemical reactions. Note that today chemists use the word "base" instead of "alkali," the reason being that the latter term has a narrower meaning: all alkalies are bases, but not all bases are alkalies.

Originally, "alkali" referred only to the ashes of burned plants, such as seawort, that contained either sodium or potassium, and from which the oxides of sodium and potassium could be obtained. Eventually, alkali came to mean the soluble hydroxides of the alkali and alkaline earth metals. This includes sodium hydroxide, the active ingredient in drain and oven cleaners; magnesium hydroxide, used for instance in milk of magnesia; potassium hydroxide, found in soaps and other substances; and other compounds. Broad as this range of substances is, it fails to encompass the wide array of materials known today as bases—compounds which react with acids to form salts and water.

TOWARD A STRUCTURAL DEFINITION

The reaction to form salts and water is, in fact, one of the ways that acids and bases can be defined. In an aqueous solution, hydrochloric acid and sodium hydroxide react to form sodium chloride—which, though it is suspended in an aqueous solution, is still common table salt—along with water. The equation for this reaction is $HCl(aq) + NaOH(aq) \rightarrow H_2O + NaCl(aq)$. In other words, the sodium (Na) ion in sodium hydroxide switches places with the hydrogen ion in hydrochloric acid, resulting in the creation of NaCl (salt) along with water.

But why does this happen? Useful as this definition regarding the formation of salts and water is, it is still not structural—in other words, it does not delve into the molecular structure and behavior of acids and bases. Credit for the first truly structural definition of the difference goes to the Swedish chemist Svante Arrhenius (1859-1927). It was Arrhenius who, in his doctoral dissertation in 1884, introduced the concept of an ion, an atom possessing an electric charge.

His understanding was particularly impressive in light of the fact that it was 13 more years before the discovery of the electron, the subatomic particle responsible for the creation of ions. Atoms have a neutral charge, but when an electron or electrons depart, the atom becomes a positive ion or cation. Similarly, when an electron or electrons join a previously uncharged

SVANTE ARRHENIUS. *(Hulton-Deutsch Collection/Corbis. Reproduced by permission.)*

atom, the result is a negative ion or anion. Not only did the concept of ions greatly influence the future of chemistry, but it also provided Arrhenius with the key necessary to formulate his distinction between acids and bases.

THE ARRHENIUS DEFINITION

Arrhenius observed that molecules of certain compounds break into charged particles when placed in liquid. This led him to the Arrhenius acid-base theory, which defines an acid as any compound that produces hydrogen ions (H^+) when dissolved in water, and a base as any compound that produces hydroxide ions (OH^-) when dissolved in water.

This was a good start, but two aspects of Arrhenius's theory suggested the need for a definition that encompassed more substances. First of all, his theory was limited to reactions in aqueous solutions. Though many acid-base reactions do occur when water is the solvent, this is not always the case.

Second, the Arrhenius definition effectively limited acids and bases only to those ionic compounds, such as hydrochloric acid or sodium hydroxide, which produced either hydrogen or

THE TARTNESS OF LEMONS AND LIMES IS A FUNCTION OF THEIR ACIDITY—SPECIFICALLY THE WAY IN WHICH CITRIC ACID MOLECULES FIT INTO THE TONGUE'S "SWEET" RECEPTORS. (*Orion Press/Corbis. Reproduced by permission.*)

hydroxide ions. However, ammonia, or NH_3, acts like a base in aqueous solutions, even though it does not produce the hydroxide ion. The same is true of other substances, which behave like acids or bases without conforming to the Arrhenius definition.

These shortcomings pointed to the need for a more comprehensive theory, which arrived with the formulation of the Brønsted-Lowry definition by English chemist Thomas Lowry (1874-1936) and Danish chemist J. N. Brønsted (1879-1947). Nonetheless, Arrhenius's theory represented an important first step, and in 1903, he was awarded the Nobel Prize in Chemistry for his work on the dissociation of molecules into ions.

THE BRØNSTED-LOWRY DEFINITION

The Brønsted-Lowry acid-base theory defines an acid as a proton (H^+) donor, and a base as a proton acceptor, in a chemical reaction. Protons are represented by the symbol H^+, and in representing acids and bases, the symbols HA and A^-, respectively, are used. These symbols indicate that an acid has a proton it is ready to give away, while a base, with its negative charge, is ready to receive the positively charged proton.

Though it is used here to represent a proton, it should be pointed out that H^+ is also the hydrogen ion—a hydrogen atom that has lost its sole electron and thus acquired a positive charge.

It is thus really nothing more than a lone proton, but this is the one and only case in which an atom and a proton are exactly the same thing. In an acid-base reaction, a molecule of acid is "donating" a proton, in the form of a hydrogen ion. This should not be confused with a far more complex process, nuclear fusion, in which an atom gives up a proton to another atom.

AN ACID-BASE REACTION IN BRØNSTED-LOWRY THEORY. The most fundamental type of acid-base reaction in Brønsted-Lowry theory can be symbolized thus $HA(aq) + H_2O(l) \rightarrow H_3O^+(aq) + A^-(aq)$. The first acid shown—which, like three of the four "players" in this equation, is dissolved in an aqueous solution—combines with water, which can serve as either an acid or a base. In the present context, it functions as a base.

Water molecules are polar, meaning that the negative charges tend to congregate on one end of the molecule with the oxygen atom, while the positive charges remain on the other end with the hydrogen atoms. The Brønsted-Lowry model emphasizes the role played by water, which pulls the proton from the acid, resulting in the creation of H_3O^+, known as the hydronium ion.

The hydronium ion produced here is an example of a conjugate acid, an acid formed when a base accepts a proton. At the same time, the acid has lost its proton, becoming A^-, a conjugate base—that is, the base formed when an acid releases a proton. These two products of the reaction are called a conjugate acid-base pair, a term that refers to two substances related to one another by the donating of a proton.

Brønsted and Lowry's definition represents an improvement over that of Arrhenius, because it includes all Arrhenius acids and bases, as well as other chemical species not encompassed in Arrhenius theory. An example, mentioned earlier, is ammonia. Though it does not produce OH^- ions, ammonia does accept a proton from a water molecule, and the reaction between these two (with water this time serving the function of acid) produces the conjugate acid-base pair of NH_4^+ (an ammonium ion) and OH^-. Note that the latter, the hydroxide ion, was not produced by ammonia, but is the conjugate base that resulted when the water molecule lost its H^+ atom or proton.

THE LEWIS DEFINITION

Despite the progress offered to chemists by the Brønsted-Lowry model, it was still limited to describing compounds that contain hydrogen. As American chemist Gilbert N. Lewis (1875-1946) recognized, this did not encompass the full range of acids and bases; what was needed, instead, was a definition that did not involve the presence of a hydrogen atom.

Lewis is particularly noted for his work in the realm of chemical bonding. The bonding of atoms is the result of activity on the part of the valence electrons, or the electrons at the "outside" of the atom. Electrons are arranged in different ways, depending on the type of bonding, but they always bond in pairs.

According to the Lewis acid-base theory, an acid is the reactant that accepts an electron pair from another reactant in a chemical reaction, while a base is the reactant that donates an electron pair to another reactant. As with the Brønsted-Lowry definition, the Lewis definition is reaction-dependant, and does not define a compound as an acid or base in its own right. Instead, the manner in which the compound reacts with another serves to identify it as an acid or base.

AN IMPROVEMENT OVER ITS PREDECESSORS. The beauty of the Lewis definition lies in the fact that it encompasses all the situations covered by the others—and more. Just as Brønsted-Lowry did not disprove Arrhenius, but rather offered a definition that covered more substances, Lewis expanded the range of substances beyond those covered by Brønsted-Lowry. In particular, Lewis theory can be used to differentiate the acid and base in bond-producing chemical reactions where ions are not produced, and in which there is no proton donor or acceptor. Thus it represents an improvement over Arrhenius and Brønsted-Lowry respectively.

An example is the reaction of boron trifluoride (BF_3) with ammonia (NH_3), both in the gas phases, to produce boron trifluoride ammonia complex (F_3BNH_3). In this reaction, boron trifluoride accepts an electron pair and is therefore a Lewis acid, while ammonia donates the electron pair and is thus a Lewis base. Though hydrogen is involved in this particular reaction, Lewis theory also addresses reactions involving no hydrogen.

REAL-LIFE APPLICATIONS

pH AND ACID-BASE INDICATORS

Though chemists apply the sophisticated structural definitions for acids and bases that we have discussed, there are also more "hands-on" methods for identifying a particular substance (including complex mixtures) as an acid or base. Many of these make use of the pH scale, developed by Danish chemist Søren Sørensen (1868-1939) in 1909.

The term pH stands for "potential of hydrogen," and the pH scale is a means of determining the acidity or alkalinity of a substance. (Though, as noted, the term "alkali" has been replaced by "base," alkalinity is still used as an adjectival term to indicate the degree to which a substance displays the properties of a base.) There are theoretically no limits to the range of the pH scale, but figures for acidity and alkalinity are usually given with numerical values between 0 and 14.

THE MEANING OF pH VALUES.

A rating of 0 on the pH scale indicates a substance that is virtually pure acid, while a 14 rating represents a nearly pure base. A rating of 7 indicates a neutral substance. The pH scale is logarithmic, or exponential, meaning that the numbers represent exponents, and thus an increased value of 1 represents not a simple arithmetic addition of 1, but an increase of 1 power. This, however, needs a little further explanation.

The pH scale is actually based on negative logarithms for the values of H_3O^+ (the hydronium ion) or H^+ (protons) in a given substance. The formula is thus $pH = -\log[H_3O^+]$ or $-\log[H^+]$, and the presence of hydronium ions or protons is measured according to their concentration of moles per liter of solution.

pH VALUES OF VARIOUS SUBSTANCES.

The pH of a virtually pure acid, such as the sulfuric acid in car batteries, is 0, and this represents 1 mole (mol) of hydronium per liter (l) of solution. Lemon juice has a pH of 2, equal to 10^{-2} mol/l. Note that the pH value of 2 translates to an exponent of -2, which, in this case, results in a figure of 0.01 mol/l.

Distilled water, a neutral substance with a pH of 7, has a hydronium equivalent of 10^{-7} mol/l. It is interesting to observe that most of the fluids in the human body have pH values in the neutral range blood (venous, 7.35; arterial, 7.45); urine (6.0—note the higher presence of acid); and saliva (6.0 to 7.4).

At the alkaline end of the scale is borax, with a pH of 9, while household ammonia has a pH value of 11, or 10^{-11} mol/l. Sodium hydroxide, or lye, an extremely alkaline chemical with a pH of 14, has a value equal to 10^{-14} moles of hydronium per liter of solution.

LITMUS PAPER AND OTHER INDICATORS.

The most precise pH measurements are made with electronic pH meters, which can provide figures accurate to 0.001 pH. However, simpler materials are also used. Best known among these is litmus paper (made from an extract of two lichen species), which turns blue in the presence of bases and red in the presence of acids. The term "litmus test" has become part of everyday language, referring to a make-or-break issue—for example, "views on abortion rights became a litmus test for Supreme Court nominees."

Litmus is just one of many materials used for making pH paper, but in each case, the change of color is the result of the neutralization of the substance on the paper. For instance, paper coated with phenolphthalein changes from colorless to pink in a pH range from 8.2 to 10, so it is useful for testing materials believed to be moderately alkaline. Extracts from various fruits and vegetables, including red cabbages, red onions, and others, are also applied as indicators.

SOME COMMON ACIDS AND BASES

The tables below list a few well-known acids and bases, along with their formulas and a few applications

Common Acids

- Acetic acid (CH_3COOH): vinegar, acetate
- Acetylsalicylic acid ($HOOCC_6H_4OOCCH_3$): aspirin
- Ascorbic acid ($H_2C_6H_6O_6$): vitamin C
- Carbonic acid (H_2CO_3): soft drinks, seltzer water
- Citric acid ($C_6H_8O_7$): citrus fruits, artificial flavorings
- Hydrochloric acid (HCl): stomach acid
- Nitric acid (HNO_3): fertilizer, explosives
- Sulfuric acid (H_2SO_4): car batteries

Common Bases

- Aluminum hydroxide ($Al[OH]_3$): antacids, deodorants
- Ammonium hydroxide (NH_4OH): glass cleaner
- Calcium hydroxide ($Ca[OH]_2$): caustic lime, mortar, plaster
- Magnesium hydroxide ($Mg[OH]_2$): laxatives, antacids
- Sodium bicarbonate/sodium hydrogen carbonate ($NaHCO_3$): baking soda
- Sodium carbonate (Na_2CO_3): dish detergent
- Sodium hydroxide ($NaOH$): lye, oven and drain cleaner
- Sodium hypochlorite ($NaClO$): bleach

Of course these represent only a few of the many acids and bases that exist. Selected substances listed above are discussed briefly below.

ACIDS

ACIDS IN THE HUMAN BODY AND FOODS. As its name suggests, citric acid is found in citrus fruits—particularly lemons, limes, and grapefruits. It is also used as a flavoring agent, preservative, and cleaning agent. Produced commercially from the fermentation of sugar by several species of mold, citric acid creates a taste that is both tart and sweet. The tartness, of course, is a function of its acidity, or a manifestation of the fact that it produces hydrogen ions. The sweetness is a more complex biochemical issue relating to the ways that citric acid molecules fit into the tongue's "sweet" receptors.

Citric acid plays a role in one famous stomach remedy, or antacid. This in itself is interesting, since antacids are more generally associated with alkaline substances, used for their ability to neutralize stomach acid. The fizz in Alka-Seltzer, however, comes from the reaction of citric acids (which also provide a more pleasant taste) with sodium bicarbonate or baking soda, a base. This reaction produces carbon dioxide gas. As a preservative, citric acid prevents metal ions from reacting with, and thus hastening the degradation of, fats in foods. It is also used in the production of hair rinses and low-pH shampoos and toothpastes.

The carboxylic acid family of hydrocarbon derivatives includes a wide array of substances—not only citric acids, but amino acids. Amino acids combine to make up proteins, one of the principal components in human muscles, skin, and hair. Carboxylic acids are also applied industrially, particularly in the use of fatty acids for making soaps, detergents, and shampoos.

SULFURIC ACID. There are plenty of acids found in the human body, including hydrochloric acid or stomach acid—which, in large quantities, causes indigestion, and the need for neutralization with a base. Nature also produces acids that are toxic to humans, such as sulfuric acid.

Though direct exposure to sulfuric acid is extremely dangerous, the substance has numerous applications. Not only is it used in car batteries, but sulfuric acid is also a significant component in the production of fertilizers. On the other hand, sulfuric acid is damaging to the environment when it appears in the form of acid rain. Among the impurities in coal is sulfur, and this results in the production of sulfur dioxide and sulfur trioxide when the coal is burned. Sulfur trioxide reacts with water in the air, creating sulfuric acid and thus acid rain, which can endanger plant and animal life, as well as corrode metals and building materials.

BASES

The alkali metal and alkaline earth metal families of elements are, as their name suggests, bases. A number of substances created by the reaction of these metals with nonmetallic elements are taken internally for the purpose of settling gastric trouble or clearing intestinal blockage. For instance, there is magnesium sulfate, better known as Epsom salts, which provide a powerful laxative also used for ridding the body of poisons.

Aluminum hydroxide is an interesting base, because it has a wide number of applications, including its use in antacids. As such, it reacts with and neutralizes stomach acid, and for that reason is found in commercial antacids such as Di-Gel™, Gelusil™, and Maalox™. Aluminum hydroxide is also used in water purification, in dyeing garments, and in the production of certain kinds of glass. A close relative, aluminum hydroxychloride or $Al_2(OH)_5Cl$, appears in many commercial antiperspirants, and helps to close pores, thus stopping the flow of perspiration.

SODIUM HYDROGEN CARBONATE (BAKING SODA). Baking soda, known by chemists both as sodium bicar-

KEY TERMS

ACID: A substance that, in its edible form, is sour to the taste, and in non-edible forms, is often capable of dissolving metals. Acids and bases react to form salts and water. These are all phenomenological definitions, however, in contrast to the three structural definitions of acids and bases—the Arrhenius, Brønsted-Lowry, and Lewis acid-base theories.

ALKALI: A term referring to the soluble hydroxides of the alkali and alkaline earth metals. Once "alkali" was used for the class of substances that react with acids to form salts; today, however, the more general term base is preferred.

ALKALINITY: An adjectival term used to identify the degree to which a substance displays the properties of a base.

ANION: The negatively charged ion that results when an atom gains one or more electrons. "Anion" is pronounced "AN-ie-un".

AQUEOUS SOLUTION: A substance in which water constitutes the solvent. A large number of chemical reactions take place in an aqueous solution.

ARRHENIUS ACID-BASE THEORY: The first of three structural definitions of acids and bases. Formulated by Swedish chemist Svante Arrhenius (1859-1927), the Arrhenius theory defines acids and bases according to the ions they produce in an aqueous solution: an acid produces hydrogen ions (H^+), and a base hydroxide ions (OH^-).

BASE: A substance that, in its edible form, is bitter to the taste. Bases tend to be slippery to the touch, and in reaction with acids they produce salts and water. Bases and acids are most properly defined, however, not in these phenomenological terms, but by the three structural definitions of acids and bases—the Arrhenius, Brønsted-Lowry, and Lewis acid-base theories.

BRØNSTED-LOWRY ACID-BASE THEORY: The second of three structural definitions of acids and bases. Formulated by English chemist Thomas Lowry (1874-1936) and Danish chemist J. N. Brønsted (1879-1947), Brønsted-Lowry theory defines an acid as a proton (H^+) donor, and a base as a proton acceptor.

CATION: The positively charged ion that results when an atom loses one or more electrons. "Cation" is pronounced "KAT-ie-un".

CHEMICAL SPECIES: A generic term used for any substance studied in chemistry—whether it be an element, compound, mixture, atom, molecule, ion, and so forth.

CONJUGATE ACID: An acid formed when a base accepts a proton (H^+).

bonate and sodium hydrogen carbonate, is another example of a base with multiple purposes. As noted earlier, it is used in Alka-Seltzer™, with the addition of citric acid to improve the flavor; in fact, baking soda alone can perform the function of an antacid, but the taste is rather unpleasant.

Baking soda is also used in fighting fires, because at high temperatures it turns into carbon dioxide, which smothers flames by obstructing

KEY TERMS CONTINUED

CONJUGATE ACID-BASE PAIR: The acid and base produced when an acid donates a single proton to a base. In the reaction that produces this pair, the acid and base switch identities. By donating a proton, the acid becomes a conjugate base, and by receiving the proton, the base becomes a conjugate acid.

CONJUGATE BASE: A base formed when an acid releases a proton.

ION: An atom or atoms that has lost or gained one or more electrons, and thus has a net electric charge. There are two types of ions: anions and cations.

IONIC BONDING: A form of chemical bonding that results from attractions between ions with opposite electric charges.

IONIC COMPOUND: A compound in which ions are present. Ionic compounds contain at least one metal and nonmetal joined by an ionic bond.

LEWIS ACID-BASE THEORY: The third of three structural definitions of acids and bases. Formulated by American chemist Gilbert N. Lewis (1875-1946), Lewis theory defines an acid as the reactant that accepts an electron pair from another reactant in a chemical reaction, and a base as the reactant that donates an electron pair to another reactant.

PH SCALE: A logarithmic scale for determining the acidity or alkalinity of a substance, from 0 (virtually pure acid) to 7 (neutral) to 14 (virtually pure base).

PHENOMENOLOGICAL: A term describing scientific definitions based purely on experimental phenomena. These convey only part of the picture, however—primarily, the part a chemist can perceive either through measurement or through the senses, such as sight. A structural definition is therefore usually preferable to a phenomenological one.

REACTANT: A substance that interacts with another substance in a chemical reaction, resulting in the creation of a product.

SALTS: Ionic compounds formed by the reaction between an acid and a base. In this reaction, one or more of the hydrogen ions of an acid is replaced with another positive ion. In addition to producing salts, acid-base reactions produce water.

SOLUTION: A homogeneous mixture in which one or more substances (the solute) is dissolved in one or more other substances (the solvent)—for example, sugar dissolved in water.

SOLVENT: A substance that dissolves another, called a solute, in a solution.

STRUCTURAL: A term describing scientific definitions based on aspects of molecular structure and behavior rather than purely phenomenological data.

the flow of oxygen to the fire. Of course, baking soda is also used in baking, when it is combined with a weak acid to make baking powder. The reaction of the acid and the baking soda produces carbon dioxide, which causes dough and batters to rise. In a refrigerator or cabinet, baking soda can absorb unpleasant odors, and additionally, it can be applied as a cleaning product.

SODIUM HYDROXIDE (LYE). Another base used for cleaning is sodium hydroxide, known commonly as lye or caustic soda. Unlike baking soda, however, it is not to be taken internally, because it is highly damaging to human tissue—particularly the eyes. Lye appears in drain cleaners, such as Drano™, and oven cleaners, such as Easy-Off™, which make use of its ability to convert fats to water-soluble soap.

In the process of doing so, however, relatively large amounts of lye may generate enough heat to boil the water in a drain, causing the water to shoot upward. For this reason, it is not advisable to stand near a drain being treated with lye. In a closed oven, this is not a danger, of course; and after the cleaning process is complete, the converted fats (now in the form of soap) can be dissolved and wiped off with a sponge.

WHERE TO LEARN MORE

"Acids and Bases Frequently Asked Questions." *General Chemistry Online* (Web site). <http://antoine.fsu. umd.edu/chem/senese/101/acidbase/faq.shtml> (June 7, 2001).

"Acids, Bases, and Salts." *Chemistry Coach* (Web site). <http://www.chemistrycoach.com/acids.htm> (June 7, 2001).

"Acids, Bases, and Salts." *University of Akron, Department of Chemistry* (Web site). <http://ull.chemistry. uakron.edu/genobc/Chapter_09/title.html> (June 7, 2001).

ChemLab. Danbury, CT: Grolier Educational, 1998.

Ebbing, Darrell D.; R. A. D. Wentworth; and James P. Birk. *Introductory Chemistry.* Boston: Houghton Mifflin, 1995.

Haines, Gail Kay. *What Makes a Lemon Sour?* Illustrated by Janet McCaffery. New York: Morrow, 1977.

Oxlade, Chris. *Acids and Bases.* Chicago: Heinemann Library, 2001.

Patten, J. M. *Acids and Bases.* Vero Beach, FL: Rourke Book Company, 1995.

Walters, Derek. *Chemistry.* Illustrated by Denis Bishop and Jim Robins. New York: F. Watts, 1982.

Zumdahl, Steven S. *Introductory Chemistry A Foundation,* 4th ed. Boston: Houghton Mifflin, 2000.

ACID-BASE REACTIONS

CONCEPT

To an extent, acids and bases can be defined in terms of factors that are apparent to the senses: edible acids taste sour, for instance, while bases are bitter-tasting and slippery to the touch. The best way to understand these two types of substances, however, is in terms of their behavior in chemical reactions. Not only do the reactions of acids and bases result in the creation of salts and water, but acids and bases can be defined by the ways in which they participate in a reaction—for instance, by donating or accepting electron pairs. The reaction of acids and bases to form water and salts is called neutralization, and it has a wide range of applications, including the promotion of plant growth in soil and the treatment of heartburn in the human stomach. Neutralization also makes it possible to test substances for their pH level, a measure of the degree to which the substance is acidic or alkaline.

HOW IT WORKS

PHENOMENOLOGICAL DEFINITIONS OF ACIDS AND BASES

Before studying the reactions of acids and bases, it is necessary to define exactly what each is. This is not as easy as it sounds, and the Acids and Bases essay discusses in detail a subject covered more briefly here: the arduous task chemists faced in developing a workable distinction. Let us start with the phenomenological differences between the two—that is, aspects relating to things that can readily be observed without referring to the molecule properties and behaviors of acids and bases.

Acids are fairly easy to understand on the phenomenological level: the name comes from the Latin term *acidus,* or "sour," and many sour substances from daily life—lemons, for instance, or vinegar—are indeed highly acidic. In fact, lemons and most citrus fruits contain citric acid ($C_6H_8O_7$), while the acidic quality of vinegar comes from acetic acid (CH_3COOH). In addition, acids produce characteristic colors in certain vegetable dyes, such as those used in making litmus paper.

The word "base," as it is used in this context, may be a bit more difficult to appreciate on a sensory level. It helps, perhaps, if the older term "alkali" is used, though even so, people tend to think of alkaline substances primarily in contrast to acids. "Alkali," which serves to indicate the basic quality of both the alkali metal and alkaline earth metal families of elements, comes from the Arabic *al-qili.* The latter refers to the ashes of the seawort plant, which usually grows in marshy areas and, in the past, was often burned to produce soda ash for making soap.

The reason chemists of today use the word "base" instead of "alkali" is that the latter term has a narrower meaning: all alkalies are bases, but not all bases are alkalies. Originally referring only to the ashes of burned plants containing either sodium or potassium, alkali was eventually used to designate the soluble hydroxides of the alkali and alkaline earth metals. Among these are sodium hydroxide or lye; magnesium hydroxide (found in milk of magnesia); potassium hydroxide, which appears in soaps; and other compounds. Because these represent only a few of the substances that react with acids in the ways discussed in this essay, the term "base" is preferred.

GILBERT N. LEWIS.

THE FORMATION OF SALTS. As chemistry evolved, and physical scientists became aware of the atomic and molecular substructures that make up the material world, they developed more fundamental distinctions between acids and bases. By the early twentieth century, chemists had applied structural distinctions between acids and bases—that is, definitions based on the molecular structures and behaviors of those substances.

An important intermediary step occurred as chemists came to the conclusion that reactions of acids and bases form salts and water. For instance, in an aqueous solution, hydrochloric acid or HCl(*aq*) reacts with the base sodium hydroxide, designated as NaOH(*aq*), to form sodium chloride, or common table salt (NaCl[*aq*]) and H_2O. What happens is that the sodium (Na) ion (an atom with an electric charge) in sodium hydroxide switches places with the hydrogen ion in hydrochloric acid, resulting in the creation of NaCl and water.

Ions themselves had yet to be defined in 1803, when the great Swedish chemist Jons Berzelius (1779-1848) added another piece to the foundation for a structural definition. Acids and bases, he suggested, have opposite electric charges. In this, he was about eight decades ahead

of his time: only in 1884 did his countryman Svante Arrhenius (1859-1927) introduce the concept of the ion. This, in turn, enabled Arrhenius to formulate the first structural distinction between acids and bases.

THE ARRHENIUS ACID-BASE THEORY

Arrhenius acid-base theory defines the two substances with regard to their behavior in an aqueous solution: an acid is any compound that produces hydrogen ions (H^+), and a base is one that produces hydroxide ions (OH^-) when dissolved in water. This occurred, for instance, in the reaction discussed above: the hydrochloric acid produced a hydrogen ion, while the sodium hydroxide produced a hydroxide ion, and these two ions bonded to form water.

Though it was a good start, Arrhenius's theory was limited to reactions in aqueous solutions. In addition, it confined its definition of acids and bases only to those ionic compounds, such as hydrochloric acid or sodium hydroxide, that produced either hydrogen or hydroxide ions. But ammonia, or NH_3, acts like a base in aqueous solutions, even though it does not produce the hydroxide ion. These shortcomings pointed to the need for a more comprehensive theory, which came with the formulation of the Brønsted-Lowry definition.

THE BRØNSTED-LOWRY ACID-BASE THEORY

Developed by English chemist Thomas Lowry (1874-1936) and Danish chemist J. N. Brønsted (1879-1947), the Brønsted-Lowry acid-base theory defines an acid as a proton (H^+) donor, and a base as a proton acceptor, in a chemical reaction. Protons are represented by the symbol H^+, a cation (positively charged ion) of hydrogen.

Elemental hydrogen, called protium to distinguish it from its isotopes, has just one proton and one electron—no neutrons. Therefore, the hydrogen cation, which has to lose its sole electron to gain a positive charge, is essentially nothing but a proton. It is thus at once an atom, an ion, and a proton, but the ionization of hydrogen constitutes the only case in which this is possible.

Thus when the term "proton donor" or "proton acceptor" is used, it does not mean that a proton is splitting off from an atom or joining another, as in a nuclear reaction. Rather, when an acid behaves as a proton donor, this means that the hydrogen proton/ion/atom is separating from an acidic compound; conversely, when a base acts as a proton acceptor, the positively charged hydrogen ion is bonding with the basic compound.

REACTIONS IN BRØNSTED-LOWRY ACID-BASE THEORY. In representing Brønsted-Lowry acids and bases, the symbols HA and A⁻, respectively, are used. These appear in the equation representing the most fundamental type of Brønsted-Lowry acid-base reaction: $HA(aq) + H_2O(l) \rightarrow H_3O^+(aq) + A^-(aq)$. The symbols (aq), (l), and $\rightarrow$ are explained in the Chemical Reactions essay. In plain English, this equation states that when an acid in an aqueous solution reacts with liquid water, the result is the creation of H_3O^+, known as the hydronium ion, along with a base. Both products of the reaction are dissolved in an aqueous solution.

Because water molecules are polar, the negative charges tend to congregate on one end of the molecule with the oxygen atom, while the positive charges remain on the other end with the hydrogen atoms. The Brønsted-Lowry model emphasizes the role played by water, which pulls the proton from the acid, resulting in the creation of the hydronium ion.

The hydronium ion, in this equation, is an example of a conjugate acid, an acid formed when a base accepts a proton. At the same time, the acid has lost its proton, becoming A⁻, a conjugate base—that is, the base formed when an acid releases a proton. These two products of the reaction are called a conjugate acid-base pair, a term that refers to two substances related to one another by the donating of a proton.

Brønsted and Lowry's definition includes all Arrhenius acids and bases, as well as other chemical species not encompassed in Arrhenius theory. As mentioned earlier, ammonia is a base, yet it does not produce OH⁻ ions; however, it does accept a proton from a water molecule. Water can serve either as an acid or base; in this instance, it is an acid, and in reaction with ammonia, it produces the conjugate acid-base pair of NH_4^+ (an ammonium ion) and OH⁻. Ammonia did not produce the hydroxide ion here; rather, OH⁻ is the conjugate base that resulted when the water molecule lost its H⁺ atom (i.e., a proton.)

THE LEWIS ACID-BASE THEORY

The Brønsted-Lowry model still had its limitations, in that it only described compounds containing hydrogen. American chemist Gilbert N. Lewis (1875-1946), however, developed a theory of acids and bases that makes no reference to the presence of hydrogen. Instead, it relates to something much more fundamental: the fact that chemical bonding always involves pairs of electrons.

Lewis acid-base theory defines an acid as the reactant that accepts an electron pair from another reactant in a chemical reaction, while a base is the reactant that donates an electron pair to another reactant. Note that, as with the Brønsted-Lowry definition, the Lewis definition is reaction-dependant. Instead of defining a compound as an acid or base in its own right, it identifies these in terms of how the compound reacts with another.

The Lewis definition encompasses all the situations covered by the others, as well as many other reactions not described in the theories of either Arrhenius or Brønsted-Lowry. In particular, Lewis theory can be used to differentiate the acid and base in chemical reactions where ions are not produced, something that takes it far beyond the scope of Arrhenius theory. Also, Lewis theory addresses situations in which there is no proton donor or acceptor, thus offering an improvement over Brønsted-Lowry.

When boron trifluoride (BF_3) and ammonia (NH_3), both in the gas phases, react to produce boron trifluoride ammonia complex (F_3BNH_3), boron trifluoride accepts an electron pair. Therefore, it is a Lewis acid, while ammonia—which donates the electron pair—can be defined as a Lewis base. This particular reaction involves hydrogen, but since the operative factor in Lewis theory relates to electron pairs and not hydrogen, the theory can be used to address reactions in which that element is not present.

REAL-LIFE APPLICATIONS

DISSOCIATION

Dissociation is the separation of a molecule into ions, and it is a key factor for evaluating the "strength" of acids and bases. The more a substance is prone to dissociation, the better it can conduct an electric current, because the separation of charges provides a "pathway" for the current's flow. A substance that dissociates completely, or almost completely, is called a strong electrolyte, whereas one that dissociates only slightly (or not at all) is designated as a weak electrolyte.

The terms "weak" and "strong" are also applied to acids and bases. For instance, vinegar is a weak acid, because it dissociates only slightly, and therefore conducts little electric current. By contrast, hydrochloric acid (HCl) is a strong acid, because it dissociates almost completely into positively charged hydrogen ions and negatively charged chlorine ones. Represented symbolically, this is: $HCl \rightarrow H^+ + Cl^-$.

A REACTION INVOLVING A STRONG ACID. It may seem a bit backward that a strong acid or base is one that "falls apart," while the weak one stays together. To understand the difference better, let us return to the reaction described earlier, in which an acid in aqueous solution reacts with water to produce a base in aqueous solution, along with hydronium: $HA(aq) + H_2O(l) \rightarrow H_3O^+(aq) + A^-(aq)$. Instead of using the generic symbols HA and A⁻, however, let us substitute hydrochloric acid (HCl) and chloride (Cl⁻) respectively.

The reaction $HCl(aq) + H_2O(l) \rightarrow H_3O^+(aq) + Cl^-(aq)$ is a reversible one, and for that reason, the symbol for chemical equilibrium ($\rightleftharpoons$) can be inserted in place of the arrow pointing to the right. In other words, the substances on the right can just as easily react, producing the substances on the left. In this reverse reaction, the reactants of the forward reaction would become products, and the products of the forward reaction serve as the reactants.

However, the reaction described here is not perfectly reversible, and in fact the most proper chemical symbolism would show a longer arrow pointing to the right, with a shorter arrow pointing to the left. Due to the presence of a strong electrolyte, there is more forward "thrust" to this reaction.

Because it is a strong acid, the hydrogen chloride in solution is not a set of molecules, but a collection of H^+ and Cl^- ions. In the reaction, the weak Cl^- ions to the right side of the equilibrium symbol exert very little attraction for the H^+ ions. Instead of bonding with the chloride, these

hydrogen ions join the water (a stronger base) to form hydronium.

The chloride, incidentally, is the conjugate base of the hydrochloric acid, and this illustrates another principal regarding the "strength" of electrolytes: a strong acid produces a relatively weak conjugate base. Likewise, a strong base produces a relatively weak conjugate acid.

THE STRONG ACIDS AND BASES. There are only a few strong acids and bases, which are listed below:

Strong Acids
- Hydrobromic acid (HBr)
- Hydrochloric acid (HCl)
- Hydroiodic acid (HI)
- Nitric acid (HNO_3)
- Perchloric acid ($HClO_4$)
- Sulfuric acid (H_2SO_4)

Strong Bases
- Barium hydroxide (Ba[OH]$_2$)
- Calcium hydroxide (Ca[OH]$_2$)
- Lithium hydroxide (LiOH)
- Potassium hydroxide (KOH)
- Sodium hydroxide (NaOH)
- Strontium hydroxide (Sr[OH]$_2$)

Virtually all others are weak acids or bases, meaning that only a small percentage of molecules in these substances ionize by dissociation. The concentrations of the chemical species involved in the dissociation of weak acids and bases are mathematically governed by the equilibrium constant K..

NEUTRALIZATION

Neutralization is the process whereby an acid and base react with one another to form a salt and water. The simplest example of this occurs in the reaction discussed earlier, in which hydrochloric acid or HCl(aq) reacts with the base sodium hydroxide, designated as NaOH(aq), in an aqueous solution. The result is sodium chloride, or common table salt (NaCl[aq]) and H_2O. This equation is written thus: $HCl(aq) + NaOH(aq) \rightarrow NaCl(aq) + H_2O$.

The human stomach produces hydrochloric acid, commonly known as "stomach acid." It is generated in the digestion process, but when a person eats something requiring the stomach to work overtime in digesting it—say, a pizza—the stomach may generate excess hydrochloric acid, and the result is "heartburn." When this happens, people often take antacids, which contain a base such as aluminum hydroxide ($Al[OH]_3$) or magnesium hydroxide ($Mg[OH]_2$).

When a person takes an antacid, the reaction leads to the creation of *a* salt, but not the salt with which most people are familiar—NaCl. As shown above, that particular salt is the product of a reaction between hydrochloric acid and sodium hydroxide, but a person who ingested sodium hydroxide (a substance used to unclog drains and clean ovens) would have much worse heartburn than before! In any case, the antacid reacts with the stomach acid to produce a salt, as well as water, and thus the acid is neutralized.

When land formerly used for mining is reclaimed, the acidic water in the area must be neutralized, and the use of calcium oxide (CaO) as a base is one means of doing so. Acidic soil, too, can be neutralized by the introduction of calcium carbonate ($CaCO_3$) or limestone, along with magnesium carbonate ($MgCO_3$). If soil is too basic, as for instance in areas where there has been too little precipitation, acid-like substances such as calcium sulfate or gypsum ($SaSO_4$) can be used. In either case, neutralization promotes plant growth.

TITRATION AND pH. One of the most important applications of neutralization is in titration, the use of a chemical reaction to determine the amount of a chemical substance in a sample of unknown purity. In a typical form of neutralization titration, a measured amount of an acid is added to a solution containing an unknown amount of a base. Once enough of the acid has been added to neutralize the base, it is possible to determine how much base exists in the solution.

Titration can also be used to measure pH ("power of hydrogen") level by using an acid-base indicator. The pH scale assigns values ranging from 0 (a virtually pure acid) to 14 (a virtually pure base), with 7 indicating a neutral substance. An acid-base indicator such as litmus paper changes color when it neutralizes the solution.

The transition interval (the pH at which the color of an indicator changes) is different for different types of indicators, and thus various indicators are used to measure substances in specific pH ranges. For instance, methyl red changes

from red to yellow across a pH range of 4.4 to 6.2, so it is most useful for testing a substance suspected of being moderately acidic.

BUFFERED SOLUTIONS. A buffered solution is one that resists a change in pH even when a strong acid or base is added to it. This buffering results from the presence of a weak acid and a strong conjugate base, and it can be very important to organisms whose cells can endure changes only within a limited range of pH values. Human blood, for instance, contains buffering systems, because it needs to be at pH levels between 7.35 and 7.45.

The carbonic acid-bicarbonate buffer system is used to control the pH of blood. The most important chemical equilibria (that is, reactions involving chemical equilibrium) for this system are: $H^+ + HCO_3^- \leftrightarrows H_2CO_3 \leftrightarrows H_2O + CO_2$. In other words, the hydrogen ion (H^+) reacts with the hydrogen carbonate ion (HCO_3^-) to produce carbonic acid (H_2CO_3). The latter is in equilibrium with the first set of reactants, as well as with water and carbon dioxide in the forward reaction.

The controls the pH level by changing the concentration of carbon dioxide by exhalation. In accordance with Le Châtelier's principle, this shifts the equilibrium to the right, consuming H^+ ions. In normal blood plasma, the concentration of HCO_3^- is about 20 times as great as that of H_2CO_3, and this large concentration of hydrogen carbonate gives the buffer a high capacity to neutralize additional acid. The buffer has a much lower capacity to neutralize bases because of the much smaller concentration of carbonic acid.

WATER: BOTH ACID AND BASE

Water is an amphoteric substance; in other words, it can serve either as an acid or a base. When water experiences ionization, one water molecule serves as a Brønsted-Lowry acid, donating a proton to another water molecule— the Brønsted-Lowry base. This results in the production of a hydroxide ion and a hydronium ion: $H_2O(l) + H_2O(l) \leftrightarrows H_3O^+(aq) + OH^-(aq)$.

This equilibrium equation is actually one in which the tendency toward the reverse reaction is much greater; therefore the equilibrium symbol, if rendered in its most proper form, would show a much shorter arrow pointing toward the right. In water purified by distillation, the concentra-

KEY TERMS

ACID: A substance that in its edible form is sour to the taste, and in non-edible forms is often capable of dissolving metals. Acids and bases react to form salts and water. These are all phenomenological definitions, however, in contrast to the three structural definitions of acids and bases—the Arrhenius, Brønsted-Lowry, and Lewis acid-base theories.

ALKALI: A term referring to the soluble hydroxides of the alkali and alkaline earth metals. Once "alkali" was used for the class of substances that react with acids to form salts; today, however, the more general term base is preferred.

ALKALINITY: An adjectival term used to identify the degree to which a substance displays the properties of a base.

AMPHOTERIC: A term describing a substance that can serve either as an acid or a base. Water is the most significant amphoteric substance.

AQUEOUS SOLUTION: A substance in which water constitutes the solvent. A large number of chemical reactions take place in an aqueous solution.

ARRHENIUS ACID-BASE THEORY: The first of three structural definitions of acids and bases. Formulated by Swedish chemist Svante Arrhenius (1859-1927), the Arrhenius theory defines acids and bases according to the ions they produce in an aqueous solution an acid produces hydrogen ions (H^+), and a base hydroxide ions (OH^-).

BASE: A substance that in its edible form is bitter to the taste. Bases tend to be slippery to the touch, and in reaction with acids they produce salts and water. Bases and acids are most properly defined, however, not in these phenomenological terms, but by the three structural definitions of acids and bases—the Arrhenius, Brønsted-Lowry, and Lewis acid-base theories.

BASIC: In the context of acids and bases, the word is the counterpart to "acidic," identifying the base-like quality of a substance.

BRØNSTED-LOWRY ACID-BASE THEORY: The second of three structural definitions of acids and bases. Formulated by English chemist Thomas Lowry (1874-1936) and Danish chemist J. N. Brønsted (1879-1947), Brønsted-Lowry theory defines an acid as a proton (H^+) donor, and a base as a proton acceptor.

CHEMICAL SPECIES: A generic term used for any substance studied in chemistry—whether it be an element, compound, mixture, atom, molecule, ion, and so forth.

CONJUGATE ACID: An acid formed when a base accepts a proton (H^+).

CONJUGATE ACID-BASE PAIR: The acid and base produced when an acid donates a single proton to a base. In the reaction that produces this pair, the acid and base switch identities. By donating a proton, the acid becomes a conjugate base, and by receiving the proton, the base becomes a conjugate acid.

CONJUGATE BASE: A base formed when an acid releases a proton.

DISSOCIATION: The separation of molecules into ions.

ION: An atom or atoms that has lost or gained one or more electrons, and thus has a net electric charge. There are two types of ions: anions and cations.

IONIC BONDING: A form of chemical bonding that results from attractions between ions with opposite electric charges.

IONIC COMPOUND: A compound in which ions are present. Ionic compounds contain at least one metal and nonmetal joined by an ionic bond.

LEWIS ACID-BASE THEORY: The third of three structural definitions of acids and bases. Formulated by American chemist Gilbert N. Lewis (1875-1946), Lewis theory defines an acid as the reactant that accepts an electron pair from another reactant in a chemical reaction, and a base as the reactant that donates an electron pair to another reactant.

NEUTRALIZATION: The process whereby an acid and base react with one another to form a salt and water.

PH SCALE: A logarithmic scale for determining the acidity or alkalinity of a substance, from 0 (virtually pure acid) to 7 (neutral) to 14 (virtually pure base).

PHENOMENOLOGICAL: A term describing scientific definitions based purely on experimental phenomena. These only convey part of the picture, however—primarily, the part a chemist can perceive either through measurement or through the senses, such as sight. A structural definition is therefore usually preferable to a phenomenological one.

REACTANT: A substance that interacts with another substance in a chemical reaction, resulting in the creation of a product.

SALTS: Ionic compounds formed by the reaction between an acid and a base. In this reaction, one or more of the hydrogen ions of an acid is replaced with another positive ion. In addition to producing salts, acid-base reactions produce water.

SOLUTION: A homogeneous mixture in which one or more substances (the solute) is dissolved in another substance (the solvent)—for example, sugar dissolved in water.

SOLVENT: A substance that dissolves another, called a solute, in a solution.

STRONG ELECTROLYTE: A substance highly prone to dissociation. The terms "strong acid" or "strong base" refers to those acids or bases which readily dissociate.

STRUCTURAL: A term describing scientific definitions based on aspects of molecular structure and behavior rather than purely phenomenological data.

TITRATION: The use of a chemical reaction to determine the amount of a chemical substance in a sample of unknown purity. Testing pH levels is an example of titration.

TRANSITION INTERVAL: The pH level at which the color of an acid-base indicator changes.

WEAK ELECTROLYTE: A substance that experiences little or no dissociation. The terms "weak acid" or "weak base" refer to those acids or bases not prone to dissociation.

tions of hydronium (H_3O^+) and hydroxide (OH^-) ions are equal. When multiplied by one another, these yield the constant figure $1.0 \cdot 10^{-14}$, which is the equilibrium constant for water. In fact, this constant—denoted as K_w—is called the ion-product constant for water.

Because the product of these two concentrations is always the same, this means that if one of them goes up, the other one must go down in order to yield the same constant. This explains the fact, noted earlier, that water can serve either as an acid or base—or, if the concentrations of hydronium and hydroxide ions are equal—as a neutral substance. In situations where the concentration of hydronium is higher, and the hydroxide concentration automatically decreases, water serves as an acid. Conversely, when the hydroxide concentration is high, the hydronium concentration decreases correspondingly, and the water is a base.

WHERE TO LEARN MORE

"*Acids and Bases.*" *Vision Learning* (Web site). <http://www.visionlearning.com/library/science/chemistry-2/CHE2.2-acid_base.htm> (June 7, 2001).

"*Acids, Bases, and Chemical Reactions*" *Open Access College* (Web site). <http://oac.schools.sa.edu.au/8-10science/acids.htm> (June 7, 2001).

"*Acids, Bases, pH.*" *About.com* (Web site). <http://chemistry.about.com/science/chemistry/cs/acidsbasesph/> (June 7, 2001).

"*Acids, Bases, and Salts*" (Web site). <http://edie.cprost.sfu.ca/~rhlogan/ionic_eq.html> (June 7, 2001).

"*Junior Part: Acids, Bases, and Salts*" (Web site). <http://www.rjclarkson.demon.co.uk/junior/junior4.htm> (June 7, 2001).

Knapp, Brian J. *Acids, Bases, and Salts.* Danbury, CT: Grolier Educational, 1998.

Moje, Steven W. *Cool Chemistry: Great Experiments with Simple Stuff.* New York: Sterling Publishing Company, 1999.

Walters, Derek. *Chemistry.* Illustrated by Denis Bishop and Jim Robins. New York: F. Watts, 1982.

Zumdahl, Steven S. *Introductory Chemistry: A Foundation,* 4th ed. Boston: Houghton Mifflin, 2000.

SOLUTIONS AND MIXTURES

MIXTURES

SOLUTIONS

OSMOSIS

DISTILLATION AND FILTRATION

MIXTURES

CONCEPT

Elements and compounds are pure substances, but much of the material around us—including air, wood, soil, and even (in most cases) water—appears in the form of a mixture. Unlike a pure substance, a mixture has variable composition: in other words, it cannot be reduced to a single type of atom or molecule. Mixtures can be either homogeneous—that is, uniform in appearance, and having only one phase—or heterogeneous, meaning that the mixture is separated into various regions with differing properties. Most homogeneous mixtures are also known as solutions, and examples of these include air, coffee, and even metal alloys. As for heterogeneous mixtures, these occur, for instance, when sand or oil are placed in water. Oil can, however, be dispersed in water as an emulsion, with the aid of a surfactant or emulsifier, and the result is an almost homogeneous mixture. Another interesting example of a heterogeneous mixture occurs when colloids are suspended in fluid, whether that fluid be air or water.

HOW IT WORKS

MIXTURES AND PURE SUBSTANCES

A mixture is a substance with a variable composition, meaning that it is composed of molecules or atoms of differing types. These may have intermolecular bonds, or bonds between molecules, but such bonds are not nearly as strong as those formed when elements bond to form a compound.

Examples of mixtures include milk, coffee, tea, and soft drinks—indeed, they include virtually any drink except for pure water, which is a rarity, because water is highly soluble and tends to contain minerals and other impurities. Air, too, is a mixture, because it contains oxygen, nitrogen, noble gases, carbon dioxide, and water vapor. Likewise metal alloys such as bronze, brass, and steel are mixtures.

STRUCTURAL AND PHENOME-NOLOGICAL DISTINCTIONS. Before chemists accepted the atomic theory of matter, it was often difficult to distinguish a mixture, such as air or steel, from pure substances, which have the same composition throughout. Pure substances are either elements such as oxygen or iron, or compounds—for example, carbon dioxide or iron oxide. An element is made up of only one type of atom, while a compound can be reduced to one type of bonded chemical species—usually either a molecule or a formation of ions.

A distinction between mixtures and pure substances based on atomic and molecular structures is called, fittingly enough, a structural definition. Prior to the development of atomic theory by English chemist John Dalton (1766-1844), and the subsequent identification of molecules by Italian physicist Amedeo Avogadro (1776-1856), chemists defined compounds and mixtures on the basis of phenomenological data garnered through observation and measurement. This could and did lead to incorrect suppositions.

PROUST AND THE DEBATE OVER CONSTANT COMPOSITION. Around the time of Dalton and Avogadro, French

THE WONDROUS INVENTION KNOWN AS SOAP IS A MIXTURE WITH SURFACTANT QUALITIES. *(Chad Weckler/Corbis. Reproduced by permission.)*

chemist Antoine Lavoisier (1743-1794) correctly identified an element as a substance that could not be broken down into a simpler substance. Later, his countryman Joseph-Louis Proust (1754-1826) clarified the definition of a compound, stating it in much the same terms that we have used: the composition of a compound is the same throughout. But with the very rudimentary knowledge of atoms and molecules that chemists of the early nineteenth century possessed, the distinctions between compounds and mixtures remained elusive.

Proust's contemporary and intellectual rival Claude Louis Berthollet (1748-1822) observed that when metals such as iron are heated, they form oxides in which the percentage of oxygen increases with temperature. If constant composition is a fact, he challenged Proust, then how is this possible? Proust maintained that the proportions of elements in a compound are always the same, a hypothesis Berthollet countered by observing that metals can form variable alloys. Bronze, for instance, is an alloy of tin and copper at a ratio of about 25:75, but if the ratio were changed to, say, 30:70, it would still be called bronze.

At the time, the equipment—both in terms of knowledge and tools for analysis—simply did not exist whereby Berthollet's assertions could be successfully proven wrong. Indeed, Berthollet, a student of Lavoisier who contributed to progress in the study of acids and reactants, was no anti-scientific villain: he was simply responding to the evidence as he saw it. Only with the discovery of subatomic particles in the late nineteenth and early twentieth centuries, discoveries that changed the entire character of chemistry as a discipline, did it truly become possible to answer Berthollet's challenges.

Chemists now understand that the oxide formed on the surface of a metal is a compound separate from the metal itself, and that alloys (discussed later in this essay) are not compounds at all: they are mixtures.

MIXTURES AND COMPOUNDS

With our modern knowledge of atomic and molecular properties, we can more fully distinguish between a compound and a mixture. First, a mixture can exist in virtually any proportion among its constituent parts, whereas a compound has a definite and constant composition. Coffee, whether weak or strong, is still coffee, but if a second oxygen atom chemically bonds to the oxygen in carbon monoxide (CO), the resulting

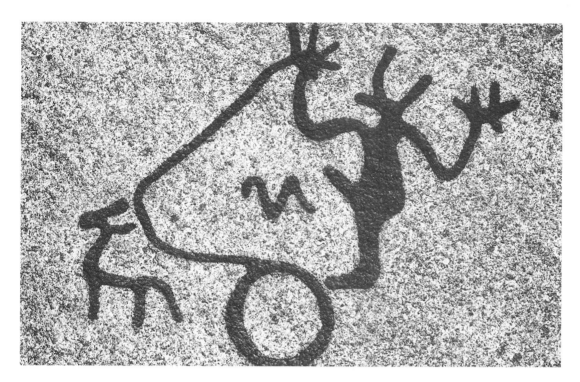

compound is not simply "strong carbon monoxide" or even "oxygenated carbon monoxide": it is carbon dioxide (CO_2), an entirely different substance. The difference is enormous: carbon dioxide is a part of the interactions between plant and animal life that sustains the natural environment, whereas carbon monoxide is a toxic gas produced, for instance, by automobiles.

Second, the parts of a mixture retain most of their properties when they join together, but elements lose their elemental characteristics once they form a compound. Sugar is still sweet, regardless of the substance into which it is dissolved. In coffee or another drink, it may no longer be dry and granular, though it could be returned to that state by evaporating the coffee. In any case, dryness and granularity are physical properties, whereas sweetness is a chemical one. On the other hand, when elemental carbon bonds with hydrogen and oxygen to form sugar itself, the resulting compound is nothing like the elements from which it is made.

Third, a mixture can usually be created simply by the physical act of stirring items together, and/or by applying heat. As discussed below, relatively high temperatures are sometimes needed to dissolve sugar in tea, for instance; likewise, alloys are usually formed by heating metals to

much higher temperatures. On the other hand, the formation of a compound involves a chemical reaction, a vastly more complex interaction than the one required to create a mixture. Carbon, a black solid found in coal, can be mixed with hydrogen and oxygen—both colorless and odorless gases—and the resulting mess would be something; but it would not be sugar. Only the chemical reaction between the three elements, in specific proportions and under specific conditions, results in their chemical bonding to form table sugar, known to chemists as sucrose.

HOMOGENEOUS AND HETEROGENEOUS MIXTURES

HOMOGENEOUS MIXTURES AND SOLUTIONS. As we have seen, the composition of a mixture is variable. Coffee, for instance, can be weak or strong, and milk can be "whole" or low-fat. Beer can be light or dark in color, as well as "light" in terms of calorie content; furthermore, its alcohol content can be varied, as with all alcoholic drinks. Though their molecular composition is variable, each of the mixtures described here is the same throughout: in other words, there is no difference between one region and another in a glass of beer. Such a mixture is described as homogeneous.

A homogeneous mixture is one that is the same throughout, but this is not the same as saying that a compound has a constant composition. Rather, when we say that a homogeneous mixture is the same in every region, we are speaking phenomenologically rather than structurally. From the standpoint of ordinary observation, a glass of milk is uniform, but as we shall see, milk is actually a type of emulsion, in which one substance is dispersed in another. There are no milk "molecules," and structural analysis would reveal that milk does not have a constant molecular composition.

Coffee is an example of a solution, a specific type of homogeneous mixture. Actually, most homogeneous mixtures can be considered solutions, but since a solution is properly defined as a homogeneous mixture in which one or more substances is dissolved in another substance, a 50:50 homogeneous mixture is technically not a solution. When particles are perfectly dissolved in a solution, the composition is uniform, but again, this is not a compound, because that composition can always be varied.

HETEROGENEOUS MIXTURES.

Anyone who has ever made iced tea has observed that it is easy to sweeten when it is hot, because temperature affects the ability of a solvent such as water—in this case, water with particles from tea leaves filtered into it—to dissolve a solute, such as sugar. Cold tea, on the other hand, is much harder to sweeten with ordinary sugar. Usually what happens is that, instead of obtaining a homogeneous mixture, the result is heterogeneous.

Whereas every region within a homogeneous mixture is more or less the same as every other region, heterogeneous mixtures contain regions that differ from one another. In the example we have used, a glass of cold tea with undissolved sugar at the bottom is a heterogeneous mixture: the tea at the top is unsweetened or even bitter, whereas at the bottom, there is an overly sweet sludge of tea and sugar.

FROM HOMOGENEOUS TO HETEROGENEOUS AND BACK AGAIN.

The distinction between homogeneous and heterogeneous solutions only serves to further highlight the difference between compounds and mixtures. Whereas the composition of a compound is definite and quantitative (so many atoms of element x bonded with so many atoms of element y in such-and-such a structure), there is less of a sharp dividing line between homogeneous and heterogeneous solutions.

When too much of a solute is added to the solvent, the solvent becomes saturated, and is no longer truly homogeneous. Such is the case when the sugar drifts to the bottom of the tea, or when coffee grounds form in the bottom of a coffee pot containing an otherwise uniform mixture of coffee and water.

Milk comes to us as a homogeneous substance, thanks to the process of homogenization, but when it comes out of the cow, it is a more heterogeneous mixture of milk fat and water. In fact, milk is an example of an emulsion, the result of a process whereby two substances that would otherwise form a heterogeneous mixture are made to form a homogeneous one.

REAL-LIFE APPLICATIONS

Colloids

In 1827, Scottish naturalist Robert Brown (1773-1858) was studying pollen grains under a microscope when he noticed that the grains underwent a curious zigzagging motion in the water. At first, he assumed that the motion had a biological explanation—in other words, that it resulted from life processes within the pollen—but later, he discovered that even pollen from long-dead plants behaved in the same way.

What Brown had observed was the suspension of a colloid—a particle intermediate in size between a molecule and a speck of dust—in the water. When placed in water or another fluid (both liquids and gases are called "fluids" in the physical sciences), a colloid forms a mixture similar both to a solution and a suspension. As with a solution, the composition remains homogeneous—the particles never settle to the bottom of the container. At the same time, however, the particles remain suspended rather than dissolved, rendering a cloudy appearance to the dispersion.

Almost everyone, especially as a child, has been fascinated by the colloidal dispersion of dust particles in a beam of sunlight. They seem to be continually in motion, as indeed they are, and this movement is called "Brownian motion" in honor of the man who first observed it. Yet Brown did not understand what he was seeing; only later did scientists recognize that Brownian motion is the result of movement on the part of molecules in a fluid. Even though the molecules are much smaller than the colloid, there are enough of them producing enough collisions to cause the colloid to be in constant motion.

Another remarkable aspect of dust particles floating in sunlight is the way that they cause a column of sunlight to become visible. This phenomenon, called the Tyndall effect after English physicist John Tyndall (1820-1893), makes it seem as though we are actually seeing a beam of light. In fact, what we are seeing is the reflection of light off the colloidal dispersion. Another example of a colloidal dispersion occurs when a puff of smoke hangs in the air; furthermore, as we shall see, milk is a substance made up of colloids—in this case, particles of fat and protein suspended in water.

EMULSIONS

Miscibility is a qualitative term identifying the relative ability of two substances to dissolve in one another. Generally, water and water-based substances have high miscibility with regard to one another, as do oil and oil-based substances with one another. Oil and water, on the other hand (or substances based in either) have very low miscibility.

The reason is that water molecules are polar, meaning that the positive electric charge is in one region of the molecule, while the negative charge is in another region. Oil, on the other hand, is nonpolar, meaning that charges are more evenly distributed throughout the molecule. Trying to mix oil and water is thus rather like attempting to use a magnet to attract a nonmagnetic metal such as gold.

HOW SURFACTANTS AID THE EMULSION PROCESS. An emulsion is a mixture of two immiscible liquids such as oil and water—we will use these simple terms, with the idea that these stand for all oil- and water-soluble substances—by dispersing microscopic droplets of one liquid in another. In order to achieve this dispersion, there needs to be a "middle man," known as an emulsifier or surfactant. The surfactant, made up of molecules that are both water- and oil-soluble, acts as an agent for joining other substances in an emulsion.

All liquids have a certain degree of surface tension. This is what causes water on a hard surface to bead up instead of lying flat. (Mercury has an extremely high surface tension.) Surfactants break down this surface tension, enabling the marriage of two formerly immiscible liquids.

In an emulsion, millions of surfactants surround the dispersed droplets (known as the internal phase), shielding them from the other liquid (the external phase). Supposing oil is the internal phase, then the oil-soluble end of the surfactant points toward the oils, while the water-soluble side joins with the water of the external phase.

PHYSICAL CHARACTERISTICS OF AN EMULSION. The resulting emulsion has physical (as opposed to chemical) properties different from those of the substances that make it up. Water and most oils are transparent, whereas emulsions are typically opaque and may have a lustrous, pearly gleam. Oil and water flow freely, but emulsions can be made to form thick, non-flowing creams.

Emulsions are inherently unstable: they are, as it were, only temporarily homogeneous. A good example of this is an oil-and-vinegar salad dressing, which has to be shaken before putting it on salad; otherwise, what comes out of the bottle is likely to be mainly oil, which floats to the top of the water-based vinegar. It is characteristic of emulsions that eventually gravity pulls them apart, so that the lighter substances float to the top. This process may take only a few minutes, as with the salad dressing; on the other hand, it can take thousands of years.

APPLICATIONS. Surfactants themselves are often used in laundry or dish detergent, because most stains on plates or clothes are oil-based, whereas the detergent itself is applied to the clothes in an aqueous solution. As for emulsions, these are found in a wide variety of products, from cosmetics to paint to milk.

In the pharmaceutical industry, for instance, emulsions are used as a means of delivering the active ingredients of drugs, many of which are not water soluble, in a medium that is. Pharmaceutical manufacturers also use emulsions to make medicines more palatable (easier to take); to control dosage and improve effectiveness; and to improve the usefulness of topical drugs such as ointments, which must contend with both oily and water-soluble substances on the skin.

The oils and other conditioning agents used in hair-care products would leave hair limp and sticky if applied by themselves. Instead, these products are applied in emulsions that dilute the oils in other materials. Emulsions are also used in color photography, as well as in the production of pesticides and fungicides.

Paints and inks, too, are typically emulsified substances, and these sometimes come in the form of dispersions akin to those of colloids. As we have seen, in a dispersion fine particles of solids are suspended in a liquid. Surfactants are used to bond the particles to the liquid, though paint must be shaken, usually with the aid of a high-speed machine, before it is consistent enough to apply.

As mentioned earlier, milk is a form of emulsion that contains particles of milk fat or cream dispersed in water. Light strikes the microscopic fat and protein colloids, resulting in the familiar white color. Other examples of emulsified foods include not only salad dressings, but gravies and various sauces, peanut butter, ice cream, and other items. Not only does emulsion affect the physical properties of foods, but it may enhance the taste by coating the tongue with emulsified oils.

SOAP

The wondrous invention called soap, which has made the world a cleaner place, exhibits several of the properties we have discussed. It is a mixture with surfactant qualities, and in powdered form, it can form something close to a true solution with water. Discovered probably by the Phoenicians around 600 B.C., it was made in ancient times by boiling animal fat (tallow) or vegetable oils with an alkali or base of wood ashes.

This was a costly process, and for many centuries, only the wealthy could afford soap—a fact that contributed to the notorious lack of personal hygiene that prevailed among Europeans of the Middle Ages. In fact, the situation did not change until late in the eighteenth century, when the work of two French chemists yielded a more economical manufacturing method.

In 1790, Nicholas Leblanc (1742-1806) developed a process for creating sodium hydroxide (called caustic soda at the time) from sodium chloride, or ordinary table salt. This made for an inexpensive base or alkali, with which soap manufacturers could combine natural fats and oils.

Then in 1823, Michel Eugène Chevreul (1786-1889) showed that fat is a compound of glycerol with organic acids, a breakthrough that led to the use of fatty acids in producing soaps.

THE CHEMISTRY OF SOAP. Soap as it is made today comes from a salt of an alkali metal, such as sodium or potassium, combined with a mixture of carboxylic acids ("fatty acids"). These carboxylic acids are the result of a reaction called saponification, involving triglycerides and a base, such as sodium hydroxide. Saponification breaks the triglycerides into their component fatty acids; then, the base neutralizes these to salts. This reaction produces glycerin, a hydrocarbon bonded to a hydroxyl (-OH) group. (Hydrocarbons are discussed in the essays on Organic Chemistry; Polymers.)

In general, the formula for soap is RCOOX, where X stands for the alkali metals, and R for the hydrocarbon chain. Because it is a salt (meaning that it is formed from the reaction of an acid with a base), soap partially separates into its component ions in water. The active ion is RCOO-, whose two ends behave in different fashions, making it a surfactant. The hydrocarbon end (R-) is said to be lipophilic, or "oil-loving," which is appropriate, since most oils are hydrocarbons. On the other hand, the carboxylate end (-COO-) is hydrophilic, or "water-loving." As a result, soap can dissolve in water, but can also clean greasy stains.

When soap is mixed with water, it does not form a true homogeneous mixture or solution due to the presence of hydrocarbons. These attract one another, forming spherical aggregates called micelles. The lipophilic "tails" of the hydrocarbons are turned toward the interior of the micelle, while the hydrophilic heads remain facing toward the water that forms the external phase.

ALLOYS

We have primarily discussed liquid mixtures, but in fact mixtures can be gaseous or even solid—as for example in the case of an alloy, a mixture of two or more metals. Alloys are usually created by melting the component metals, then mixing them together in specific proportions, but an alloy can also be created by bonding metal powders.

The structure of an elemental metal includes tight "electron sea" bonds, which are discussed in the essay on Metals. These, combined with the

KEY TERMS

ALLOY: A mixture of two or more metals.

AQUEOUS: A term describing a solution in which a substance or substances are dissolved in water.

ATOM: The smallest particle of an element.

CHEMICAL SPECIES: A generic term used for any substance studied in chemistry—whether it be an element, compound, mixture, atom, molecule, ion, and so forth.

COMPOUND: A substance made up of atoms of more than one element, which are chemically bonded and usually joined in molecules. The composition of a compound is always the same, unless it is changed chemically.

DISPERSION: A term describing the distribution of particles in a fluid.

ELEMENT: A substance made up of only one kind of atom. Unlike compounds, elements cannot be chemically broken down into other substances.

EMULSIFIER: A surfactant.

EMULSION: A mixture of two immiscible liquids such as oil and water, made by dispersing microscopic droplets of one liquid in another. In creating an emulsion, surfactants act as bridges between the two liquids.

FLUID: A substance with the ability to flow. In physical sciences such as chemistry and physics, "fluid" refers both to liquids and gases.

HETEROGENEOUS: A term describing a mixture that is not the same through-

out; rather, it has various regions possessing different properties. An example of a mixture is sand in a container of water. Rather than dissolving to form a homogeneous mixture, as sugar would, the sand sinks to the bottom.

HOMOGENEOUS: A term describing a mixture that is the same throughout, as for example when sugar is fully dissolved in water. A solution is a homogenous mixture.

IMMISCIBLE: Possessing a negligible value of miscibility.

ION: An atom or group of atoms that has lost or gained one or more electrons, and thus has a net electric charge.

MISCIBILITY: A qualitative term identifying the relative ability of two substances to dissolve in one another.

MIXTURE: A substance with a variable composition, meaning that it is composed of molecules or atoms of differing types. Compare with pure substance or compound.

MOLECULE: A group of atoms, usually but not always representing more than one element, joined in a structure. Compounds are typically made of up molecules.

PHENOMENOLOGICAL: A term describing scientific definitions based purely on experimental phenomena. These only convey part of the picture, however—primarily, the part a chemist can perceive either through measurement or through the senses, such as sight. A structural definition is therefore usually preferable to a phenomenological one.

KEY TERMS CONTINUED

PURE SUBSTANCE: A substance that has the same composition throughout. Pure substances are either elements or compounds, and should not be confused with mixtures. A homogeneous mixture is not the same as a pure substance, because although it has an unvarying composition, it cannot be reduced to a single type of atom or molecule.

SOLUTION: A homogeneous mixture in which one or more substances is dissolved in another substances—for example, sugar dissolved in water.

STRUCTURAL: A term describing scientific definitions based on aspects of molecular structure rather than purely phenomenological data.

SURFACTANT: A substance made up of molecules that are both water- and oil-soluble, which acts as an agent for joining other substances in an emulsion.

SUSPENSION: A term that refers to a mixture in which solid particles are suspended in a fluid.

metal's crystalline structure, create a situation in which internal bonding is very strong, but nondirectional. As a result, most metals form alloys, but again, this is not the same as a compound: the elemental metals retain their identity in these metal composites, forming characteristic fibers, beads, and other shapes.

SOME FAMOUS ALLOYS. One of the most important alloys in early human history was a 25:75 mixture of tin and copper that gave its name to an entire stage of technological development: the Bronze Age (c. 3300-1200 B.C.). This combination formed a metal much stronger than either copper or tin, and bronze remained dominant until the discovery of new iron-smelting methods ushered in the Iron Age in about 1200 B.C.

Another important alloy of the ancient world was brass, which is about one-third zinc to two-thirds copper. Introduced in the period c. 1400-c. 1200 B.C, brass was one of the metals that defined the world in which the Bible was written. Biblical references to it abound, as in the famous passage from I Corinthians 13 (read at many weddings), in which the Apostle Paul proclaims that "without love, I am as sounding brass." Some biblical scholars, however, maintain that some of the references to brass in the Old Testament are actually mistranslations of "bronze."

Tin, copper, and antimony form pewter, a soft mixture that can be molded when cold and beaten regularly without turning brittle. Though used in Roman times, it became most popular in England from the fourteenth to the eighteenth centuries, when it offered a cheap substitute for silver in plates, cups, candelabra, and pitchers. The pewter turned out by colonial American metalsmiths is still admired for its beauty and functionality.

As for iron, it appears in nature as an impure ore, but even when purified, it is typically alloyed with other elements. Among the well-known forms of iron are cast iron, a variety of mixtures containing carbon and/or silicon; wrought iron, which contains small amounts of various other elements, such as nickel, cobalt, copper, chromium, and molybdenum; and steel. Steel is a mixture of iron with manganese and chromium, purified with a blast of hot air. Steel is also sometimes alloyed with aluminum in a one-third/two-thirds mixture to make duraluminum, developed for the superstructures of German zeppelins during World War I.

WHERE TO LEARN MORE

"All About Colloids." Synthashield (Web site). <http://www.synthashield.net/vault/colloids.html> (June 6, 2001).

Carona, Phillip B. *Magic Mixtures: Alloys and Plastics.* Englewood Cliffs, NJ: Prentice Hall, 1963.

ChemLab. Danbury, CT: Grolier Educational, 1998.

"*Corrosion of Alloys.*" *Corrosion Doctors* (Web site). <http://www.corrosion-doctors.org/MatSelect/corralloys.htm> (June 6, 2001).

"*Emulsions.*" *Pharmweb* (Web site). <http://pharmweb.usc.edu/phar306/Handouts/emulsions/> (June 6, 2001).

Hauser, Jill Frankel. *Super Science Concoctions: 50 Mysterious Mixtures for Fabulous Fun.* Charlotte, VT: Williamson Publishing, 1996.

Knapp, Brian J. *Elements, Compounds, and Mixtures.* Danbury, CT: Grolier Educational, 1998.

Maton, Anthea. *Exploring Physical Science.* Upper Saddle River, NJ: Prentice Hall, 1997.

Patten, J. M. *Elements, Compounds, and Mixtures.* Vero Beach, FL: Rourke Book Company, 1995.

The Soap and Detergent Association (Web site). <http://www.sdahq.org/> (June 6, 2001).

SOLUTIONS

CONCEPT

We are most accustomed to thinking of solutions as mixtures of a substance dissolved in water, but in fact the meaning of the term is broader than that. Certainly there is a special place in chemistry for solutions in which water—"the universal solvent"—provides the solvent medium. This is also true in daily life. Coffee, tea, soft drinks, and even water itself (since it seldom appears in pure form) are solutions, but the meaning of the term is not limited to solutions involving water. Indeed, solutions do not have to be liquid; they can be gaseous or solid as well. One of the most important solutions in the world, in fact, is the air we breathe, a combination of nitrogen, oxygen, noble gases, carbon dioxide, and water vapor. Nonetheless, aqueous or water-based substances are the focal point of study where solutions are concerned, and reactions that take place in an aqueous solution provide an important area of study.

HOW IT WORKS

Mixtures

Solutions belong to the category of substances identified as mixtures, substances with variable composition. This may seem a bit confusing, since among the defining characteristics of a solution are its uniformity and consistency. But this definition of a solution involves external qualities—the things we can observe with our senses—whereas a mixture can be distinguished from a pure substance not only in terms of external characteristics, but also because of its internal and structural properties.

At one time, chemists had a hard time differentiating between mixtures and pure substances (which are either elements or compounds), primarily because they had little or no knowledge of those "internal and structural properties." Externally, a mixture and compound can be difficult to distinguish, but the "variable composition" mentioned above is a key factor. Coffee can be almost as weak as water, or so strong it seems to galvanize the throat as it goes down—yet in both cases, though its composition has varied, it is still coffee.

By contrast, if a compound is altered by the variation of just one atom in its molecular structure, it becomes something else entirely. When an oxygen atom bonds with a carbon and oxygen atom, carbon monoxide—a poisonous gas produced by the burning of fossil fuels—is turned into carbon dioxide, an essential component of the interaction between plants and animals in the natural environment.

MIXTURES VS. COMPOUNDS: THREE DISTINCTIONS. Air (a mixture) can be distinguished from a compound, such as carbon dioxide, in three ways. First, a mixture can exist in virtually any proportions among its constituent parts, whereas a compound has a definite and constant composition. Air can be oxygen-rich or oxygen-poor, but if we vary the numbers of oxygen atoms chemically bonded to form a compound, the result is an entirely different substance.

Second, the parts of a mixture retain most of their properties when they join together, but elements—once they form a compound—lose their elemental characteristics. If pieces of carbon were floating in the air, their character would be quite

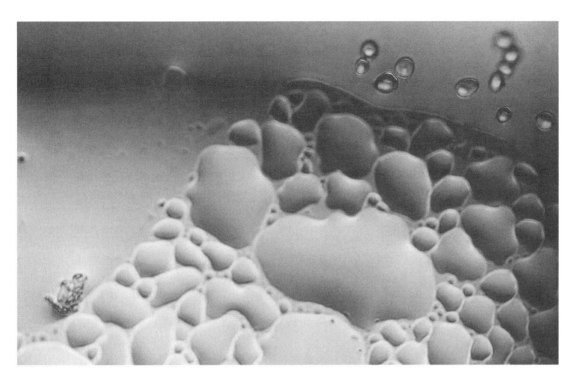

WATER AND OIL DROPLETS EXPOSED TO POLARIZED LIGHT. WATER AND OIL HAVE VERY LITTLE MISCIBILITY WITH REGARD TO ONE ANOTHER, MEANING THAT THEY TEND TO REMAIN SEPARATE WHEN COMBINED. *(Science Pictures Limited/Corbis. Reproduced by permission.)*

different from that of the carbon in carbon dioxide. Carbon, in its natural form, is much like coal or graphite (graphite is pure carbon, and coal an impure form of the element), but when a carbon atom combines with an oxygen atom, the two form a gas. Pieces of carbon in the air, on the other hand, would be solid, as carbon itself is.

Third, a mixture can usually be created simply by the physical act of stirring items together, and/or by applying heat. Certainly this is true of many solutions, such as brewed coffee. On the other hand, a compound such as carbon dioxide can only be formed by the chemical reaction of atoms in specific proportions and at specific temperatures. Some such reactions consume heat, whereas others produce it. These interactions of heat are highly complex; on the other hand, in making coffee, heat is simply produced by an external source (the coffee maker) and transferred to the coffee in the brewing process.

HETEROGENEOUS AND HOMO-GENEOUS MIXTURES. There are two basic types of mixtures. One of these occurs, for instance, when sand is added to a beaker of water: the sand sinks to the bottom, and the composition of the sand-water mixture cannot be said to be the same throughout. It is thus a heterogeneous mixture. The same is true of cold tea when table sugar (sucrose) is added to it: the sugar drifts to the bottom, and as a result, the first sip of tea from the glass is not nearly as sweet as the last.

Wood is a mixture that typically appears in heterogeneous form, a fact that can be easily confirmed by studying wood paneling. There are knots, for instance—areas where branches once grew, and where the concentrations of sap are higher. Striations of color run through the boards of paneling, indicating complex variations in the composition of the wood. Much the same is true of soil, with a given sample varying widely depending on a huge array of factors: the concentrations of sand, rock, or decayed vegetation, for instance, or the activities of earthworms and other organisms in processing the soil.

In a homogeneous mixture, by contrast, there is no difference between one area of the mixture and another. If coffee has been brewed properly, and there are no grounds at the bottom of the pot, it is homogeneous. If sweetener is added to tea while it is hot, it too should yield a homogeneous mixture.

WATER AND ETHANOL ARE COMPLETELY MISCIBLE, WHETHER THE SOLUTION BE 99% WATER AND 1% ETHANOL, 50% OF BOTH, OR 99% ETHANOL AND 1% WATER. SHOWN HERE IS A GLASS OF BEER—AN EXAMPLE OF A WATER-ETHANOL SOLUTION. *(Owen Franken/Corbis. Reproduced by permission.)*

SOLUTIONS AND SOLUBILITY

Virtually all varieties of homogeneous mixtures can be classified as a solution—that is, a homogeneous mixture in which one or more substances is dissolved in another substance. If two or more substances were combined in exactly equal proportions, this would technically not be a solution; hence the distinction between solutions and the larger class of homogeneous mixtures.

A solution is made up of two parts: the solute, the substance or substances that are dissolved; and the solvent, the substance that dissolves the solute. The solvent typically determines the physical state of the solution: in other words, if the solvent is a liquid, the solution will most likely be a liquid. Likewise if the solvent is a solid such as a metal, melted at high temperatures to form a metallic solution called an alloy, then when the solution is cooled to its normal temperature, it will be solid again.

MISCIBILITY AND SOLUBILITY. Miscibility is a term describing the relative ability of two substances to dissolve in one another. Generally, water and water-based substances have high miscibility with regard to one another, as do oil and oil-based substances. Oil and water, on the other hand (or substances based in either) have very low miscibility. The reasons for this will be discussed below.

Miscibility is a qualitative term like "fast" or "slow"; on the other hand, solubility—the maximum amount of a substance that dissolves in a given amount of solvent at a specific temperature—can be either qualitative or quantitative. In a general sense, the word is qualitative, referring to the property of being soluble, or able to dissolve in a solution.

At some points in this essay, "solubility" will be used in this qualitative sense; however, within the realm of chemistry, it also has a quantitative meaning. In other words, instead of involving an imprecise quality such as "fast" or "slow," solubility in this sense relates to precise quantities more along the lines of "100 MPH (160.9 km/h)." Solubility is usually expressed in grams (g) (0.0022 lb) of solute per 100 g (0.22 lb) of solvent.

OTHER QUANTITATIVE TERMS. Another quantitative means of describing a solution is in terms of mass percent: the mass of the solute divided by the mass of the solution, and multiplied by 100%. If there are 25 g of solute in a solution of 200 g, for instance, 25 would be divided by 200 and multiplied by 100% to yield a 12.5% figure of solution composition.

Molarity also provides a quantitative means of showing the concentration of solute to solution. Whereas mass percent is, as its name indicates, a comparison of the mass of the solute to that of the solvent, molarity shows the amount of solute in a given volume of solution. This is measured in moles of solute per liter of solution, abbreviated mol/l.

REAL-LIFE APPLICATIONS

SATURATION AND DILUTION

The quantitative terms for describing solubility that we have reviewed are useful to a professional chemist, or to anyone performing work in a laboratory. For the most part, however, qualitative expressions will suffice for the present discussion. Among the most useful qualitative terms is saturation.

When it is possible to dissolve solute in a solvent, the solution is said to be unsaturated—rather like a sponge that has not been filled to capacity with liquid. By contrast, a solution that contains as much solute as it can at a given temperature is like a sponge that has been filled with all the liquid it can possibly contain. It can no longer absorb more of the liquid; rather, it can only push liquid particles along, and the sponge is said to be saturated. In the same way, if tea is hot, it is easier to introduce sugar to it, but when the temperature is low, sugar will not dissolve as easily.

SATURATION AND TEMPERATURE.

With tea and many other substances, higher temperatures mean that a greater amount of solute can be added before the solution is fully saturated. The reason for this relationship between saturation and temperature, according to generally accepted theory, is that heated solvent particles move more quickly than cold ones, and as a result, create more space into which the solvent can fit. Indeed, "space" is a prerequisite for a solution: the molecules of solute need to find a "hole" between the molecules of solvent into which they can fit. Thus the molecules in a solution can be compared to a packed crowd: if a crowd is suddenly dispersed, it is easier to walk through it.

Not all substances respond to rises in temperature in the manner described, however. As a rule, gases (with the exception of helium) are more soluble at lower temperatures than at higher ones. Higher temperatures mean an increase in kinetic energy, with more molecules colliding with one another, and this makes it easier for the gases to escape the liquid in which they are dissolved.

Carbonated soft drinks get their "fizz" from carbon dioxide gas dissolved, along with sugar and other flavorings, in a solution of water. When the soft-drink container is opened, much of the carbon dioxide quickly departs, while a certain proportion stays dissolved in the cola. The remaining carbon dioxide will eventually escape as well, but it will do so more quickly if left in a warm environment rather than a cold one, such as the inside of a refrigerator.

DILUTION AND CONCENTRATION.

Another useful set of qualitative terms for solutions also relates to the relative amount of solute that has been dissolved. But whereas saturation is an absolute condition at a certain temperature (in other words, the solution is definitely saturated at such-and-such a temperature), concentration and dilution are much more relative terms.

When a solution contains a relatively small amount of solute, it is said to be dilute; on the other hand, a solution with a relatively large amount of solute is said to be concentrated. For instance, hydrogen peroxide (H_2O_2) as sold over the counter in drug stores, to be used in homes as a disinfectant of bleaching agent, is a dilute 3% solution in water. In higher concentrations, hydrogen peroxide can cause burns to human skin, and is used to power rockets.

Chemists often keep on hand in their laboratories concentrated versions of various frequently used solutions. Maintaining them in concentrated form saves space; when more of a solution is needed, the chemist dilutes it in water according to desired molarity values. To do so, the chemist determines how much water needs to be added to a particular "stock solution" (as these concentrated solutions in a laboratory are called) to create a solution of a particular concentration. In the dilution with water, the number of moles of the solute does not change; rather, the particles of solute are dispersed over a larger volume.

"LIKE DISSOLVES LIKE"

In the molecules of a compound, the atoms are held together by powerful attractions. The molecules within a compound, however, also have an intermolecular attraction, but this is much weaker than the interatomic bonds in a molecule. It is thus quite possible for the molecules in a solution to exert a greater attractive force than the one holding the molecules of solute together with one another. When this happens, the solute dissolves.

A useful rule of thumb when studying solutions is "like dissolves like." In other words, water dissolves water-based solutes, whereas oil dissolves oily substances. Or to put it another way, water and water-based substances are miscible with one another, as are oil and oil-based substances. Water and oil together, however, are largely immiscible.

POLAR WATER, NONPOLAR OIL. Water and oil and immiscible due to their respective molecule structures, and hence their inherent characteristics of intermolecular bonding. Water molecules are polar, meaning that the positive electric charge is at one end of the molecule, while the negative charge is at the other end. Oil, on the other hand, is nonpolar—charges are more evenly distributed throughout the molecule.

The oxygen atom in a water molecule has a much greater electronegativity than the hydrogen atoms, so it pulls negatively charged electrons to its end of the molecule. Conversely, most oils are composed of carbon and hydrogen, which have similar values of electronegativity. As a result, positive and negative charges are distributed evenly throughout the oil molecule. Whereas a water molecule is like a magnet with a north and south pole, an oil molecule is like a nonmagnetic metal. Thus, as most people know, "oil and water don't mix."

The immiscible quality of oil and water in relation to one another explains a number of phenomena from the everyday world. It is easy enough to wash mud from your hands under running water, because the water and the mud (which, of course, contains water) are highly miscible. But motor oil or oil-based paint will leave a film on the hands, as these substances bond to oily molecules in the skin itself, and no amount of water will wash the film off. To remove it, one needs mineral spirits, paint thinner, or soap manufactured to bond with the oils.

Everyone has had the experience of eating spicy foods and then trying to cool down the mouth with cold water—and just about everyone has discovered that this does not work. This is because most spicy substances contain emulsified oils that coat the tongue, and these oils are not attracted to water. A much better "solution" (in more ways than one) is milk, and in particular, whole milk. Though a great deal of milk is composed of water, it also contains tiny droplets of fats and proteins that join with the oils in spicy substances, thus cooling down the mouth.

CROSSING THE OIL/WATER BARRIER

The active ingredient in a dry-cleaning chemical usually has nonpolar molecules, because the toughest stains are made by oils and greases. But what about ordinary soap, which we use in water to wash both water- and oil-based substances from our bodies? Though, as noted earlier, certain kinds of oil stains require special cleaning solutions, ordinary soap is fine for washing off the natural oils secreted by the skin. Likewise, it can wash off small concentrations of oil that come from the environment and attach to the skin—for instance, from working over a deep fryer in a fast-food restaurant.

How does the soap manage to "connect" both with oils and water? Likewise, how does milk—which is water-soluble and capable of dilution in water—wash away the oils associated with spicy food? To answer these questions, we need to briefly consider a subject examined in more depth within the Mixtures essay: emulsions, or mixtures of two immiscible liquids.

EMULSIFIERS AND SURFACTANTS. The dispersion of two substances in an emulsion is achieved through the use of an emulsifier or surfactant. Made up of molecules that are both water- and oil-soluble, an emulsifier or surfactant acts as an agent for joining other substances in an emulsion. The two words are virtually synonymous, but "emulsifier" is used typically in reference to foods, whereas "surfactant" most often refers to an ingredient in detergents and related products.

In an emulsion, millions of surfactants surround the dispersed droplets of solute, known as the internal phase, shielding them from the solvent, or external phase. Surfactants themselves are often used in laundry or dish detergent, because most stains on plates or clothes are oil-based, whereas the detergent itself is applied to the clothes in a water-based external phase. The emulsifiers in milk help to bond oily particles of milk fat (cream) and protein to the external phase of water that comprises the majority of milk's volume.

As for soap, it is a mixture of an acid and a base—specifically, carboxylic acids joined with a base such as sodium hydroxide. Carboxylic acids, chains of hydrogen and carbon atoms, are just some of many hydrocarbons that form the chemical backbone of a vast array of organic substances. Because it is a salt, meaning that it is formed from the reaction of an acid with a base, soap partially separates into its component ions in water. The active ion is RCOO- (R stands for the hydrocarbon chain), whose two ends behave in different fashions, making it a surfactant.

The hydrocarbon end (R-) is said to be lipophilic, or "oil-loving"; on the other hand, the carboxylate end (-COO-) is hydrophilic, or "water-loving." As a result, soap can dissolve in water, but can also clean greasy stains. When soap is mixed with water, it does not form a true solution, due to the presence of the hydrocarbons, which attract one another to form spherical aggregates called micelles. The lipophilic "tails" of the hydrocarbons are turned toward the interior of the micelle, while the hydrophilic "heads" remain facing toward the water that forms the external phase.

ETHANOL. When a hydrocarbon joins with other substances in one of the hydrocarbon functional groups, these form numerous hydrocarbon derivatives. Among the hydrocarbon derivatives is alcohol, and within this grouping is ethyl alcohol or ethanol, which includes a hydrocarbon bonded to the –OH functional group. The formula for ethanol is C_2H_6O, though this is sometimes rendered as C_2H_5OH to show that the alcohol functional group (–OH) is bonded to the hydroxyl radical (C_2H_5).

Ethanol, the same alcohol found in alcoholic beverages, is obviously miscible with water. Beer, after all, is mostly water, and if ethanol and water were not miscible, it would be impossible to mix alcohol with water or a water-based substance to make Scotch and water or a Bloody Mary (vodka and tomato juice.) In fact, water and ethanol are completely miscible, whether the solution be 99% water and 1% ethanol, 50% of both, or 99% ethanol and 1% water.

How can this be, given the fact that ethanol is a hydrocarbon derivative—in other words, a cousin of petroleum? The addition of the term "derivative" gives us a clue: note that ethanol contains oxygen, just as water does. As in water, the oxygen and hydrogen form a polar bond, giving that portion of the ethanol molecule a high affinity for water. Molecules of water and the alcohol functional group are joined through an intermolecular force called hydrogen bonding.

AQUEOUS SOLUTIONS

The term aqueous refers to water, and because water has extraordinary solvent qualities (discussed in the Osmosis essay) it serves as the medium for numerous solutions. Furthermore, reactions in aqueous solutions—though we can only touch on them briefly here—constitute a large and significant body of reactions studied by chemists.

Most of the solutions we have discussed up to this point are aqueous. We have referred, at least in an external or phenomenological way, to the means by which sugar dissolves in an aqueous solution, such as tea. Now let us consider, from a molecular or structural standpoint, the means by which salt does so as well.

SALTWATER. Salt is formed of positively charged sodium ions, firmly bonded to negatively charged chlorine ions. To break these ionic bonds and dissolve the sodium chloride, water must exert a strong attraction. However, salt is not really composed of molecules but simply repeating series of sodium and chlorine atoms joined together in a simple face-centered cubic lattice which has each each sodium ion (Na^+) surrounded by six chlorine ions (Cl^-); at the same time, each chlorine ion is surrounded by six sodium ions.

In dissolving the salt, water molecules surround ions that make up the salt, holding them in place through the electrical attraction of opposite charges. Due to the differences in electronegativity that give the water molecule its high polarity, the hydrogen atoms are more positively charged, and thus these attract the negatively charged chlorine ion. Similarly, the negatively charged oxygen end of the water molecule attracts the positively charged sodium ion. As a result, the salt is "surrounded" or dissolved.

PLASMA AND AQUEOUS-SOLUTION REACTIONS IN THE BODY. Whereas saltwater is an aqueous solution that can eventually kill someone who drinks it, plasma is an essential component of the life process—and in fact, it contains a small quantity of sodium chloride. Not to be confused with the phase of matter also called plasma, this plasma is the liquid portion of blood. Blood itself is about 55% plasma, with red and white blood cells and platelets suspended in it.

Plasma is in turn approximately 90% water, with a variety of other substances suspended or dissolved in it. Prominent among these substances are proteins, of which there are about 60 in plasma, and these serve numerous important functions—for instance, as antibodies. Plasma

KEY TERMS

AQUEOUS: A term describing a solution in which a substance or substances are dissolved in water.

COMPOUND: A substance made up of atoms of more than one element, which are chemically bonded and usually joined in molecules. The composition of a compound is always the same, unless it is changed chemically.

CONCENTRATED: A qualitative term describing a solution with a relatively large ratio of solute to solvent.

DILUTE: A qualitative term describing a solution with a relatively small ratio of solute to solvent.

EMULSIFIER: A surfactant. (Usually the term "emulsifier" is applied to foods, and "surfactant" to detergents and other inedible products.)

EMULSION: A mixture of two immiscible liquids, such as oil and water, made by dispersing microscopic droplets of one liquid in another. In creating an emulsion, surfactants act as bridges between the two liquids.

HETEROGENEOUS: A term describing a mixture that is not the same through-

out; rather, it has various regions possessing different properties. One example of this is sand in a container of water. Rather than dissolving to form a homogeneous mixture as sugar does, the sand sinks to the bottom.

HOMOGENEOUS: A term describing a mixture that is the same throughout, as for example when sugar is fully dissolved in water. A solution is a homogenous mixture.

IMMISCIBLE: Possessing a negligible value of miscibility.

MASS PERCENT: A quantitative means of measuring solubility in terms of the mass of the solute divided by the mass of the solution, multiplied by 100%.

MISCIBILITY: A term identifying the relative ability of two substances to dissolve in one another. Miscibility is qualitative, like "fast" or "slow"; solubility, on the other hand, can be a quantitative term.

MIXTURE: A substance with a variable composition, composed of molecules or atoms of differing types. Compare with pure substance.

---

also contains ions, which prevent red blood cells from taking up excess water in osmosis. Prominent among these ions are those of salt, or sodium chloride. Plasma also transports nutrients such as amino acids, glucose, and fatty acids, as well as waste products such as urea and uric acid, which it passes on to the kidneys for excretion.

Much of the activity that sustains life in the body of a living organism can be characterized as a chemical reaction in an aqueous solution. When we breathe in oxygen, it is taken to the lungs and fed into the bloodstream, where it

associates with the iron-containing hemoglobin in red blood cells and is transported to the organs. In the stomach, various aqueous-solution reactions process food, turning part of it into fuel that the blood carries to the cells, where the oxygen engages in complex reactions with nutrients.

A MISCELLANY OF SOLUTIONS

Aqueous-solution reactions can lead to the formation of a solid, as when a solution of potassi-

KEY TERMS CONTINUED

MOLARITY: A quantitative means of showing the concentration of solute to solution. This is measured in moles of solute per liter of solution, abbreviated mol/L.

PURE SUBSTANCE: A substance—either an element or compound—that has an unvarying composition. This means that by changing the proportions of atoms, the result would be an entirely different substance. Compare with mixture.

QUALITATIVE: Involving a comparison between qualities that are not precisely defined, such as "fast" and "slow" or "warm" and "cold."

QUANTITATIVE: Involving a comparison between precise quantities—for instance, 10 lb vs. 100 lb, or 50 MPH vs. 120 MPH.

SATURATED: A qualitative term describing a solution that contains as much solute as it can dissolve at a given temperature.

SOLUBILITY: In a broad, qualitative sense, solubility refers to the property of being soluble. Chemists, however, usually

apply the word in a quantitative sense, to indicate the maximum amount of a substance that dissolves in a given amount of solvent at a specific temperature. Solubility is usually expressed in grams of solute per 100 g of solvent.

SOLUBLE: Capable of dissolving in a solvent.

SOLUTE: The substance or substances that are dissolved in a solvent to form a solution.

SOLUTION: A homogeneous mixture in which one or more substances (the solute) is dissolved in a solvent—for example, sugar dissolved in water.

SOLVENT: The substance or substances that dissolve a solute to form a solution.

SURFACTANT: A substance made up of molecules that are both water- and oil-soluble, which acts as an agent for joining other substances in an emulsion.

UNSATURATED: A term describing a solution that is capable of dissolving additional solute, if that solute is introduced to it.

um chromate (K_2CrO_4) is added to an aqueous solution of barium nitrate ($Ba[NO_3]_2$ to form solid barium chromate ($BaCrO_4$) and a solution of potassium nitrate (KNO_3). This reaction is described in the essay on Chemical Reactions.

We have primarily discussed liquid solutions, and in particular aqueous solutions. It should be stressed, however, that solutions can also exist in the gaseous or solid phases. The air we breathe is a solution, not a compound: in other words, there is no such thing as an "air molecule." Instead, it is made up of diatomic ele-

ments (those in which two atoms join to form a molecule of a single element); monatomic elements (those elements that exist as single atoms); one element in a triatomic molecule; and two compounds.

The "solvent" in air is nitrogen, a diatomic element that accounts for 78% of Earth's atmosphere. Oxygen, also diatomic, constitutes an additional 21%. Argon, which like all noble gases is monatomic, ranks a distant third, with 0.93%. The remaining 0.07% is made up of traces of other noble gases; the two compounds mentioned, carbon dioxide and water (in vapor form); and, high in the atmosphere, the triatomic form of oxygen known as ozone (O_3).

The most significant solid solutions are alloys of metals, discussed in the essay on Mixtures, as well as in essays on various metal families, particularly the Transition Metals. Some well-known alloys include bronze (three-quarters copper, one-quarter tin); brass (two-thirds copper, one-third zinc); pewter (a mixture of tin and copper with traces of antimony); and numerous alloys of iron—particularly steel—as well as alloys involving other metals.

WHERE TO LEARN MORE

"*Aqueous Solutions*" (Web site). <http://www.tannerm. com/aqueous.htm> (June 6, 2001).

Gibson, Gary. *Making Things Change.* Brookfield, CT: Copper Beech Books, 1995.

Hauser, Jill Frankel. *Super Science Concoctions: 50 Mysterious Mixtures for Fabulous Fun.* Charlotte, VT: Williamson Publishing, 1996.

"*Introduction to Solutions*" (Web site). <http://edie. cprost.sfu.ca/~rhlogan/solintro.html> (June 6, 2001).

Knapp, Brian J. *Air and Water Chemistry.* Illustrated by David Woodroffe. Danbury, CT: Grolier Educational, 1998.

Seller, Mick. *Elements, Mixtures, and Reactions.* New York: Shooting Star Press, 1995.

"*Solubility.*" *Spark Notes* (Web site). <http://www. sparknotes.com/chemistry/solutions/solubility/terms. html> (June 6, 2001).

"*Solutions and Solubility*" (Web site). <http://educ. queensu.ca/~science/main/concept/chem/c10/c101a mo1.htm> (June 6, 2001).

Watson, Philip. *Liquid Magic.* Illustrated by Elizabeth Wood and Ronald Fenton. New York: Lothrop, Lee, and Shepard Books, 1982.

Zumdahl, Steven S. *Introductory Chemistry: A Foundation,* 4th ed. Boston: Houghton Mifflin, 2000.

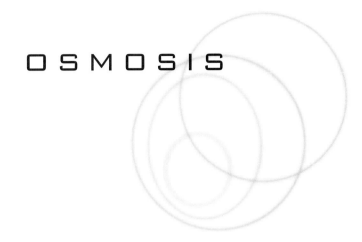

OSMOSIS

CONCEPT

The term osmosis describes the movement of a solvent through a semipermeable membrane from a less concentrated solution to a more concentrated one. Water is sometimes called "the perfect solvent," and living tissue (for example, a human being's cell walls) is the best example of a semipermeable membrane. Osmosis has a number of life-preserving functions: it assists plants in receiving water, it helps in the preservation of fruit and meat, and is even used in kidney dialysis. In addition, osmosis can be reversed to remove salt and other impurities from water.

HOW IT WORKS

If you were to insert a hollow tube of a certain diameter into a beaker of water, the water would rise inside the tube and reach the same level as the water outside it. But suppose you sealed the bottom end of the tube with a semipermeable membrane, then half-filled the tube with salt water and again inserted it into the beaker. Over a period of time, the relative levels of the salt water in the tube and the regular water in the beaker would change, with the fresh water gradually rising into the beaker.

This is osmosis at work; however, before investigating the process, it is necessary to understand at least three terms. A solvent is a liquid capable of dissolving or dispersing one or more other substances. A solute is the substance that is dissolved, and a solution is the resulting mixture of solvent and solute. Hence, when you mix a packet of sugar into a cup of hot coffee, the coffee—which is mostly water—acts as a solvent for the sugar, a solute, and the resulting sweetened coffee is a solution. (Indeed, people who need a cup of coffee in the morning might say that it is a "solution" in more ways than one!) The relative amount of solute in the solution determines whether it can be described as more or less concentrated.

WATER AND OIL: MOLECULAR DIFFERENCES

In the illustrations involving the beaker and the hollow tube, water plays one of its leading roles, as a solvent. It is possible to use a number of other solvents for osmosis, but most of the ones that will be discussed here are water-based substances. In fact, virtually everything people drink is either made with water as its central component (soft drinks, coffee, tea, beer and spirits), or comes from a water-based plant or animal life form (fruit juices, wine, milk.) Then of course there is water itself, still the world's most popular drink.

By contrast, people are likely to drink an oily product only in extreme circumstances: for instance, to relieve constipation, holistic-health practitioners often recommend a mixture of olive oil and other compounds for this purpose. Oil, unlike water, has a tendency to pass straight through a person's system, without large amounts of it being absorbed through osmosis. In fact, oil and water differ significantly at the molecular level.

Water is the best example of a polar molecule, sometimes called a dipole. As everyone knows, water is a name for the chemical H_2O, in which two relatively small hydrogen atoms bond with a large atom of oxygen. You can visualize a water molecule by imagining oxygen as a basketball with hydrogen as two baseballs fused to the

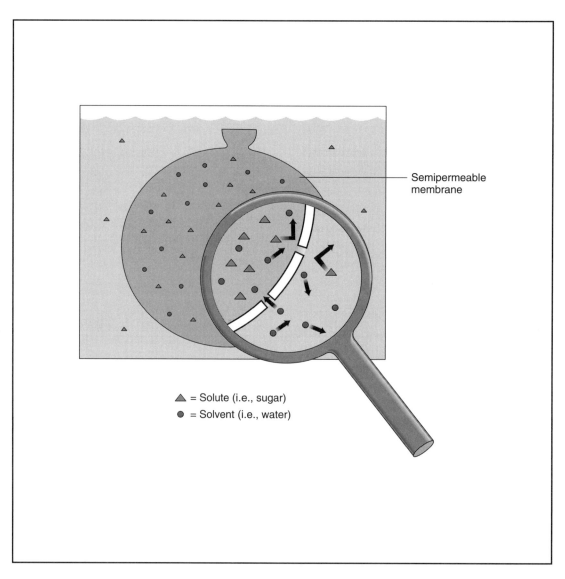

A SEMIPERMEABLE MEMBRANE PERMITS ONLY CERTAIN COMPONENTS OF SOLUTIONS, USUALLY THE SOLVENT, TO PASS THROUGH.

basketball's surface. Bonded together as they are, the oxygen tends to pull electrons from the hydrogen atoms, giving it a slight negative charge and the hydrogen a slight positive charge.

As a result, one end of a water molecule has a positive electrical charge, and the other end a negative charge. This in turn causes the positive end of one molecule to attract the negative side of its neighbor, and vice versa. Though the electromagnetic force is weak, even in relative terms, it is enough to bond water molecules tightly to one another.

By contrast, oily substances—whether the oil is animal-, vegetable-, or petroleum-based—are typically nonpolar, meaning that the positive and negative charges are distributed evenly

across the surface of the molecule. Hence, the bond between oil molecules is much less tight than for water molecules. Clean motor oil in a car's crankshaft behaves as though it were made of millions of tiny ball-bearings, each rolling through the engine without sticking. Water, on the other hand, has a tendency to stick to surfaces, since its molecules are so tightly bonded to one another.

This tight bond gives water highly unusual properties compared to other substances close to its molecular weight. Among these are its high boiling point, its surprisingly low density when frozen, and the characteristics that make osmosis possible. Thanks to its intermolecular structure, water is not only an ideal solvent, but its closely

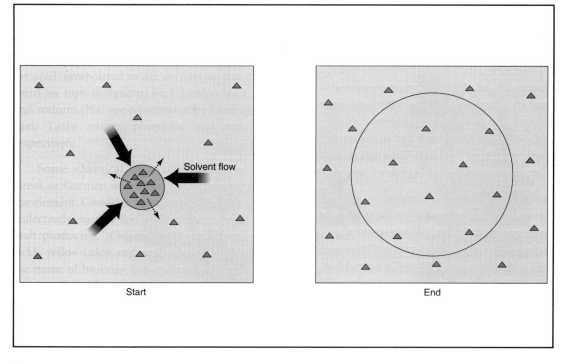

Solvent flow

Start

End

OSMOSIS EQUALIZES CONCENTRATION.

packed structure enables easy movement, as, for instance, from an area of low concentration to an area of high concentration.

In the beaker illustration, the "pure" water is almost pure solvent. (Actually, because of its solvent qualities, water seldom appears in a pure state unless one distills it: even water flowing through a "pure" mountain stream carries all sorts of impurities, including microscopic particles of the rocks over which it flows.) In any case, the fact that the water in the beaker is almost pure makes it easy for it to flow through the semipermeable membrane in the bottom of the tube. By contrast, the solute particles in the salt-water solution have a much harder time passing through, and are much more likely to block the openings in the membrane. As a result, the movement is all in one direction: water in the beaker moves through the membrane, and into the tube.

A few points of clarification are in order here. A semipermeable membrane is anything with a structure somewhere between that of, say, plastic on the one hand and cotton on the other. Were the tube in the beaker covered with Saran wrap, for instance, no water would pass through. On the other hand, if one used a piece of cotton in the bottom of the tube, the water would pass straight through without osmosis taking place. In contrast to cotton, Gore-tex is a fabric containing a very thin layer of plastic with billions of tiny pores which let water vapor flow out without allowing liquid water to seep in. This accounts for the popularity of Gore-tex for outdoor gear—it keeps a person dry without holding in their sweat. So Gore-tex would work well as a semipermeable membrane.

Also, it is important to consider the possibilities of what can happen during the process of osmosis. If the tube were filled with pure salt, or salt with only a little water in it, osmosis would reach a point and then stop due to osmotic pressure within the substance. Osmotic pressure results when a relatively concentrated substance takes in a solvent, thus increasing its pressure until it reaches a point at which the solution will not allow any more solvent to enter.

REAL-LIFE APPLICATIONS

CELL BEHAVIOR AND SALT WATER

Cells in the human body and in the bodies of all living things behave like microscopic bags of solution housed in a semipermeable membrane. The health and indeed the very survival of a person, animal, or plant depends on the ability of

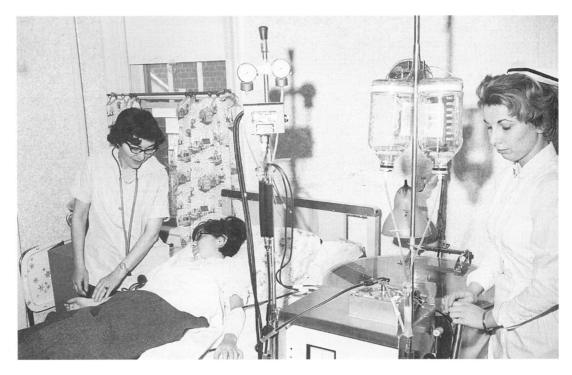

IN THIS 1966 PHOTO, A GIRL IS UNDERGOING KIDNEY DIALYSIS, A CRUCIAL MODERN MEDICAL APPLICATION THAT RELIES ON OSMOSIS. (*Bettmann/Corbis. Reproduced by permission.*)

the cells to maintain their concentration of solutes.

Two illustrations involving salt water demonstrate how osmosis can produce disastrous effects in living things. If you put a carrot in salty water, the salt water will "draw" the water from inside the carrot—which, like the human body and most other forms of life, is mostly made up of water. Within a few hours, the carrot will be limp, its cells shriveled.

Worse still is the process that occurs when a person drinks salt water. The body can handle a little bit, but if you were to consume nothing but salt water for a period of a few days, as in the case of being stranded on the proverbial desert island, the osmotic pressure would begin drawing water from other parts of your body. Since a human body ranges from 60% water (in an adult male) to 85% in a baby, there would be a great deal of water available—but just as clearly, water is the essential ingredient in the human body. If you continued to ingest salt water, you would eventually experience dehydration and die.

How, then, do fish and other forms of marine life survive in a salt-water environment? In most cases, a creature whose natural habitat is the ocean has a much higher solute concentration in its cells than does a land animal. Hence,

for them, salt water is an isotonic solution, or one that has the same concentration of solute—and hence the same osmotic pressure—as in their own cells.

OSMOSIS IN PLANTS

Plants depend on osmosis to move water from their roots to their leaves. The further toward the edge or the top of the plant, the greater the solute concentration, which creates a difference in osmotic pressure. This is known as osmotic potential, which draws water upward. In addition, osmosis protects leaves against losing water through evaporation.

Crucial to the operation of osmosis in plants are "guard cells," specialized cells dispersed along the surface of the leaves. Each pair of guard cells surrounds a stoma, or pore, controlling its ability to open and thus release moisture.

In some situations, external stimuli such as sunlight may cause the guard cells to draw in potassium from other cells. This leads to an increase in osmotic potential: the guard cell becomes like a person who has eaten a dry biscuit, and is now desperate for a drink of water to wash it down. As a result of its increased osmotic potential, the guard cell eventually takes on water through osmosis. The guard cells then swell with water, opening the stomata and increasing the rate of gas exchange through them. The outcome of this action is an increase in the rate of photosynthesis and plant growth.

When there is a water shortage, however, other cells transmit signals to the guard cells that cause them to release their potassium. This decreases their osmotic potential, and water passes out of the guard cells to the thirsty cells around them. At the same time, the resultant shrinkage in the guard cells closes the stomata, decreasing the rate at which water transpires through them and preventing the plant from wilting.

OSMOSIS AND MEDICINE

Osmosis has several implications where medical care is concerned, particularly in the case of the storage of vitally important red blood cells. These are normally kept in a plasma solution which is isotonic to the cells when it contains specific proportions of salts and proteins. However, if red blood cells are placed in a hypotonic solution, or one with a lower solute concentration than in the cells themselves, this can be highly detrimental.

Hence water, a life-giving and life-preserving substance in most cases, is a killer in this context. If red blood cells were stored in pure water, osmosis would draw the water into the cells, causing them to swell and eventually burst. Similarly, if the cells were placed in a solution with a higher solute concentration, or hypertonic solution, osmosis would draw water out of the cells until they shriveled.

In fact, the plasma solution used by most hospitals for storing red blood cells is slightly hypertonic relative to the cells, to prevent them from drawing in water and bursting. Physicians use a similar solution when injecting a drug intravenously into a patient. The active ingredient of the drug has to be suspended in some kind of medium, but water would be detrimental for the reasons discussed above, so instead the doctor uses a saline solution that is slightly hypertonic to the patient's red blood cells.

One vital process closely linked to osmosis is dialysis, which is critical to the survival of many victims of kidney diseases. Dialysis is the process by which an artificial kidney machine removes waste products from a patients' blood—performing the role of a healthy, normally function-

ing kidney. The openings in the dialyzing membrane are such that not only water, but salts and other waste dissolved in the blood, pass through to a surrounding tank of distilled water. The red blood cells, on the other hand, are too large to enter the dialyzing membrane, so they return to the patient's body.

PRESERVING FRUITS AND MEATS

Osmosis is also used for preserving fruits and meats, though the process is quite different for the two. In the case of fruit, osmosis is used to dehydrate it, whereas in the preservation of meat, osmosis draws salt into it, thus preventing the intrusion of bacteria.

Most fruits are about 75% water, and this makes them highly susceptible to spoilage. To preserve fruit, it must be dehydrated, which—as in the case of the salt in the meat— presents bacteria with a less-than-hospitable environment. Over the years, people have tried a variety of methods for drying fruit, but most of these have a tendency to shrink and harden the fruit. The reason for this is that most drying methods, such as heat from the Sun, are relatively quick and drastic; osmosis, on the other hand, is slower, more moderate—and closer to the behavior of nature.

Osmotic dehydration techniques, in fact, result in fruit that can be stored longer than fruit dehydrated by other methods. This in turn makes it possible to provide consumers with a wider variety of fruit throughout the year. Also, the fruit itself tends to maintain more of its flavor and nutritional qualities while keeping out microorganisms.

Because osmosis alone can only remove about 50% of the water in most ripe fruits, however, the dehydration process involves a secondary method as well. First the fruit is blanched, or placed briefly in scalding water to stop enzymatic action. Next it is subjected to osmotic dehydration by dipping it in, or spreading it with, a specially made variety of syrup whose sugar draws out the water in the fruit. After this, air drying or vacuum drying completes the process. The resulting product is ready to eat; can be preserved on a shelf under most climatic conditions; and may even be powdered for making confectionery items.

Whereas osmotic dehydration of fruit is currently used in many parts of the world, the salt-

KEY TERMS

HYPERTONIC: Of higher osmotic pressure.

HYPOTONIC: Of lower osmotic pressure.

ISOTONIC: Of equal osmotic pressure.

OSMOSIS: The movement of a solvent through a semipermeable membrane from a less concentrated solution to a more concentrated one.

OSMOTIC POTENTIAL: A difference in osmotic pressure that draws water from an area of less osmotic pressure to an area of greater osmotic pressure.

OSMOTIC PRESSURE: The pressure that builds in a substance as it experiences osmosis, and eventually stops that process.

SOLVENT: A liquid capable of dissolving or dispersing one or more other substances.

SOLUTE: A substance capable of being dissolved in a solvent.

SOLUTION: A mixture of solvent and solute.

curing of meat in brine is largely a thing of the past, due to the introduction of refrigeration. Many poorer families, even in the industrialized world, however, remained without electricity long after it spread throughout most of Europe and North America. John Steinbeck's *Grapes of Wrath* (1939) offers a memorable scene in which a contemporary family, the Joads, kill and cure a pig before leaving Oklahoma for California. And a Web site for Walton Feed, an Idaho company specializing in dehydrated foods, offers reminiscences by Canadians whose families were still salt-curing meats in the middle of the twentieth century. Verla Cress of southern Alberta, for instance, offered a recipe from which the following details are drawn.

First a barrel is filled with a solution containing 2 gal (7.57 l) of hot water and 8 oz (.2268 kg) of salt, or 32 parts hot water to one part salt, as well as a small quantity of vinegar. The pig or cow, which would have just been slaughtered, should then be cut up into what Cress called "ham-sized pieces (about 10-15 lb [5-7 kg]) each." The pieces are then soaked in the brine barrel for six days, after which the meat is removed, dried, "and put... in flour or gunny sacks to keep the flies away. Then hang it up in a cool dry place to dry. It will keep like this for perhaps six weeks if stored in a cool place during the Summer. Of course, it will keep much longer in the Winter. If it goes bad, you'll know it!"

Cress offered another method, one still used on ham today. Instead of salt, sugar is used in a mixture of 32 oz (.94 l) to 3 gal (11.36 l) of water. After being removed, the meat is smoked—that is, exposed to smoke from a typically aromatic wood such as hickory, in an enclosed barn—for three days. Smoking the meat tends to make it last much longer: four months in the summer, according to Cress.

The Walton Feeds Web page included another brine-curing recipe, this one used by the women of the Stirling, Alberta, Church of the Latter-Day Saints in 1973. Also included were reminiscences by Glenn Adamson (born 1915): "...When we butchered a pig, Dad filled a wooden 45-gal (170.34 l) barrel with salt brine. We cut up the pig into maybe eight pieces and put it in the brine barrel. The pork soaked in the barrel for several days, then the meat was taken out, and the water was thrown away.... In the hot summer days after they [the pieces of meat] had dried, they were put in the root cellar to keep them cool. The meat was good for eating two or three months this way."

For thousands of years, people used salt to cure and preserve meat: for instance, the sailing ships that first came to the New World carried on board barrels full of cured meat, which fed sailors on the voyage over. Meat was not the only type of food preserved through the use of salt or brine, which is hypertonic—and thus lethal—to bacteria cells. Among other items packed in brine were fish, olives, and vegetables.

Even today, some types of canned fish come to the consumer still packed in brine, as do olives. Another method that survives is the use of sugar—which can be just as effective as salt for keeping out bacteria—to preserve fruit in jam.

REVERSE OSMOSIS

Given the many ways osmosis is used for preserving food, not to mention its many interactions with water, it should not be surprising to discover that osmosis can also be used for desalination, or turning salt water into drinking water. Actually, it is not osmosis, strictly speaking, but rather reverse osmosis that turns salt water from the ocean—97% of Earth's water supply—into water that can be used for bathing, agriculture, and in some cases even drinking.

When you mix a teaspoon of sugar into a cup of coffee, as mentioned in an earlier illustration, this is a non-reversible process. Short of some highly complicated undertaking—for instance, using ultrasonic sound waves—it would be impossible to separate solute and solvent.

Osmosis, on the other hand, can be reversed. This is done by using a controlled external pressure of approximately 60 atmospheres, an atmosphere being equal to the air pressure at sea level—14.7 pounds-per-square-inch (1.013×10^5 Pa.) In reverse osmosis, this pressure is applied to the area of higher solute concentration—in this case, the seawater. As a result, the pressure in the seawater pushes water molecules into a reservoir of pure water.

If performed by someone with a few rudimentary tools and a knowledge of how to provide just the right amount of pressure, it is possible that reverse osmosis could save the life of a shipwreck victim stranded in a location without a fresh water supply. On the other hand, a person in such a situation may be able to absorb sufficient water from fruits and plant life, as Tom Hanks's character did in the 2001 film *Cast Away*.

Companies such as Reverse Osmosis Systems in Atlanta, Georgia, offer a small unit for home or business use, which actually performs the reverse-osmosis process on a small scale. The unit makes use of a process called crossflow, which continually cleans the semipermeable membrane of impurities that have been removed from the water. A small pump provides the pressure necessary to push the water through the membrane. In addition to an under-the-sink model, a reverse osmosis water cooler is also available.

Not only is reverse osmosis used for making water safe, it is also applied to metals in a variety of capacities, not least of which is its use in treating wastewater from electroplating. But there are other metallurgical methods of reverse osmosis that have little to do with water treatment: metal finishing, as well as recycling of metals and chemicals. These processes are highly complicated, but they involve the same principle of removing impurities that governs reverse osmosis.

WHERE TO LEARN MORE

Francis, Frederick J., editor-in-chief. *Encyclopedia of Food Science and Technology.* New York: Wiley, 2000.

Gardner, Robert. *Science Project Ideas About Kitchen Chemistry.* Berkeley, N.J.: Enslow Publishers, 2002.

Laschish, Uri. *"Osmosis, Reverse Osmosis, and Osmotic Pressure: What They Are"* (Web site). <http://members.tripod.com/~urila/> (February 20, 2001).

"Lesson 5: Osmosis" (Web site). <http://www.biologylessons.sdsu.edu/classes/lab5/semnet/> (February 20, 2001).

Rosenfeld, Sam. *Science Experiments with Water.* Illustrated by John J. Floherty, Jr. Irvington-on-Hudson, NY: Harvey House, 1965.

"Salt-Curing Meat in Brine." Walton Feed (Web site). <http://waltonfeed.com/old/brine.html> (February 20, 2001).

DISTILLATION AND FILTRATION

CONCEPT

When most people think of chemistry, they think about joining substances together. Certainly, the bonding of elements to form compounds through chemical reactions is an integral component of the chemist's study; but chemists are also concerned with the separation of substances. Some forms of separation, in which compounds are returned to their elemental form, or in which atoms split off from molecules to yield a compound and a separated element, are complex phenomena that require chemical reactions. But chemists also use simpler physical methods, distillation and filtration, to separate mixtures. Distillation can be used to purify water, make alcohols, or separate petroleum into various components that range from natural gas to gasoline to tar. Filtration also makes possible the separation of gases or liquids from solids, and sewage treatment—a form of filtration—separates liquids, solids, and gases through a series of chemical and physical processes.

HOW IT WORKS

MIXTURES: HOMOGENEOUS AND HETEROGENEOUS

In contrast to a pure substance, a class that includes only elements and compounds, mixtures constitute a much broader category. Usually, a mixture consists of many compounds mixed together, but it is possible to have a mixture (air is an example) in which elemental substances are combined with compounds. Elements alone can be combined in a mixture, as when copper and zinc are alloyed to make brass. It is also possible to have a mixture of mixtures, as for example

when milk (a particular type of mixture called an emulsion) is added to coffee.

It is sometimes difficult, without studying the chemical aspects involved, to recognize the difference between a mixture and a pure substance. A mixture, however, can be defined as a substance with a variable composition, meaning that if more of one component is added, it does not change the essential character of the mixture. People unconsciously recognize this when they joke, "Would you like a little coffee with your milk?"

This comment, made when someone puts a great quantity of milk into a cup of coffee, implies that we generally regard coffee and milk as a solution in which the coffee dissolves the milk. But even if someone made a cup of coffee that was half milk, so that the resulting substance had a vanilla-like color, we would still call it coffee. If the components of a pure substance are altered, however, it becomes something entirely different.

Two hydrogen atoms bonded to an oxygen atom create water, but when two hydrogen atoms bond to two oxygen atoms, this is hydrogen peroxide, an altogether different substance. Water at ordinary temperatures does not bubble, whereas hydrogen peroxide does. Hydrogen peroxide boils at a temperature more than 1.5 times the boiling point of water, and potatoes boiled in hydrogen peroxide would be nothing like potatoes boiled in water. Eating them, in fact, could very well be fatal.

HOMOGENEOUS VS. HETEROGENEOUS. Unlike a compound or element, a mixture can never be broken down to a single type of molecule or atom, yet there are mixtures that appear to be the same throughout. These are called homogeneous mixtures, in which the properties and characteristics are the same throughout the mixture. In a heterogeneous mixture, on the other hand, the composition is not the same throughout, and in fact the mixture is separated into regions with varying properties.

Chemists of the past sometimes had difficulty distinguishing between homogeneous mixtures—virtually all of which can be described as solutions—and pure substances. Air, for example, appears to be a pure substance; yet there is no such thing as an "air molecule." Instead, air is composed primarily of nitrogen and oxygen, which appear in molecular rather than atomic form, along with separated atoms of noble gases and two important compounds: carbon dioxide and water. (Water exists in air as a vapor.)

There is less probability of confusing a heterogeneous mixture with a pure substance, because one of the defining aspects of a heterogeneous substance is the fact that it is clearly a mixture. When sand is added to a container of water, and the sand sinks to the bottom, it is obvious that there is no "sand-and-water" molecule pervading the entire mixture. Clearly, the upper portion of the container is mostly water, and the lower portion mostly sand.

SEPARATING MIXTURES

There are two basic processes for separating mixtures, distillation and filtration. In general, these are applied for the separation of homogeneous and heterogeneous mixtures, respectively. Distillation is the use of heat to separate the components of a liquid and/or gas, while filtration is the separation of solids from a fluid (either a gas or a liquid) by allowing the fluid to pass through a filter.

In a solution such as salt water, there are two components: the solvent (the water) and the solute (the salt). These can be separated by distillation in a laboratory using a burner placed under a beaker containing the salt water. As the water is heated, it passes out of the beaker in the form of steam, and travels through a tube cooled by a continual flow of cold water. Inside the tube, the steam condenses to form liquid water, which passes into a second beaker. Eventually, all of the solvent will be distilled from the first beaker, leaving behind the salt that constituted the solute of the original solution.

Distillation may seem like a chemical process, but in fact it is purely physical, because the composition of neither compound—salt or water—has been changed. As for filtration, it is clearly a physical process, just as heterogeneous mixtures are more obviously mixtures and not compounds. Suppose we wanted to separate the heterogeneous mixture we described earlier, of sand in water: all we would need would be a mesh screen through which the liquid could be strained, leaving behind the sand.

Despite the apparent simplicity of filtration, it can become rather complicated, as we shall see, in the treatment of sewage to turn it into water that poses no threat to the environment. Furthermore, it should be noted that sometimes these processes—distillation and filtration—are used in tandem. Suppose we had a mixture of sand and salt water taken from the beach, and we wanted to separate the mixture. The first step would involve the simpler process of filtration, separating the sand from the salt water. This would be followed by the separation of pure water and salt through the distillation process.

REAL-LIFE
APPLICATIONS

DISTILLATION

In distillation, the more volatile component of the mixture—that is, the part that is more easily vaporized—is separated from the less volatile portion. With regard to the illustration used above, of separating water from salt, clearly the water is the more volatile portion: its boiling point is much, much lower than that of salt, and the heat required to vaporize the salt is so great that the water would be long since vaporized by the time the salt was affected.

Once the volatile portion has been vaporized, or turned into distillate, it experiences condensation—the cooling of a gas to form a liquid, after which the liquid or condensate is collected. Note that in all these stages of distillation, heat is the agent of separation. This is especially important in the process of fractional distillation, discussed below.

The process of removing water vapor from salt water, described earlier, is an example of what is called "single-stage differential distillation." Sometimes, however—especially when a

A WATER FILTRATION PLANT IN CONTRA COSTA COUNTY, CALIFORNIA. *(James A. Sugar/Corbis. Reproduced by permission.)*

high level of purity in the resulting condensate is desired—it may be necessary to subject the condensate to multiple processes of distillation to remove additional impurities. This process is called rectification.

PURIFIED WATER AND ETHA-NOL. If distillation can be used to separate water from salt, obviously it can also be used to separate water from microbes and other impurities through a more detailed process of rectification. Distilled water, purchased for drinking and other purposes, is just one of the more common applications of the distillation process. Another well-known use of this process is the production of ethyl alcohol, made famous (or perhaps infamous) by stories of bootleggers producing

homemade whiskey from "stills" or distilleries in the woods.

Ethanol, the alcohol in alcoholic beverages, is produced by the fermentation of glucose, a sugar found in various natural substances. The choice of substances is a function of the desired product: grapes for wine; barley and hops for beer; juniper berries for gin; potatoes for vodka, and so on. Glucose reacts with yeast, fermenting to yield ethanol, and this reaction is catalyzed by enzymes in the yeast. The reaction continues until the alcohol content reaches a point equal to about 13%.

After that point, the yeast enzymes can no longer survive, and this is where the production of beer or wine usually ends. Fortified wine or

malt liquor—variations of wine and beer, respectively, that are more than 13% alcohol, or 26 proof—are exceptions. These two products are the result of mixtures between undistilled beer and wine and products of distillation having a higher alcohol content.

Distillation constitutes the second stage in the production of alcohol. Often, the fermentation products are subjected to a multistage rectification process, which yields a mixture of up to 95% ethanol and 5% water. This mixture is called an azeotrope, meaning that it will not change composition when distilled further. Usually the production of whiskey or other liquor stops well below the point of yielding an azeotrope, but in

some states, brands of grain alcohol at 190 proof (95%) are sold.

INDUSTRIAL DISTILLATION. In contrast to ethanol, methanol or "wood alcohol" is a highly toxic substance whose ingestion can lead to blindness or death. It is widely used in industry for applications in adhesives, plastics, and other products, and was once produced by the distillation of wood. In this old production method, wood was heated in the absence of air, such that it did not burn, but rather decomposed into a number of chemicals, a fraction of which were methanol and other alcohols.

The distillation of wood alcohol was abandoned for a number of reasons, not least of which was the environmental concern: thousands of trees had to be cut down for use in an inefficient process yielding only small portions of the desired product. Today, methanol is produced from synthesis gas, itself a product of coal gasification; nonetheless, numerous industrial processes use distillation.

Distillation is employed, for instance, to separate the hydrocarbons benzene and toluene, as well as acetone from acetic acid. In nuclear power plants that require quantities of the hydrogen isotope deuterium, the lighter protium isotope, or plain hydrogen, is separated from the heavier deuterium by means of distillation. Few industries, however, employ distillation to a greater degree than the petroleum industry.

Through a process known as fractional distillation, oil companies separate hydrocarbons from petroleum, allowing those of lower molecular mass to boil off and be collected first. For instance, at temperatures below 96.8°F (36°C), natural gas separates from petroleum. Other substances, including petroleum ether and naphtha, separate before the petroleum reaches the 156.2°F-165.2°F (69°C-74°C) range, at which point gasoline separates.

Fractional distillation continues, with the separation of other substances, all the way up to 959°F (515°C), above which tar becomes the last item to be separated. Petroleum and petroleum-product companies account for a large portion of the more than 40,000 distillation units in operation across the country. About 95% of all industrial separation processes involve distillation, which consumes large amounts of energy; for that reason, ever more efficient forms of distillation are continually being researched.

FILTRATION

In the simplest form of filtration, described earlier, a suspension (solid particles floating in a liquid) is allowed to pass through filter paper supported in a glass funnel. The filter paper traps the solid particles, allowing a clear solution (the filtrate) to pass through the funnel and into a receiving container.

There are two basic purposes for filtration: either to capture the solid material suspended in the fluid, or to clarify the fluid in which the solid is suspended. An example of the former occurs, for instance, when panning for gold: the water is allowed to pass through a sieve, leaving behind rocks that—the prospector hopes—contain gold.

An example of the latter occurs in sewage treatment. Here the solid left behind is feces—as undesirable as gold is desirable.

Filtration can further be classified as either gaseous or liquid, depending on which of the fluids constitutes the filtrate. Furthermore, the force that moves the fluid through the filter—whether gravity, a vacuum, or pressure—is another defining factor. The type of filter used can also serve to distinguish varieties of filtration.

LIQUID FILTRATION. In liquid filtration, such as that applied in sewage treatment, a liquid can be pulled through the filter by gravitational force, as in the examples given. On the other hand, some sort of applied pressure, or a pressure differential created by the existence of a vacuum, can force the liquid through the filter.

Water filtration for purification purposes is often performed by means of gravitation. Usually, water is allowed to run down through a thick layer of granular material such as sand, gravel, or charcoal, which removes impurities. These layers may be several feet thick, in which case the filter is known as a deep-bed filter. Water purification plants may include fine particles of charcoal, known as activated carbon, in the deep-bed filter to absorb unpleasant-smelling gases.

For separating smaller volumes of solution, a positive-pressure system—one that uses external pressure to push the liquid through the filter—may be used. Fluid may be introduced under pressure at one end of a horizontal tank, then forced through a series of vertical plates covered with thick filtering cloths that collect solids. In a vacuum filter, a vacuum (an area virtually devoid of matter, including air) is created beneath the solution to be filtered, and as a result, atmospheric pressure pushes the liquid through the filter.

When the liquid is virtually freed of impurities, it may be passed through a second variety of filter, known as a clarifying filter. This type of filter is constructed of extremely fine-mesh wires or fibers designed to remove the smallest particles suspended in the liquid. Clarifying filters may also use diatomaceous earth, a finely powdered material produced from the decay of marine organisms.

GAS FILTRATION. Gas filtration is used in a common appliance, the vacuum cleaner, which passes a stream of dust-filled air

KEY TERMS

AZEOTROPE: A solution that has been subjected through distillation and is still not fully separated, but cannot be separated any further by means of distillation. An example is ethanol, which cannot be distilled beyond the point at which it reaches a 95% concentration with water.

CONDENSATE: The liquid product that results from distillation.

CONDENSATION: The cooling of a gas to form a liquid or condensate.

DISTILLATE: The vapor collected from the volatile material or materials in distillation. This vapor is then subjected to condensation.

DISTILLATION: The use of heat to separate the components of a liquid and/or gas. Generally, distillation is used to separate mixtures that are homogeneous.

FILTRATE: The liquid or gas separated from a solid by means of filtration.

FILTRATION: The separation of solids from a fluid (either gas or liquid) by allowing the fluid to pass through a filter. Generally, filtration is used to separate mixtures that are heterogeneous.

HETEROGENEOUS: A term describing a mixture that is not the same throughout; rather, it has various regions possessing different properties. An example is sand in a container of water.

HOMOGENEOUS: A term describing a mixture that is the same throughout, as for example when sugar is fully dissolved in water. A solution is a homogenous mixture.

MIXTURE: A substance with a variable composition, composed of molecules or atoms of differing types. Compare with pure substance.

PURE SUBSTANCE: A substance—either an element or compound—that has an unvarying composition. This means that by changing the proportions of atoms, the result would be an entirely different substance. Compare with mixture.

SOLUTE: The substance or substances that are dissolved in a solvent to form a solution.

SOLUTION: A homogeneous mixture in which one or more substances is dissolved in an another substance—for example, sugar dissolved in water.

SOLVENT: The substance that dissolves a solute to form a solution.

SUSPENSION: A term that refers to a mixture in which solid particles are suspended in a fluid.

VOLATILE: Easily vaporized.

through a filtering bag inside the machine. The bag traps solid particles, while allowing clean air to pass back out into the room. This is essentially the same principle applied in air filters and even air conditioning and heating systems, which, in addition to regulating temperature, also remove dust, pollen, and other impurities from the air.

Various industries use gas filtration, not only to purify the products released into the atmosphere, but also to filter the air in the workplace, as for instance in a factory room where fine airborne particles are a hazard of production. The gases from power plants that burn coal and oil are often purified by passing them through filtering systems that collect particles.

Sewage treatment is a complex, multistage process whereby waste water is freed of harmful contents and rendered safe, so that it can be returned to the environment. This is a critical part of modern life, because when people congregate in cities, they generate huge amounts of wastes that contain disease-causing bacteria, viruses, and other microorganisms. Lack of proper waste management in ancient and medieval times led to devastating plagues in cities such as Athens, Rome, and Constantinople.

Indeed, one of the factors contributing to the end of the Athenian golden age of the fifth century B.C. was a plague, caused by lack of proper sewage-disposal methods, that felled many of the city's greatest figures. The horrors created by improper waste management reached their apogee in 1347-51, when microorganisms carried by rats brought about the Black Death, in which Europe's population was reduced by one-third.

AEROBIC AND ANAEROBIC DECAY. Small amounts of sewage can be dealt with through aerobic (oxygen-consuming) decay, in which microorganisms such as bacteria and fungi process the toxins in human waste. With larger amounts of waste, however, oxygen is depleted and the decay mechanism must be anaerobic, or carried out in the absence of oxygen.

For people who are not connected to a municipal sewage system, and therefore must rely on a septic tank for waste disposal, waste products pass through a tank in which they are subjected to anaerobic decay by varieties of microorganisms that do not require oxygen to survive. From there, the waste goes through the drain field, a network of pipes that allow the water to seep into a deep-bed filter. In the drain field, the waste is subjected to aerobic decay by oxygen-dependant microorganisms before it either filters through the drain pipes into the ground, or is evaporated.

MUNICIPAL WASTE TREATMENT. Municipal waste treatment is a much more involved process, and will only be described in the most general terms here. Essentially, large-scale sewage treatment requires the separation of larger particles first, followed by the separation of smaller particles, as the process continues. Along the way, the sewage is chemically treated as well. The separation of smaller solid particles is aided by aluminum sulfate, which causes these particles to settle to the bottom more rapidly.

Through continual processing by filtration, water is eventually clarified of biosolids (that is, feces), but it is still far from being safe. At that point, all the removed biosolids are dried and incinerated; or they may be subjected to additional processes to create fertilizer. The water, or effluent, is further clarified by filtration in a deep-bed filter, and additional air may be introduced to the water to facilitate aerobic decay—for instance, by using algae.

Through a long series of such processes, the water is rendered safe to be released back into the environment. However, wastes containing extremely high levels of toxins may require additional or different forms of treatment.

WHERE TO LEARN MORE

"Case Study: Petroleum Distillation" (Web site). <http://www.pafko.com/history/h_distill.html> (June 6, 2001).

"Classification of Matter" (Web site). <http://www.cards.anoka.k12.mn.us/projects/html>_98/smith/smith2.ht ml (June 6, 2001).

"Fractional Distillation" (Web site). <http://www.geocities.com/chemforum/fdistillation.htm> (June 6, 2001).

"Introduction to Distillation" (Web site). <http://lorien.ncl.ac.uk/ming/distil/distil0.htm> (June 6, 2001).

Kellert, Stephen B. and Matthew Black. The Encyclopedia of the Environment. New York: Franklin Watts, 1999.

Oxlade, Chris. Chemistry. Illustrated by Chris Fairclough. Austin, TX: Raintree Steck-Vaughn, 1999.

Seuling, Barbara. Drip! Drop! How Water Gets to Your Tap. Illustrated by Nancy Tobin. New York: Holiday House, 2000.

"Sewage Treatment" (Web site). <http://www.dcs.exeter.ac.uk/water/sewage.htm> (June 6, 2001).

Zumdahl, Steven S. Introductory Chemistry: A Foundation, 4th ed. Boston: Houghton Mifflin, 2000.

ORGANIC CHEMISTRY

ORGANIC CHEMISTRY

POLYMERS

ORGANIC CHEMISTRY

CONCEPT

There was once a time when chemists thought "organic" referred only to things that were living, and that life was the result of a spiritual "life force." While there is nothing wrong with viewing life as having a spiritual component, spiritual matters are simply outside the realm of science, and to mix up the two is as silly as using mathematics to explain love (or vice versa). In fact, the "life force" has a name: carbon, the common denominator in all living things. Not everything that has carbon is living, nor are all the areas studied in organic chemistry—the branch of chemistry devoted to the study of carbon and its compounds—always concerned with living things. Organic chemistry addresses an array of subjects as vast as the number of possible compounds that can be made by strings of carbon atoms. We can thank organic chemistry for much of what makes life easier in the modern age: fuel for cars, for instance, or the plastics found in many of the products used in an average day.

HOW IT WORKS

INTRODUCTION TO CARBON

As the element essential to all of life, and hence the basis for a vast field of study, carbon is addressed in its own essay. The Carbon essay, in addition to examining the chemical properties of carbon (discussed below), approaches a number of subjects, such as the allotropes of carbon. These include three crystalline forms (graphite, diamond, and buckminsterfullerene), as well as amorphous carbon. In addition, two oxides of carbon—carbon dioxide and carbon monoxide—are important, in the case of the former, to

the natural carbon cycle, and in the case of the latter, to industry. Both also pose environmental dangers.

The purpose of this summary of the carbon essay is to provide a hint of the complexities involved with this sixth element on the periodic table, the 14th most abundant element on Earth. In the human body, carbon is second only to oxygen in abundance, and accounts for 18% of the body's mass. Capable of combining in seemingly endless ways, carbon, along with hydrogen, is at the center of huge families of compounds. These are the hydrocarbons, present in deposits of fossil fuels: natural gas, petroleum, and coal.

A PROPENSITY FOR LIMIT-LESS BONDING. Carbon has a valence electron configuration of $2s^2 2p^2$. Likewise all the members of Group 4 on the periodic table (Group 14 in the IUPAC version of the table)—sometimes known as the "carbon family"—have configurations of $ns^2 np^2$, where n is the number of the period or row the element occupies on the table. There are two elements noted for their ability to form long strings of atoms and seemingly endless varieties of molecules: one is carbon, and the other is silicon, directly below it on the periodic table.

Just as carbon is at the center of a vast network of organic compounds, silicon holds the same function in the inorganic realm. It is found in virtually all types of rocks, except the calcium carbonates—which, as their name implies, contain carbon. In terms of known elemental mass, silicon is second only to oxygen in abundance on Earth. Silicon atoms are about one and a half times as large as those of carbon; thus not even silicon can compete with carbon's ability to form

"I JUST WANT TO SAY ONE WORD TO YOU...PLASTICS." SIGNIFYING THAT PLASTICS WERE THE WAVE OF THE FUTURE, THESE WORDS WERE UTTERED TO DUSTIN HOFFMAN'S CHARACTER IN THE 1967 FILM THE GRADUATE. (*The Kobal Collection. Reproduced by permission.*)

an almost limitless array of molecules in various shapes and sizes, and with various chemical properties.

ELECTRONEGATIVITY

Carbon is further distinguished by its high value of electronegativity, the relative ability of an atom to attract valence electrons. To mention a few basic aspects of chemical bonding, developed at considerably greater length in the Chemical Bonding essay, if two atoms have an electric charge and thus are ions, they form strong ionic bonds. Ionic bonding occurs when a metal bonds with a nonmetal. The other principal type of bond is a covalent bond, in which two uncharged atoms share eight valence electrons. If the electronegativity values of the two elements involved are equal, then they share the electrons equally; but if one element has a higher electronegativity value, the electrons will be more drawn to that element.

The electronegativity of carbon is certainly not the highest on the periodic table. That distinction belongs to fluorine, with an electronegativity value of 4.0, which makes it the most reactive of all elements. Fluorine is at the head of Group 7, the halogens, all of which are highly

reactive and most of which have high electronegativity values. If one ignores the noble gases, which are virtually unreactive and occupy the extreme right-hand side of the periodic table, electronegativity values are highest in the upper right-hand side of the table—the location of fluorine—and lowest in the lower left. In other words, the value increases with group or column number (again, leaving out the noble gases in Group 8), and decreases with period or row number.

With an electronegativity of 2.5, carbon ties with sulfur and iodine (a halogen) for sixth place, behind only fluorine; oxygen (3.5); nitrogen and chlorine (3.0); and bromine (2.8). Thus its electronegativity is high, without being too high. Fluorine is not likely to form the long chains for which is carbon is known, simply because its electronegativity is so high, it overpowers other elements with which it comes into contact. In addition, with four valence electrons, carbon is ideally suited to find other elements (or other carbon atoms) for forming covalent bonds according to the octet rule, whereby most elements bond so that they have eight valence electrons.

CARBON'S MULTIPLE BONDS

Carbon—with its four valence electrons—happens to be tetravalent, or capable of bonding to four other atoms at once. It is not necessarily the case that an element has the ability to bond with as many other elements as it has valence electrons; in fact, this is rarely the case. Additionally, carbon is capable of forming not only a single bond, with one pair of shared valence electrons, but a double bond (two pairs) or even a triple bond (three pairs.)

Another special property of carbon is its ability to bond in long chains that constitute strings of carbons and other atoms. Furthermore, though sometimes carbon forms a typical molecule (for example, carbon dioxide, or CO_2, is just one carbon atom with two oxygens), it is also capable of forming "molecules" that are really not molecules in the way that the word is typically used in chemistry. Graphite, for instance, is just a series of "sheets" of carbon atoms bonded tightly in a hexagonal, or six-sided, pattern, while a diamond is simply a huge "molecule" composed of carbon atoms strung together by covalent bonds.

ORGANIC CHEMISTRY

Organic chemistry is the study of carbon, its compounds, and their properties. The only major carbon compounds considered inorganic are carbonates (for instance, calcium carbonate, alluded to above, which is one of the major forms of mineral on Earth) and oxides, such as carbon dioxide and carbon monoxide. This leaves a huge array of compounds to be studied, as we shall see.

The term "organic" in everyday language connotes "living," but organic chemistry is involved with plenty of compounds not part of living organisms: petroleum, for instance, is an organic compound that ultimately comes from the decayed bodies of organisms that once were alive. It should be stressed that organic compounds do not have to be produced by living things, or even by things that once were alive; they can be produced artificially in a laboratory.

The breakthrough event in organic chemistry came in 1828, when German chemist Friedrich Wöhler (1800-1882) heated a sample of ammonium cyanate (NH_4OCN) and converted it to urea (H_2N-CO-NH_2). Ammonium cyanite is an inorganic material, whereas urea, a waste product in the urine of mammals, is an organic one. "Without benefit of a kidney, a bladder, or a

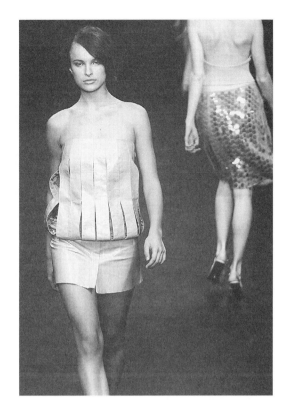

IT IS VIRTUALLY IMPOSSIBLE FOR A PERSON IN MODERN AMERICA TO SPEND AN ENTIRE DAY WITHOUT COMING INTO CONTACT WITH AT LEAST ONE, AND MORE LIKELY DOZENS, OF PLASTIC PRODUCTS, ALL MADE POSSIBLE BY ORGANIC CHEMISTRY. HERE, A FASHION MODEL PARADES A DRESS MADE OF PVC. *(Reuters NewMedia Inc./Corbis. Reproduced by permission.)*

dog," as Wöhler later said, he had managed to transform an inorganic substance into an organic one.

Ammonium cyanate and urea are isomers: substances having the same formula, but possessing different chemical properties. Thus they have exactly the same numbers and proportions of atoms, yet these atoms are arranged in different ways. In urea, the carbon forms an organic chain, and in ammonium cyanate, it does not. Thus, to reduce the specifics of organic chemistry even further, this discipline can be said to constitute the study of carbon chains, and ways to rearrange them to create new substances.

REAL-LIFE APPLICATIONS

ORGANIC CHEMISTRY AND MODERN LIFE

At first glance, the term "organic chemistry" might sound like something removed from

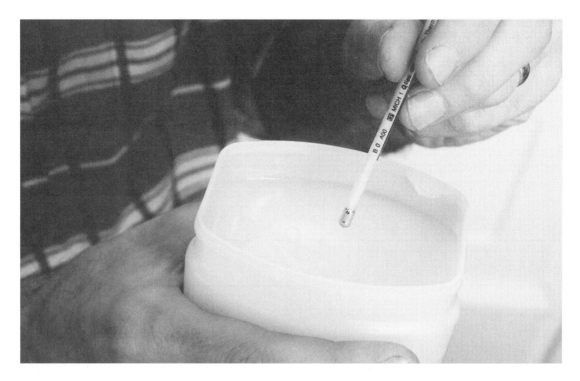

ANOTHER BYPRODUCT OF ORGANIC CHEMISTRY: PETROLEUM JELLY. *(Laura Dwight/Corbis. Reproduced by permission.)*

everyday life, but this could not be further from the truth. The reality of the role played by organic chemistry in modern existence is summed up in a famous advertising slogan used by E. I. du Pont de Nemours and Company (usually referred to as "du Pont"): "Better Things for Better Living Through Chemistry."

Often rendered simply as "Better Living Through Chemistry," the advertising campaign made its debut in 1938, just as du Pont introduced a revolutionary product of organic chemistry: nylon, the creation of a brilliant young chemist named Wallace Carothers (1896-1937). Nylon, an example of a polymer (discussed below), started a revolution in plastics that was still unfolding three decades later, in 1967. That was the year of the film *The Graduate*, which included a famous interchange between the character of Benjamin Braddock (Dustin Hoffman) and an adult named Mr. McGuire (Walter Brooke):

- Mr. McGuire: I just want to say one word to you... just one word.
- Benjamin Braddock: Yes, sir.
- Mr. McGuire: Are you listening?
- Benjamin Braddock: Yes, sir, I am.
- Mr. McGuire: Plastics.

The meaning of this interchange was that plastics were the wave of the future, and that an intelligent young man such as Ben should invest his energies in this promising new field. Instead, Ben puts his attention into other things, quite removed from "plastics," and much of the plot revolves around his revolt against what he perceives as the "plastic" (that is, artificial) character of modern life.

In this way, *The Graduate* spoke for a whole generation that had become ambivalent concerning "better living through chemistry," a phrase that eventually was perceived as ironic in view of concerns about the environment and the many artificial products that make up modern life. Responding to this ambivalence, du Pont dropped the slogan in the late 1970s; yet the reality is that people truly do enjoy "better living through chemistry"—particularly organic chemistry.

APPLICATIONS OF ORGANIC CHEMISTRY. What would the world be like without the fruits of organic chemistry? First, it would be necessary to take away all the various forms of rubber, vitamins, cloth, and paper made from organically based compounds. Aspirins and all types of other drugs; preservatives that keep food from spoiling; perfumes and toiletries; dyes and flavorings—all these things would have to go as well.

Synthetic fibers such as nylon—used in everything from toothbrushes to parachutes—would be out of the picture if it were not for the enormous progress made by organic chemistry. The same is true of plastics or polymers in general, which have literally hundreds upon hundreds of applications. Indeed, it is virtually impossible for a person in twenty-first century America to spend an entire day without coming into contact with at least one, and more likely dozens, of plastic products. Car parts, toys, computer housings, Velcro fasteners, PVC (polyvinyl chloride) plumbing pipes, and many more fixtures of modern life are all made possible by plastics and polymers.

Then there is the vast array of petrochemicals that power modern civilization. Best-known among these is gasoline, but there is also coal, still one of the most significant fuels used in electrical power plants, as well as natural gas and various other forms of oil used either directly or indirectly in providing heat, light, and electric power to homes. But the influence of petrochemicals extends far beyond their applications for fuel. For instance, the roofing materials and tar that (quite literally) keep a roof over people's heads, protecting them from sun and rain, are the product of petrochemicals—and ultimately, of organic chemistry.

Hydrocarbons

Carbon, together with other elements, forms so many millions of organic compounds that even introductory textbooks on organic chemistry consist of many hundreds of pages. Fortunately, it is possible to classify broad groupings of organic compounds. The largest and most significant is that class of organic compounds known as hydrocarbons—chemical compounds whose molecules are made up of nothing but carbon and hydrogen atoms.

Every molecule in a hydrocarbon is built upon a "skeleton" composed of carbon atoms, either in closed rings or in long chains. The chains may be straight or branched, but in each case—rings or chains, straight chains or branched ones—the carbon bonds not used in tying the carbon atoms together are taken up by hydrogen atoms.

Theoretically, there is no limit to the number of possible hydrocarbons. Not only does carbon form itself into apparently limitless molecular shapes, but hydrogen is a particularly good partner. It has the smallest atom of any element on the periodic table, and therefore it can bond to one of carbon's valence electrons without getting in the way of the other three.

There are two basic varieties of hydrocarbon, distinguished by shape: aliphatic and aromatic. The first of these forms straight or branched chains, as well as rings, while the second forms only benzene rings, discussed below. Within the aliphatic hydrocarbons are three varieties: those that form single bonds (alkanes), double bonds (alkenes), and triple bonds (alkynes.)

ALKANES. The alkanes are also known as saturated hydrocarbons, because all the bonds not used to make the skeleton itself are filled to their capacity (that is, saturated) with hydrogen atoms. The formula for any alkane is C_nH_{2n+2}, where n is the number of carbon atoms. In the case of a linear, unbranched alkane, every carbon atom has two hydrogen atoms attached, but the two end carbon atoms each have an extra hydrogen.

What follows are the names and formulas for the first eight normal, or unbranched, alkanes. Note that the first four of these received common names before their structures were known; from C_5 onward, however, they were given names with Greek roots indicating the number of carbon atoms (e.g., octane, a reference to "eight.")

- Methane (CH_4)
- Ethane (C_2H_6)
- Propane (C_3H_8)
- Butane (C_4H_{10})
- Pentane (C_5H_{12})
- Hexane (C_6H_{14})
- Heptane (C_7H_{16})
- Octane (C_8H_{18})

The reader will undoubtedly notice a number of familiar names on this list. The first four, being the lowest in molecular mass, are gases at room temperature, while the heavier ones are oily liquids. Alkanes even heavier than those on this list tend to be waxy solids, an example being paraffin wax, for making candles. It should be noted that from butane on up, the alkanes have numerous structural isomers, depending on whether they are straight or branched, and these isomers have differing chemical properties.

Branched alkanes are named by indicating the branch attached to the principal chain. Branches, known as substituents, are named by taking the name of an alkane and replacing the suffix with yl—for example, methyl, ethyl, and so on. The general term for an alkane which functions as a substituent is alkyl.

Cycloalkanes are alkanes joined in a closed loop to form a ring-shaped molecule. They are named by using the names above, with cyclo- as a prefix. These start with propane, or rather cyclopropane, which has the minimum number of carbon atoms to form a closed shape: three atoms, forming a triangle.

ALKENES AND ALKYNES. The names of the alkenes, hydrocarbons that contain one or more double bonds per molecule, are parallel to those of the alkanes, but the family ending is -ene. Likewise they have a common formula: C_nH_{2n}. Both alkenes and alkynes, discussed below, are unsaturated—in other words, some of the carbon atoms in them are free to form other bonds. Alkenes with more than one double bond are referred to as being polyunsaturated.

As with the alkenes, the names of alkynes (hydrocarbons containing one or more triple bonds per molecule) are parallel to those of the alkanes, only with the replacement of the suffix -yne in place of -ane. The formula for alkenes is C_nH_{2n-2}. Among the members of this group are acetylene, or C_2H_2, used for welding steel. Plastic polystyrene is another important product from this division of the hydrocarbon family.

AROMATIC HYDROCARBONS. Aromatic hydrocarbons, despite their name, do not necessarily have distinctive smells. In fact the name is a traditional one, and today these compounds are defined by the fact that they have benzene rings in the middle. Benzene has a formula C_6H_6, and a benzene ring is usually represented as a hexagon (the six carbon atoms and their attached hydrogen atoms) surrounding a circle, which represents all the bonding electrons as though they were everywhere in the molecule at once.

In this group are products such as naphthalene, toluene, and dimethyl benzene. These last two are used as solvents, as well as in the synthesis of drugs, dyes, and plastics. One of the more famous (or infamous) products in this part of

the vast hydrocarbon network is trinitrotoluene, or TNT. Naphthalene is derived from coal tar, and used in the synthesis of other compounds. A crystalline solid with a powerful odor, it is found in mothballs and various deodorant-disinfectants.

PETROCHEMICALS. As for petrochemicals, these are simply derivatives of petroleum, itself a mixture of alkanes with some alkenes, as well as aromatic hydrocarbons. Through a process known as fractional distillation, the petrochemicals of the lowest molecular mass boil off first, and those having higher mass separate at higher temperatures.

Among the products derived from the fractional distillation of petroleum are the following, listed from the lowest temperature range (that is, the first material to be separated) to the highest: natural gas; petroleum ether, a solvent; naphtha, a solvent (used for example in paint thinner); gasoline; kerosene; fuel for heating and diesel fuel; lubricating oils; petroleum jelly; paraffin wax; and pitch, or tar. A host of other organic chemicals, including various drugs, plastics, paints, adhesives, fibers, detergents, synthetic rubber, and agricultural chemicals, owe their existence to petrochemicals.

Obviously, petroleum is not just for making gasoline, though of course this is the first product people think of when they hear the word "petroleum." Not all hydrocarbons in gasoline are desirable. Straight-chain or normal heptane, for instance, does not fire smoothly in an internal-combustion engine, and therefore disrupts the engine's rhythm. For this reason, it is given a rating of zero on a scale of desirability, while octane has a rating of 100. This is why gas stations list octane ratings at the pump: the higher the presence of octane, the better the gas is for one's automobile.

HYDROCARBON DERIVATIVES

With carbon and hydrogen as the backbone, the hydrocarbons are capable of forming a vast array of hydrocarbon derivatives by combining with other elements. These other elements are arranged in functional groups—an atom or group of atoms whose presence identifies a specific family of compounds. Below we will briefly discuss some of the principal hydrocarbon derivatives, which are basically hydrocarbons with the addition of other molecules or single atoms.

Alcohols are oxygen-hydrogen molecules wedded to hydrocarbons. The two most important commercial types of alcohol are methanol, or wood alcohol; and ethanol, which is found in alcoholic beverages, such as beer, wine, and liquor. Though methanol is still known as "wood alcohol," it is no longer obtained by heating wood, but rather by the industrial hydrogenation of carbon monoxide. Used in adhesives, fibers, and plastics, it can also be applied as a fuel. Ethanol, too, can be burned in an internal-combustion engine, when combined with gasoline to make gasohol. Another significant alcohol is cholesterol, found in most living organisms. Though biochemically important, cholesterol can pose a risk to human health.

Aldehydes and ketones both involve a double-bonded carbon-oxygen molecule, known as a carbonyl group. In a ketone, the carbonyl group bonds to two hydrocarbons, while in an aldehyde, the carbonyl group is always at the end of a hydrocarbon chain. Therefore, instead of two hydrocarbons, there is always a hydrocarbon and at least one other hydrogen bonded to the carbon atom in the carbonyl. One prominent example of a ketone is acetone, used in nail polish remover. Aldehydes often appear in nature—for instance, as vanillin, which gives vanilla beans their pleasing aroma. The ketones carvone and camphor impart the characteristic flavors of spearmint leaves and caraway seeds.

CARBOXYLIC ACIDS AND ESTERS. Carboxylic acids all have in common what is known as a carboxyl group, designated by the symbol -COOH. This consists of a carbon atom with a double bond to an oxygen atom, and a single bond to another oxygen atom that is, in turn, wedded to a hydrogen. All carboxylic acids can be generally symbolized by RCOOH, with R as the standard designation of any hydrocarbon. Lactic acid, generated by the human body, is a carboxylic acid: when a person overexerts, the muscles generate lactic acid, resulting in a feeling of fatigue until the body converts the acid to water and carbon dioxide. Another example of a carboxylic acid is butyric acid, responsible in part for the smells of rancid butter and human sweat.

When a carboxylic acid reacts with an alcohol, it forms an ester. An ester has a structure similar to that described for a carboxylic acid, with a few key differences. In addition to its bonds (one double, one single) with the oxygen atoms, the carbon atom is also attached to a hydrocarbon, which comes from the carboxylic acid. Furthermore, the single-bonded oxygen atom is attached not to a hydrogen, but to a second hydrocarbon, this one from the alcohol. One well-known ester is acetylsalicylic acid—better known as aspirin. Esters, which are a key factor in the aroma of various types of fruit, are often noted for their pleasant smell.

POLYMERS

Polymers are long, stringy molecules made of smaller molecules called monomers. They appear in nature, but thanks to Carothers—a tragic figure, who committed suicide a year before Nylon made its public debut—as well as other scientists and inventors, synthetic polymers are a fundamental part of daily life.

The structure of even the simplest polymer, polyethylene, is far too complicated to discuss in ordinary language, but must be represented by chemical symbolism. Indeed, polymers are a subject unto themselves, but it is worth noting here just how many products used today involve polymers in some form or another.

Polyethylene, for instance, is the plastic used in garbage bags, electrical insulation, bottles, and a host of other applications. A variation on polyethylene is Teflon, used not only in nonstick cookware, but also in a number of other devices, such as bearings for low-temperature use. Polymers of various kinds are found in siding for houses, tire tread, toys, carpets and fabrics, and a variety of other products far too lengthy to enumerate.

WHERE TO LEARN MORE

Blashfield, Jean F. *Carbon*. Austin, TX: Raintree Steck-Vaughn, 1999.

"*Carbon*." *Xrefer* (Web site). <http://www.xrefer.com/entry/639742> (May 30, 2001).

Chemistry Help Online for Students (Web site). <http://members.tripod.com/chemistryhelp/> May 30, 2001).

Knapp, Brian J. *Carbon Chemistry*. Illustrated by David Woodroffe. Danbury, CT: Grolier Educational, 1998.

Loudon, G. Marc. *Organic Chemistry*. Menlo Park, CA: Benjamin/Cummings, 1988.

KEY TERMS

ALKANES: Hydrocarbons that form single bonds. Alkanes are also called saturated hydrocarbons.

ALKENES: Hydrocarbons that form double bonds.

ALKYL: A general term for an alkane that functions as a substituent.

ALKYNES: Hydrocarbons that form triple bonds.

ALLOTROPES: Different versions of the same element, distinguished by molecular structure.

AMORPHOUS: Having no definite structure.

COVALENT BONDING: A type of chemical bonding in which two atoms share valence electrons.

CRYSTALLINE: A term describing a type of solid in which the constituent parts have a simple and definite geometric arrangement repeated in all directions.

DOUBLE BOND: A form of bonding in which two atoms share two pairs of valence electrons. Carbon is also capable of single bonds and triple bonds.

ELECTRONEGATIVITY: The relative ability of an atom to attract valence electrons.

FUNCTIONAL GROUPS: An atom or group of atoms whose presence identifies a specific family of compounds.

HYDROCARBON: Any chemical compound whose molecules are made up of nothing but carbon and hydrogen atoms.

HYDROCARBON DERIVATIVES: Families of compounds formed by the joining of hydrocarbons with various functional groups.

ISOMERS: Substances having the same chemical formula, but that are different

chemically due to disparities in the arrangement of atoms.

OCTET RULE: A term describing the distribution of valence electrons that takes place in chemical bonding for most elements, which end up with eight valence electrons.

ORGANIC CHEMISTRY: The study of carbon, its compounds, and their properties. (Some carbon-containing compounds, most notably oxides and carbonates, are not considered organic.)

SATURATED: A term describing a hydrocarbon in which each carbon is already bound to four other atoms. Alkanes are saturated hydrocarbons.

SINGLE BOND: A form of bonding in which two atoms share one pair of valence electrons. Carbon is also capable of double bonds and triple bonds.

SUBSTITUENTS: Branches of alkanes, named by taking the name of an alkane and replacing the suffix with yl—for example, methyl, ethyl, and so on.

TETRAVALENT: Capable of bonding to four other elements.

TRIPLE BOND: A form of bonding in which two atoms share three pairs of valence electrons. Carbon is also capable of single bonds and double bonds.

UNSATURATED: A term describing a hydrocarbon in which the carbons involved in a multiple bond (a double bond or triple bond) are free to bond with other atoms. Alkenes and alkynes are both unsaturated.

VALENCE ELECTRONS: Electrons that occupy the highest principal energy level in an atom. These are the electrons involved in chemical bonding.

"Organic Chemistry" (Web site). <http://edie.cprost.sfu.
ca/~rhlogan/organic.html> (May 30, 2001).

"Organic Chemistry." Frostburg State University Chemistry Helper (Web site). <http://www.chemhelper.
com/> (May 30, 2001).

Sparrow, Giles. Carbon. New York: Benchmark
Books, 1999.

Zumdahl, Steven S. Introductory Chemistry: A Foundation, 4th ed. Boston: Houghton Mifflin, 2000.

POLYMERS

CONCEPT

Formed from hydrocarbons, hydrocarbon derivatives, or sometimes from silicon, polymers are the basis not only for numerous natural materials, but also for most of the synthetic plastics that one encounters every day. Polymers consist of extremely large, chain-like molecules that are, in turn, made up of numerous smaller, repeating units called monomers. Chains of polymers can be compared to paper clips linked together in long strands, and sometimes cross-linked to form even more durable chains. Polymers can be composed of more than one type of monomer, and they can be altered in other ways. Likewise they are created by two different chemical processes, and thus are divided into addition and condensation polymers. Among the natural polymers are wool, hair, silk, rubber, and sand, while the many synthetic polymers include nylon, synthetic rubber, Teflon, Formica, Dacron, and so forth. It is very difficult to spend a day without encountering a natural polymer—even if hair is removed from the list—but in the twenty-first century, it is probably even harder to avoid synthetic polymers, which have collectively revolutionized human existence.

HOW IT WORKS

Polymers of Silicon and Carbon

Polymers can be defined as large, typically chain-like molecules composed of numerous smaller, repeating units known as monomers. There are numerous varieties of monomers, and since these can be combined in different ways to form polymers, there are even more of the latter.

The name "polymer" does not, in itself, define the materials that polymers contain. A handful of polymers, such as natural sand or synthetic silicone oils and rubbers, are built around silicon. However, the vast majority of polymers center around an element that occupies a position just above silicon on the periodic table: carbon.

The similarities between these two are so great, in fact, that some chemists speak of Group 4 (Group 14 in the IUPAC system) on the periodic table as the "carbon family." Both carbon and silicon have the ability to form long chains of atoms that include bonds with other elements. The heavier elements of this "family," however (most notably lead), are made of atoms too big to form the vast array of chains and compounds for which silicon and carbon are noted.

Indeed, not even silicon—though it is at the center of an enormous range of inorganic compounds—can compete with carbon in its ability to form arrangements of atoms in various shapes and sizes, and hence to participate in an almost limitless array of compounds. The reason, in large part, is that carbon atoms are much smaller than those of silicon, and thus can bond to one another and still leave room for other bonds.

Carbon is such an important element that an entire essay in this book is devoted to it, while a second essay discusses organic chemistry, the study of compounds containing carbon. In the present context, there will be occasional references to non-carbon (that is, silicon) polymers, but the majority of our attention will be devoted to hydrocarbon and hydrocarbon-derivative polymers, which most of us know simply as "plastics."

COTTON IS AN EXAMPLE OF A NATURAL POLYMER. (*Carl Corey/Corbis. Reproduced by permission.*)

ORGANIC CHEMISTRY

As explained in the essay on Organic Chemistry, chemists once defined the term "organic" as relating only to living organisms; the materials that make them up; materials derived from them; and substances that come from formerly living organisms. This definition, which more or less represents the everyday meaning of "organic," includes a huge array of life forms and materials: humans, all other animals, insects, plants, microorganisms, and viruses; all substances that make up their structures (for example, blood, DNA, and proteins); all products that come from them (a list diverse enough to encompass everything from urine to honey); and all materials derived from the bodies of organisms that were once alive (paper, for instance, or fossil fuels).

As broad as this definition is, it is not broad enough to represent all the substances addressed by organic chemistry—the study of carbon, its compounds, and their properties. All living or once-living things do contain carbon; however, organic chemistry is also concerned with carbon-containing materials—for instance, the synthetic plastics we will discuss in this essay—that have never been part of a living organism.

It should be noted that while organic chemistry involves only materials that contain carbon, carbon itself is found in other compounds not considered organic: oxides such as carbon dioxide and monoxide, as well as carbonates, most notably calcium carbonate or limestone. In other words, as broad as the meaning of "organic" is, it still does not encompass all substances containing carbon.

HYDROCARBONS

As for hydrocarbons, these are chemical compounds whose molecules are made up of nothing but carbon and hydrogen atoms. Every molecule in a hydrocarbon is built upon a "skeleton" of carbon atoms, either in closed rings or in long chains, which are sometimes straight and sometimes branched.

Theoretically, there is no limit to the number of possible hydrocarbons: not only does carbon form itself into seemingly limitless molecular shapes, but hydrogen is a particularly good partner. It is the smallest atom of any element on the periodic table, and therefore it can bond to one of carbon's four valence electrons without getting in the way of the other three.

There are many, many varieties of hydrocarbon, classified generally as aliphatic hydrocarbons (alkanes, alkenes, and alkynes) and aromatic hydrocarbons, the latter being those that con-

MODERN APPLIANCES CONTAIN NUMEROUS EXAMPLES OF SYNTHETIC POLYMERS, FROM THE FLOORING TO THE COUNTERTOPS TO VIRTUALLY ALL APPLIANCES. *(Scott Roper/Corbis. Reproduced by permission.)*

tain a benzene ring. By means of a basic alteration in the shape or structure of a hydrocarbon, it is possible to create new varieties. Thus, as noted above, the number of possible hydrocarbons is essentially unlimited.

Certain hydrocarbons are particularly useful, one example being petroleum, a term that refers to a wide array of hydrocarbons. Among these is an alkane that goes by the name of octane (C_8H_{18}), a preferred ingredient in gasoline. Hydrocarbons can be combined with various functional groups (an atom or group of atoms whose presence identifies a specific family of compounds) to form hydrocarbon derivatives such as alcohols and esters.

REAL-LIFE APPLICATIONS

TYPES OF POLYMERS AND POLYMERIZATION

Many polymers exist in nature. Among these are silk, cotton, starch, sand, and asbestos, as well as the incredibly complex polymers known as RNA (ribonucleic acid) and DNA (deoxyribonucleic acid), which hold genetic codes. The polymers discussed in this essay, however, are primarily of the synthetic kind. Artificial polymers include such plastics (defined below) as polyethylene, styrofoam, and Saran wrap; fibers such as nylon, Dacron (polyester), and rayon; and other materials such as Formica, Teflon, and PVC pipe.

As noted earlier, most polymers are formed from monomers either of hydrocarbon or hydrocarbon derivatives. The most basic synthetic monomer is ethylene (C_2H_4), a name whose -ene ending identifies it as an alkene, a hydrocarbon formed by double bonds between carbon atoms. Another alkene hydrocarbon monomer is butadiene, whose formula is C_4H_6. This is an example of the fact that the formula of a compound does not tell the whole story: on paper, the difference between these two appears to be merely a matter of two extra atoms each of carbon and hydrogen. In fact, butadiene's structure is much more complex.

Still more complex is styrene, which includes a benzene ring. Several other monomers involve other elements: chloride, in vinyl chloride; nitrogen, in acrylonitrile; and fluorine, in tetrafluoroethylene. It is not necessary, in the present context, to keep track of all of these substances, which in any case represent just some of the more prominent among a wide variety of synthetic monomers. A good high-school or college chemistry textbook (either general chemistry or organic chemistry) should provide structural representations of these common monomers. Such representations will show, for instance, the vast differences between purely hydrocarbon monomers such as ethylene, propylene, styrene, and butadiene.

When combined into polymers, the monomers above form the basis for a variety of useful and familiar products. Once the carbon double bonds in tetrafluoroethylene (C_2F_4) are broken, they form the polymer known as Teflon, used in the coatings of cooking utensils, as well as in electrical insulation and bearings. Vinyl chloride breaks its double bonds to form polyvinyl chloride, better known as PVC, a material used for everything from plumbing pipe to toys to Saran wrap. Styrene, after breaking its double bonds, forms polystyrene, used in containers and thermal insulation.

POLYMERIZATION. Note that several times in the preceding paragraph, there was a reference to the breaking of carbon double bonds. This is often part of one variety of polymerization, the process whereby monomers join to form polymers. If monomers of a single type join, the resulting polymer is called a homopolymer, but if the polymer consists of more than one type of monomer, it is known as a copolymer. This joining may take place by one of two

A SYNTHETIC POLYMER KNOWN AS KEVLAR IS USED IN THE CONSTRUCTION OF BULLETPROOF VESTS FOR LAW-ENFORCEMENT OFFICERS. *(Anna Clopet/Corbis. Reproduced by permission.)*

processes. The first of these, addition polymerization, is fairly simple: monomers add themselves to one another, usually breaking double bonds in the process. This results in the creation of a polymer and no other products.

Much more complex is the process known as condensation polymerization, in which a small molecule called a dimer is formed as monomers join. The specifics are too complicated to discuss in any detail, but a few things can be said here about condensation polymerization. The monomers in condensation polymerization must be bifunctional, meaning that they have a functional group at each end. When characteristic structures at the ends of the monomers react to one another by forming a bond, they create a dimer, which splits off from the polymer. The products of condensation polymerization are thus not only the polymer itself, but also a dimer, which may be water, hydrochloric acid (HCl), or some other substance.

A Plastic World

A DAY IN THE LIFE. Though "plastic" has a number of meanings in everyday life, and in society at large (as we shall see), the scientific definition is much more specific. Plastics are materials, usually organic, that can be caused to flow under certain conditions of heat and pressure, and thus to assume a desired shape when the pressure and temperature conditions are withdrawn. Most plastics are made of polymers.

Every day, a person comes into contact with dozens, if not hundreds, of plastics and polymers. Consider a day in the life of a hypothetical teenage girl. She gets up in the morning, brushes her teeth with a toothbrush made of nylon, then opens a shower door—which is likely to be plastic rather than glass—and steps into a molded plastic shower or bathtub. When she gets out of the shower, she dries off with a towel containing a polymer such as rayon, perhaps while standing on tile that contains plastics, or polymers.

She puts on makeup (containing polymers) that comes in plastic containers, and later blow-dries her hair with a handheld hair dryer made of insulated plastic. Her clothes, too, are likely to contain synthetic materials made of polymers. When she goes to the kitchen for breakfast, she will almost certainly walk on flooring with a plastic coating. The countertops may be of formica, a condensation polymer, while it is likely that virtually every appliance in the room will contain plastic. If she opens the refrigerator to get out a milk container, it too will be made of plastic, or of paper with a thin plastic coating. Much of the packaging on the food she eats, as well as sandwich bags and containers for storing food, is also made of plastic.

And so it goes throughout the day. The phone she uses to call a friend, the computer she sits at to check her e-mail, and the stereo in her room all contain electrical components housed in plastic. If she goes to the gym, she may work out in Gore-tex, a fabric containing a very thin layer of plastic with billions of tiny pores, so that it lets through water vapor (that is, perspiration) without allowing the passage of liquid water. On the way to the health club, she will ride in a car that contains numerous plastic molds in the steering wheel and dashboard. If she plays a compact disc—itself a thin wafer of plastic coated with metal—she will pull it out of a plastic jewel case. Finally, at night, chances are she will sleep in sheets, and with a pillow, containing synthetic polymers.

A SILENT REVOLUTION. The scenario described above—a world surrounded by polymers, plastics, and synthetic materials—represents a very recent phenomenon. "Before the 1930s," wrote John Steele Gordon in an article about plastics for American Heritage, "almost everything people saw or handled was made of materials that had been around since ancient times: wood, stone, metal, and animal and plant fibers." All of that changed in the era just before World War II, thanks in large part to a brilliant young American chemist named Wallace Carothers (1896-1937).

By developing nylon for E. I. du Pont de Nemours and Company (known simply as "DuPont" or "du Pont"), Carothers and his colleagues virtually laid the foundation for modern polymer chemistry—a field that employs more chemists than any other. These men created what Gordon called a "materials revolution" by introducing the world to polymers and plastics, which are typically made of polymers.

Yet as Gordon went on to note, "It has been a curiously silent revolution.... When we think of the scientific triumphs of [the twentieth century], we think of nuclear physics, medicine, space exploration, and the computer. But all these developments would have been much impeded, in some cases impossible, without... plastics. And yet 'plastic' remains, as often as not, a term of opprobrium."

AMBIVALENCE TOWARD PLASTICS. Gordon was alluding to a cultural attitude discussed in the essay on Organic Chemistry: the association of plastics, a physical material developed by chemical processes, with the condition—spiritual, moral, and intellectual—of being "plastic" or inauthentic. This was symbolized in a famous piece of dialogue about plastics from the 1967 movie The Graduate, in which a nonplussed Ben Braddock (Dustin Hoffman) listens as one of his parents' friends advises him to invest his future in plastics. As Gordon noted, "however intergenerationally challenged that half-drunk friend of Dustin Hoffman's parents may have been... he was right about the importance of the materials revolution in the twentieth century."

One aspect of society's ambivalence over plastics relates to very genuine concerns about the environment. Most synthetic polymers are made from petroleum, a nonrenewable resource;

but this is not the greatest environmental danger that plastics present. Most plastics are not biodegradable: though made of organic materials, they do not contain materials that will decompose and eventually return to the ground. Nor is there anything in plastics to attract microorganisms, which, by assisting in the decomposition of organic materials, help to facilitate the balance of decay and regeneration necessary for life on Earth.

Efforts are underway among organic chemists in the research laboratories of corporations and other institutions to develop biodegradable plastics that will speed up the decomposition of materials in the polymers—a process that normally takes decades. Until such replacement polymers are developed, however, the most environmentally friendly solution to the problem of plastics is recycling. Today only about 1% of plastics are recycled, while the rest goes into waste dumps, where they account for 30% of the volume of trash.

Long before environmental concerns came to the forefront, however, people had begun almost to fear plastics as a depersonalizing aspect of modern life. It seemed that in a given day, a person touched fewer and fewer things that came directly from the natural environment: the "wood, stone, metal, and animal and plant fibers" to which Gordon alluded. Plastics seemed to have made human life emptier; yet the truth of the matter—including the fact that plastics add more than they take away from the landscape of our world—is much more complex.

THE PLASTICS REVOLUTION

Though the introduction of plastics is typically associated with the twentieth century, in fact the "materials revolution" surrounding plastics began in 1865. That was the year when English chemist Alexander Parkes (1813-1890) produced the first plastic material, celluloid. Parkes could have become a rich man from his invention, but he was not a successful marketer. Instead, the man who enjoyed the first commercial success in plastics was—not surprisingly—an American, inventor John Wesley Hyatt (1837-1920).

Responding to a contest in which a billiard-ball manufacturer offered $10,000 to anyone who could create a substitute for ivory, which was extremely costly, Hyatt turned to Parkes's celluloid. Actually, Parkes had given his creation—

developed from cellulose, a substance found in the cell walls of plants—a much less appealing name, "Parkesine." Hyatt, who used celluloid to make smooth, hard, round billiard balls (thereby winning the contest) took out a patent for the process involved in making the material he had dubbed "Celluloid," with a capital C.

Though the Celluloid made by Hyatt's process was flammable (as was Parkesine), it proved highly successful as a product when he introduced it in 1869. He marketed it successfully for use in items such as combs and baby rattles, and Celluloid sales received a powerful boost after photography pioneer George Eastman (1854-1932) chose the material for use in the development of film. Eventually, Celluloid would be applied in motion-picture film, and even today, the adjective "celluloid" is sometimes used in relation to the movies. Actually, Celluloid (which can be explosive in large quantities) was phased out in favor of "safety film," or cellulose acetate, beginning in 1924.

Two important developments in the creation of synthetic polymers occurred at the turn of the century. One was the development of Galalith, an ivory-like substance made from formaldehyde and milk, by German chemist Adolf Spitteler. An even more important innovation happened in 1907, when Belgian-American chemist Leo Baekeland (1863-1944) introduced Bakelite. The latter, created in a reaction between phenol and formaldehyde, was a hard, black plastic that proved an excellent insulator. It soon found application in making telephones and household appliances, and by the 1920s, chemists had figured out how to add pigments to Bakelite, thus introducing the public to colored plastics.

SYNTHETIC RUBBER. Throughout these developments, chemists had only a vague understanding of polymers, but by the 1930s, they had come to accept the model of polymers as large, flexible, chain-like molecules. One of the most promising figures in the emerging field of polymer chemistry was Carothers, who in 1926 left a teaching post at Harvard University to accept a position as director of the polymer research laboratory at DuPont.

Among the first problems Carothers tackled was the development of synthetic rubber. Natural rubber had been known for many centuries when English chemist Joseph Priestley (1733-1804) gave it its name because he used it to rub

POLYMERS

out pencil marks. In 1839, American inventor Charles Goodyear (1800-1860) accidentally discovered a method for making rubber more durable, after he spilled a mixture of rubber and sulfur onto a hot stove. Rather than melting, the rubber bonded with the sulfur to form a much stronger but still elastic product, and Goodyear soon patented this process under the name vulcanization.

Natural rubber, nonetheless, had many undesirable properties, and hence DuPont put Carothers to the task of developing a substitute. The result was neoprene, which he created by adding a chlorine atom to an acetylene derivative. Neoprene was stronger, more durable, and less likely to become brittle in cold weather than natural rubber. It would later prove an enormous boost to the Allied war effort, after the Japanese seized the rubber plantations of Southeast Asia in 1941.

NYLON. Had neoprene, which Carothers developed in 1931, been the extent of his achievements, he would still be remembered by science historians. However, his greatest creation still lay ahead of him. Studying the properties of silk, he became convinced that he could develop a more durable compound that could replicate the properties of silk at a much lower cost.

Carothers was not alone in his efforts, as Gordon showed in his account of events at the DuPont laboratories:

One day, an assistant, Julian Hill, noticed that when he stuck a glass stirring rod into a gooey mass at the bottom of a beaker the researchers had been investigating, he could draw out threads from it, the polymers forming spontaneously as he pulled. When Carothers was absent one day, Hill and his colleagues decided to see how far they could go with pulling threads out of goo by having one man hold the beaker while another ran down the hall with the glass rod. A very long, silk-like thread was produced.

Realizing what they had on their hands, DuPont devoted $27 million to the research efforts of Carothers and his associates at the lab, and in 1937, Carothers presented his boss with the results, saying "Here is your synthetic textile fabric." DuPont introduced the material, nylon, to the American public the following year with one of the most famous advertising campaigns of all time: "Better Things for Better Living Through Chemistry."

The product got an additional boost through exposure at the 1939 World's Fair. When DuPont put 4,000 pairs of nylon stockings on the market, they sold in a matter of hours. A few months later, four million pairs sold in New York City in a single day. Women stood in line to buy stockings of nylon, a much better (and less expensive) material for that purpose than silk—but they did not have long to enjoy it. During World War II, all nylon went into making war materials such as parachutes, and nylon did not become commercially available again until 1946.

As Gordon noted, Carothers would surely have won the Nobel Prize in chemistry for his work—"but Nobel prizes go only to living recipients...." Carothers had married in 1936, and by early 1937, his wife Helen was pregnant. (Presumably, he was unaware of the fact that he was about to become a father.) Though highly enthusiastic about his work, Carothers was always shy and withdrawn, and in Gordon's words, "he had few outlets other than work." He was, however, a talented singer, as was his closest sibling, Isobel, a radio celebrity. Her death in January 1937 sent him into a bout of depression, and on April 29, he killed himself with a dose of cyanide. Seven months later, on November 27, Helen gave birth to a daughter, Jane.

HOW PLASTICS HAVE ENHANCED LIFE. Despite his tragic end, Carothers had brought much good to the world by sparking enormous interest in polymer research and plastics. Over the years that followed, polymer chemists developed numerous products that had applications in a wide variety of areas. Some, such as polyester—a copolymer of terephthalic acid and ethylene—seemed to fit the idea of "plastics" as ugly, inauthentic, and even dehumanizing. During the 1970s, clothes of polyester became fashionable, but by the early 1980s, there was a public backlash against synthetics, and in favor of natural materials.

Yet even as the public rejected synthetic fabrics for everyday wear, Gore-tex and other synthetics became popular for outdoor and workout clothing. At the same time, the polyester that many regarded as grotesque when used in clothing was applied in making safer beverage bottles. The American Plastics Council dramatized this in a 1990s commercial that showed a few seconds in the life of a mother. Her child takes a soft-

KEY TERMS

ADDITION POLYMERIZATION: A form of polymerization in which monomers having at least one double bond or triple bond simply add to one another, forming a polymer and no other products. Compare to condensation polymerization.

ALKENES: Hydrocarbons that contain double bonds.

COPOLYMER: A polymer composed of more than one type of monomer.

DIMER: A molecule formed by the joining of two monomers.

DOUBLE BOND: A form of bonding in which two atoms share two pairs of valence electrons. Carbon is noted for its ability to form double bonds, as for instance in many hydrocarbons.

FUNCTIONAL GROUPS: An atom or group of atoms whose presence identifies a specific family of compounds. When combined with hydrocarbons, various functional groups form hydrocarbon derivatives.

HOMOPOLYMER: A polymer that consists of only one type of monomer.

HYDROCARBON: Any chemical compound whose molecules are made up of nothing but carbon and hydrogen atoms.

HYDROCARBON DERIVATIVES: Families of compounds formed by the joining of hydrocarbons with various functional groups.

MONOMERS: Small, individual subunits, often built of hydrocarbons, that join together to form polymers.

ORGANIC: A term referring to any compound that contains carbon, except for oxides such as carbon dioxide, or carbonates such as calcium carbonate (i.e., limestone).

ORGANIC CHEMISTRY: The study of carbon, its compounds, and their properties.

PLASTICS: Materials, usually organic, that can be caused to flow under certain conditions of heat and pressure, and thus to assume a desired shape when the pressure and temperature conditions are withdrawn. Plastics are usually made up of polymers.

POLYMERIZATION: The process whereby monomers join to form polymers.

POLYMERS: Large, typically chain-like molecules composed of numerous smaller, repeating units known as monomers.

VALENCE ELECTRONS: Electrons that occupy the highest principal energy level in an atom. These are the electrons involved in chemical bonding.

drink bottle out of the refrigerator and drops it, and the mother cringes at what she thinks she is about to see next: glass shattering around her child. But she is remembering the way things were when she was a child, when soft drinks still came in glass bottles: instead, the plastic bottle bounces harmlessly.

Of course, such dramatizations may seem a bit self-serving to critics of plastic, but the fact remains that plastics enhance—and in some cases even preserve—life. Kevlar, for instance, enhances life when it is used in making canoes for recreation; when used to make a bulletproof vest, it can save the life of a law-enforcement officer. Mylar, a form of polyester, enhances life when used to make a durable child's balloon—but this highly nonreactive material also saves lives when it is applied to make replacement human blood vessels, or even replacement skin for burn victims.

RECYCLING

As mentioned above, plastics—for all their benefits—do pose a genuine environmental threat, due to the fact that the polymers break down much more slowly than materials from living organisms. Hence the need not only to develop biodegradable plastics, but also to work on more effective means of recycling.

One of the challenges in the recycling arena is the fact that plastics come in a variety of grades. Different catalysts are used to make polymers that possess different properties, with varying sizes of molecules, and in chains that may be linear, branched, or cross-linked. Long chains of 10,000 or more monomers can be packed closely to form a hard, tough plastic known as high-density polyethylene or HDPE, used for bottles containing milk, soft drinks, liquid soap, and other products. On the other hand, shorter, branched chains of about 500 ethylene monomers each produce a much less dense plastic, low-density polyethylene or LDPE. This is used for plastic food or garment bags, spray bottles, and so forth. There are other grades of plastic as well.

In some forms of recycling, plastics of all varieties are melted down together to yield a cheap, low-grade product known as "plastic lumber," used in materials such as landscaping timbers, or in making park benches. In order to achieve higher-grade recycled plastics, the materials need to be separated, and to facilitate this, recycling codes have been developed. Many plastic materials sold today are stamped with a recycling code number between 1 and 6, identifying specific varieties of plastic. These can be melted or ground according to type at recycling centers, and reprocessed to make more plastics of the same grade.

To meet the environmental challenges posed by plastics, polymer chemists continue to research new methods of recycling, and of using recycled plastic. One impediment to recycling, however, is the fact that most state and local governments do not make it convenient, for instance by arranging trash pickup for items that have been separated into plastic, paper, and glass products. Though ideally private recycling centers would be preferable to government-operated recycling, few private companies have the financial resources to make recycling of plastics and other materials practical.

WHERE TO LEARN MORE

Bortz, Alfred B. *Superstuff!: Materials That Have Changed Our Lives.* New York: Franklin Watts, 1990.

Ebbing, Darrell D.; R. A. D. Wentworth; and James P. Birk. *Introductory Chemistry.* Boston: Houghton Mifflin, 1995.

Galas, Judith C. *Plastics: Molding the Past, Shaping the Future.* San Diego, CA: Lucent Books, 1995.

Gordon, John Steele. "*Plastics, Chance, and the Prepared Mind.*" *American Heritage,* July-August 1998, p. 18.

Mebane, Robert C. and Thomas R. Rybolt. *Plastics and Polymers.* Illustrated by Anni Matsick. New York: Twenty-First Century Books, 1995.

Plastics.com (Web site). <http://www.plastics.com> (June 5, 2001).

"Plastics 101." *Plastics Resource* (Web site). <http://www2.plasticsresource.com/plastics_101/> (June 5, 2001).

"Polymers." *University of Illinois, Urbana/Champaign, Materials Science and Technology Department* (Web site). <http://matse1.mse.uiuc.edu/~tw/polymers/polymers.html> (June 5, 2001).

"Polymers and Liquid Crystals." *Case Western Reserve University* (Web site). <http://abalone.cwru.edu/> (June 5, 2001).

Zumdahl, Steven S. *Introductory Chemistry: A Foundation,* 4th ed. Boston: Houghton Mifflin, 2000.

GENERAL
SUBJECT INDEX

Boldface type indicates main entry page numbers. Italic type indicates photo and illustration page numbers.